"Seminar on Marine
3d, Toky"

Proceedings of
the Third NEA Seminar on

MARINE RADIOECOLOGY

Compte rendu du
Troisième Colloque de l'AEN sur la

RADIOECOLOGIE MARINE

TOKYO, 1,5 October 1979

NUCLEAR ENERGY AGENCY
ORGANISATION FOR ECONOMIC CO-OPERATION AND DEVELOPMENT
AGENCE POUR L'ÉNERGIE NUCLÉAIRE
ORGANISATION DE COOPÉRATION ET DE DÉVELOPPEMENT ÉCONOMIQUES

QH
543
.6
S44
1979

The Organisation for Economic Co-operation and Development (OECD) was set up under a Convention signed in Paris on 14th December, 1960, which provides that the OECD shall promote policies designed:
- to achieve the highest sustainable economic growth and employment and a rising standard of living in Member countries, while maintaining financial stability, and thus to contribute to the development of the world economy;
- to contribute to sound economic expansion in Member as well as non-member countries in the process of economic development;
- to contribute to the expansion of world trade on a multilateral, non-discriminatory basis in accordance with international obligations.

The Members of OECD are Australia, Austria, Belgium, Canada, Denmark, Finland, France, the Federal Republic of Germany, Greece, Iceland, Ireland, Italy, Japan, Luxembourg, the Netherlands, New Zealand, Norway, Portugal, Spain, Sweden, Switzerland, Turkey, the United Kingdom and the United States.

The OECD Nuclear Energy Agency (NEA) was established on 20th April 1972, replacing OECD's European Nuclear Energy Agency (ENEA) on the adhesion of Japan as a full Member.

NEA now groups all the European Member countries of OECD and Australia, Canada, Japan, and the United States. The Commission of the European Communities takes part in the work of the Agency.

The primary objectives of NEA are to promote co-operation between its Member governments on the safety and regulatory aspects of nuclear development, and on assessing the future role of nuclear energy as a contributor to economic progress.

This is achieved by:
- *encouraging harmonisation of governments' regulatory policies and practices in the nuclear field, with particular reference to the safety of nuclear installations, protection of man against ionising radiation and preservation of the environment, radioactive waste management, and nuclear third party liability and insurance;*
- *keeping under review the technical and economic characteristics of nuclear power growth and of the nuclear fuel cycle, and assessing demand and supply for the different phases of the nuclear fuel cycle and the potential future contribution of nuclear power to overall energy demand;*
- *developing exchanges of scientific and technical information on nuclear energy, particularly through participation in common services;*
- *setting up international research and development programmes and undertakings jointly organised and operated by OECD countries.*

In these and related tasks, NEA works in close collaboration with the International Atomic Energy Agency in Vienna, with which it has concluded a Co-operation Agreement, as well as with other international organisations in the nuclear field.

© OECD, 1980
Queries concerning permissions or translation rights should be addressed to:
Director of Information, OECD
2, rue André-Pascal, 75775 PARIS CEDEX 16, France.

L'Organisation de Coopération et de Développement Économiques (OCDE), qui a été instituée par une Convention signée le 14 décembre 1960, à Paris, a pour objectif de promouvoir des politiques visant :
- à réaliser la plus forte expansion possible de l'économie et de l'emploi et une progression du niveau de vie dans les pays Membres, tout en maintenant la stabilité financière, et contribuer ainsi au développement de l'économie mondiale ;
- à contribuer à une saine expansion économique dans les pays Membres, ainsi que non membres, en voie de développement économique ;
- à contribuer à l'expansion du commerce mondial sur une base multilatérale et non discriminatoire, conformément aux obligations internationales.

Les Membres de l'OCDE sont : la République Fédérale d'Allemagne, l'Australie, l'Autriche, la Belgique, le Canada, le Danemark, l'Espagne, les États-Unis, la Finlande, la France, la Grèce, l'Irlande, l'Islande, l'Italie, le Japon, le Luxembourg, la Norvège, la Nouvelle-Zélande, les Pays-Bas, le Portugal, le Royaume-Uni, la Suède, la Suisse et la Turquie.

L'Agence de l'OCDE pour l'Énergie Nucléaire (AEN) a été créée le 20 avril 1972, en remplacement de l'Agence Européenne pour l'Énergie Nucléaire de l'OCDE (ENEA) lors de l'adhésion du Japon à titre de Membre de plein exercice.

L'AEN groupe désormais tous les pays Membres européens de l'OCDE ainsi que l'Australie, le Canada, les États-Unis et le Japon. La Commission des Communautés Européennes participe à ses travaux.

L'AEN a pour principaux objectifs de promouvoir, entre les gouvernements qui en sont Membres, la coopération dans le domaine de la sécurité et de la réglementation nucléaires, ainsi que l'évaluation de la contribution de l'énergie nucléaire au progrès économique.

Pour atteindre ces objectifs, l'AEN :
- *encourage l'harmonisation des politiques et pratiques réglementaires dans le domaine nucléaire, en ce qui concerne notamment la sûreté des installations nucléaires, la protection de l'homme contre les radiations ionisantes et la préservation de l'environnement, la gestion des déchets radioactifs, ainsi que la responsabilité civile et les assurances en matière nucléaire ;*
- *examine régulièrement les aspects économiques et techniques de la croissance de l'énergie nucléaire et du cycle du combustible nucléaire, et évalue la demande et les capacités disponibles pour les différentes phases du cycle du combustible nucléaire, ainsi que le rôle que l'énergie nucléaire jouera dans l'avenir pour satisfaire la demande énergétique totale ;*
- *développe les échanges d'informations scientifiques et techniques concernant l'énergie nucléaire, notamment par l'intermédiaire de services communs ;*
- *met sur pied des programmes internationaux de recherche et développement, ainsi que des activités organisées et gérées en commun par les pays de l'OCDE.*

Pour ces activités, ainsi que pour d'autres travaux connexes, l'AEN collabore étroitement avec l'Agence Internationale de l'Énergie Atomique de Vienne, avec laquelle elle a conclu un Accord de coopération, ainsi qu'avec d'autres organisations internationales opérant dans le domaine nucléaire.

© OCDE 1980
Les demandes de reproduction ou de traduction doivent être adressées à:
M. le Directeur de l'Information, OCDE
2, rue André-Pascal, 75775 PARIS CEDEX 16, France.

FOREWORD

The objective of this Seminar was to bring together specialists in marine radioecology, to discuss selected topics and thereby to contribute to a better understanding of the consequences of radioactive waste sea dumping operations and of isotope releases from reprocessing and reactor operations. This Seminar was a follow-up of two earlier Seminars on the same subject area which were organised in Cherbourg in 1968 and in Hamburg in 1971.

AVANT-PROPOS

L'objet de ce Colloque était de réunir des spécialistes de la radioécologie marine pour discuter d'un certain nombre de sujets sélectionnés dans ce domaine afin de contribuer à une meilleure connaissance des conséquences des opérations d'immersion de déchets radioactifs en mer et du comportement des isotopes rejetés par suite de l'exploitation d'usines de retraitement du combustible et des réacteurs. Ce Colloque a fait suite à ceux qui ont déjà été organisés sur le même sujet à Cherbourg en 1968 et à Hambourg en 1971.

TABLE OF CONTENTS

TABLE DES MATIÈRES

OPENING OF THE SEMINAR
OUVERTURE DU COLLOQUE

    THE INTERACTION BETWEEN RADIOLOGICAL ASSESSMENTS AND RESEARCH REQUIREMENTS RELATED TO WASTE DISPOSAL IN THE DEEP SEA
    G.A.M. Webb (United Kingdom) 13

Session 1 - <u>DEEP SEA INVESTIGATIONS AND STUDIES</u>
Séance 1 - <u>RECHERCHES ET ETUDES SUR LES PROFONDEURS MARINES</u>

    Chairman - Président : Dr. R. ICHIKAWA (Japan)

    RADIOLOGICAL ASPECTS OF SEABED DUMPING IN THE DEEP OCEANS
    W.L. Templeton (United States) 23

    BIOLOGICAL STUDIES OF THE U.S. SUBSEABED DISPOSAL PROGRAM
    L.S. Gomez, R.R. Hessler, D.W. Jackson, M.G. Marietta, K.L. Smith Jr., D.M. Talbert, A.A. Yayanos (United States) 35

    CONTRIBUTION AU CONTROLE RADIOLOGIQUE DU MILIEU MARIN
    A. Ortins de Bettencourt, M.C. Vaz Carreiro, M.M. Sequeira (Portugal) 47

Session 2 - <u>DEEP SEA INVESTIGATIONS AND STUDIES</u> (continued)
Séance 2 - <u>RECHERCHES ET ETUDES SUR LES PROFONDEURS MARINES</u>
            (suite)

    Chairman - Président : Mr. A.W. VAN WEERS (The Netherlands)

    HYDROGRAPHIC SURVEYS IN THE CENTRAL WESTERN NORTH PACIFIC IN RELATION TO DEEP SEA DISPOSAL OF RADIOACTIVE WASTES
    H. Sudo (Japan) 57

    DEEP AND BOTTOM CURRENT MEASUREMENTS IN AN AREA ADJACENT TO THE PROPOSED WASTE DUMP SITE
    K. Taira, S. Imawaki, T. Teramoto (Japan) 69

    A DEEP CURRENT MEASUREMENT AT A PROPOSED WASTE DUMP SITE
    S. Imawaki, K. Takano (Japan) 79

EVALUATION ON THE DISPOSAL OF RADIOACTIVE WASTES INTO THE
NORTH PACIFIC - THE EFFECT OF STEADY FLOW AND UP-WELLING

Y. Sugiura, K. Saruhashi, Y. Miyake (Japan) 87

A PRELIMINARY ASSESSMENT OF BIOLOGICAL TRANSPORT OF
RADIONUCLIDES DUMPED AT DEEP SEA BOTTOM

T. Doi, T. Kidachi, K. Honjo, Y. Matsushita, T. Nemoto,
M. Shimizu, H. Sudo, H. Tsuruga (Japan) 95

OUTLINE OF BIOLOGICAL SURVEYS AROUND PLANNED WASTE
DUMPING SITE OFF JAPAN CONDUCTED BY TOKAI REGIONAL
FISHERIES RESEARCH LABORATORY

T. Kidachi, K. Honjo, T. Okutani, T. Watanabe (Japan) 111

DISPERSION DE POLLUANTS D'UNE SOURCE AU FOND PROFOND PAR
DES TOURBILLONS A ECHELLE INTERMEDIAIRE

K. Takano, S. Matsuyama (Japan) 117

Session 3 - DEEP SEA INVESTIGATIONS AND STUDIES (continued)
Séance 3 - RECHERCHES ET ETUDES SUR LES PROFONDEURS MARINES
            (suite)

          Chairman - Président : Mr. A. AARKROG (Denmark)

THE BEHAVIOR AND THE CHEMICAL FORMS OF METALLIC ELEMENTS
DISSOLVED IN OCEAN WATERS

Y. Sugimura, Y. Suzuki, Y. Miyake (Japan) 131

DISTRIBUTION OF $Sr-90$ AND $Cs-137$ IN DEEP WATERS AROUND
JAPAN

Y. Nagaya, K. Nakamura (Japan) 143

DISTRIBUTION ANOMALIES OF RADIO ISOTOPES IN DEEP SEA
REGIONS OF THE NORTH ATLANTIC

H. Kautsky (Federal Republic of Germany) 157

PLUTONIUM MOBILIZATION FROM SEDIMENTARY SOURCES TO
SOLUTION IN THE MARINE ENVIRONMENT

V.E. Noshkin, K.M. Wong (United States) 165

CONSIDERATIONS SUR LES CHAINES ALIMENTAIRES

M.C. Vaz Carreiro, A. Ortins de Bettencourt (Portugal) ... 179

Session 4 - STUDIES IN SHALLOW WATERS
Séance 4 - ETUDES MENEES DANS LES EAUX COTIERES

          Chairman - Président : Mr. A. AARKROG (Denmark)

DONNEES RADIOECOLOGIQUES CONCERNANT LE SITE MARIN DE
LA HAGUE

J. Ancellin, P. Bovard (France) 187

THE BEHAVIOUR OF TRANSURANIC AND OTHER LONG-LIVED RADIO-
NUCLIDES IN THE IRISH SEA AND ITS RELEVANCE TO THE DEEP
SEA DISPOSAL OF RADIOACTIVE WASTES

R.J. Pentreath, D.F. Jefferies, M.B. Lovett (United
Kingdom)
D.M. Nelson (United States) 203

Session 5 - STUDIES IN SHALLOW WATERS (continued)
Séance 5 - ETUDES MENEES DANS LES EAUX COTIERES (suite)

    Chairman - Président : Dr. R.J. PENTREATH (United Kingdom)

BIOTURBATION OF SURFICIAL SEDIMENTS ON THE CONTINENTAL SLOPE, EAST OF NEWFOUNDLAND
J.N. Smith, C.T. Schafer (Canada) 225

BEHAVIOUR OF NATURAL (Th, U) AND ARTIFICIAL (Pu, Am) ACTINIDES IN COASTAL WATERS
E. Holm, B.R.R. Persson (Sweden) 237

PLUTONIUM LEVELS IN THE MARINE ENVIRONMENT AT THULE, GREENLAND
A. Aarkrog (Denmark) 245

DIFFUSION OF TRITIATED WATER IN COASTAL AREAS
M. Fukuda, A. Kasai, T. Imai, H. Amano, N. Yanase (Japan). 253

RADIOECOLOGY OF Co-60 IN URAZOKO BAY
T. Ueda, Y. Suzuki, R. Nakamura (Japan) 265

SOME ASPECTS OF THE POSSIBLE FORMATION OF THE METAL ORGANIC COMPLEXES IN A SEAWATER BASED ON THE LABORATORY WORKS
Y. Honda, Y. Kimura, T. Ishiyama, T. Matsumura (Japan) ... 275

Session 6 - STUDIES IN SHALLOW WATERS (continued)
Séance 6 - ETUDES MENEES DANS LES EAUX COTIERES (suite)

    Chairman - Président : M. J. ANCELLIN (France)

THE NORTH SEA REGION TAKEN AS AN EXAMPLE FOR THE BEHAVIOUR OF ARTIFICIAL RADIOISOTOPES IN NEARSHORE SEA AREAS
H. Kautsky (Federal Republic of Germany) 283

A PRELIMINARY ASSESSMENT OF SOME NATURALLY-OCCURRING RADIONUCLIDES IN MARINE ORGANISMS (INCLUDING DEEP SEA FISH) AND THE ABSORBED DOSE RESULTING FROM THEM
R.J. Pentreath, D.S. Woodhead, B.R. Harvey, R.D. Ibbett (United Kingdom) 291

ACCUMULATION OF TRACE METALS IN COASTAL MARINE ORGANISMS
A.W. van Weers, J.G. van Raaphorst (The Netherlands) 303

RADIONUCLIDE ACCUMULATION BY MARINE DEMERSAL FISHES
T. Koyanagi, M. Nakahara, M. Matsuba (Japan) 313

CONCENTRATION FACTORS OF MESOPELAGIC ORGANISMS
M. Nakahara, T. Ueda, Y. Suzuki, T. Ishii, H. Suzuki (Japan) 323

CONCENTRATION FACTORS OF MARINE ORGANISMS USED FOR THE ENVIRONMENTAL DOSE ASSESSMENT
M. Kurabayashi, S. Fukuda, Y. Kurokawa (Japan) 335

Session 7 - STUDIES IN SHALLOW WATERS (continued)
Séance 7 - ETUDES MENEES DANS LES EAUX COTIERES (suite)

    Chairman - Président : Dr. M. SAIKI (Japan)

DIETARY SURVEY AROUND NUCLEAR SITE IN THE TOKAI AREA OF JAPAN AND THEIR RADIOLOGICAL SIGNIFICANCE TO THE RELEVANT POPULATION

M. Sumiya, Y. Ohmomo (Japan) 349

FINNISH STUDIES IN THE BALTIC SEA ON THE BEHAVIOUR OF RADIONUCLIDES DUE TO GLOBAL FALL-OUT AND RELEASES FROM NUCLEAR POWER PLANTS

A. Salo (Finland) 359

BEHAVIOUR OF ACTIVATION PRODUCTS RELEASED FROM A NUCLEAR POWER REACTOR INTO COASTAL WATERS

M. Nilsson, S. Mattsson (Sweden) 373

SELECTION OF CRITICAL GROUP IN RELATION TO THE RELEASE OF RADIONUCLIDES FROM NUCLEAR SPENT FUEL REPROCESSING PLANT

Y. Ohmomo (Japan) 381

SOME COMMENTS IN CONNECTION WITH JAPANESE EXPERIENCE ON PRE-ESTIMATION OF RADIATION DOSES TO MAN DUE TO THE COASTAL RELEASE OF LIQUID RADIOACTIVE WASTES

M. Saiki (Japan) 383

RADIOCESIUM IN INNER DANISH WATERS FROM WINDSCALE

A. Aarkrog (Denmark) 387

Session 8 - CONCLUSIONS OF THE SEMINAR
Séance 8 - CONCLUSIONS DU COLLOQUE

    Chairman - Président : Mr. W.L. TEMPLETON (United States)

    PANEL - TABLE RONDE 393

LIST OF PARTICIPANTS - LISTE DES PARTICIPANTS 401

OPENING OF THE SEMINAR

OUVERTURE DU COLLOQUE

THE INTERACTION BETWEEN RADIOLOGICAL ASSESSMENTS AND RESEARCH
REQUIREMENTS RELATED TO WASTE DISPOSAL IN THE
DEEP SEA

G. A. M. Webb
National Radiological Protection Board
Harwell, Didcot, United Kingdom.

ABSTRACT

Radiological assessments of the disposal of radioactive waste use mathematical models to describe the release of radionuclides from the waste package, their transport through the environment, their uptake by or irradiation of man and the predicted doses to man. The results of such assessments are used for radiological protection purposes such as comparison with the criteria recommended by the International Commission on Radiological Protection and for determining the best disposal option for a particular waste category. The results are also useful in directing research into those areas which have the greatest impact on the dose assessment.

1. INTRODUCTION

The protection of man against the harmful effects of radiation has been the subject of international recommendations and concern for more than 50 years. During this time understanding has developed both of the effects and of the appropriate means of protection. Present day considerations of one aspect of disposal of a small category of radioactive waste should be conditioned by this background of knowledge and viewed as part of the overall task of protection, rather than in isolation. The general methods and objectives of radiological protection are valid when considering waste disposal in the deep sea, just as they are for all other practices causing, or likely to cause exposure of man to radiation.

In this paper I discuss the radiological protection objectives which a waste disposal system should achieve; the use of pathway analysis in assessing whether these objectives have been attained together with the further interaction between the radiological assessment using pathway analysis and the research needs related to the waste disposal systems.

2. RADIOLOGICAL PROTECTION OBJECTIVES OF WASTE DISPOSAL

The basic radiological protection objectives of waste management must be the same as the basic radiological protection objectives for all procedures involving exposure of man to radioactivity. These have been clearly stated by the International Commission on Radiological Protection (ICRP) in their latest recommendation [1] as:

"(a) no practice shall be adopted unless its introduction produces a positive net benefit;

(b) all exposures should be kept as low as reasonably achievable, economic and social factors being taken into account; and

(c) the dose equivalent to individuals shall not exceed the limits recommended for the appropriate circumstances by the Commission."

When waste management only is considered the first requirement does not apply; this would be properly applied to the total practice giving rise to the wastes, with the doses (1) and costs of the waste disposal operation decreasing the net benefit. Within the context of a waste disposal system, given that the waste exists, the key recommendation is the second, often referred to as "optimisation of protection." The technique which is suggested by the Commission for carrying out the optimisation is that of differential cost benefit analysis. This is merely a formal procedure for establishing the level of protection beyond which efforts to provide additional protection would be wasteful of society's resources in the broadest sense. The requirements as set out by the Commission include the use of the concept of "dose commitment", the infinite time integral of the dose rate, which makes it necessary to take into account the radiation exposure of future generations when assessing practices which may eventually lead to their exposure.

The third objective of the Commission is important in that it is an overriding requirement that doses shall not exceed the limits; it is this requirement which is normally translated into statutory form. However for operations leading to doses below the limit it is the optimisation procedure which indicates whether or not further effort to reduce doses or dose commitments is warranted.

---

(1) for the remainder of this paper "dose" will be used to mean "dose equivalent"

The practical application of differential cost benefit analysis is the subject of considerable effort at present, both nationally and internationally, and many papers have been written on the subject. The basic concepts, however, are worth reiterating. The essence of the technique is to achieve that level of expenditure on protection at which the total cost of the waste disposal system is least. This total cost is made up of the direct monetary costs, eg for sea disposal ship costs, handling, packaging etc., added to the costs of the radiological detriment. In general, reduction of radiological detriment will require increased direct costs but these will reach a point of "diminishing returns" which is the optimum level of protection. A simple, albeit incomplete, measure of the radiological detriment is the collective dose commitment, made up by summing all the doses to be received over all future time. This can then be converted to a monetary equivalent by assessing and costing the consequent health effects. Although this procedure is difficult, it is necessary and commonplace in other areas of decision making involving the allocation of resources to health care, road safety etc.

It is worth emphasising that the objective of waste disposal is not to attain complete containment of the waste for ever; even if this were possible it would definitely not be the optimum level of protection in ICRP terms on radiological grounds.

In order to carry out either an optimisation procedure or a comparison with dose limits from a disposal operation, an assessment is needed of the distribution of doses, in space and time, likely to result from the disposal. This is the primary purpose of a radiological assessment which is carried out using pathway analysis.

3. PATHWAY ANALYSIS

The general technique for predictive assessment of doses to man from radionuclides released into the environment is to generate, from a knowledge of physical, chemical, biological or other characteristics of the receiving environment, a network of potential pathways by which the radioactivity may return to man. Appropriate mathematical models are then constructed for each step in each pathway to quantitatively predict the rates of return. Finally parameter values are ascribed to the various sections of the mathematical models. The appropriate set of pathways and models will differ for particular radionuclides as a function of waste form, half-life, physico-chemical behaviour, radiotoxicity to man etc. Thus the choice of appropriate models and parameters is not a simple matter but requires considerable background knowledge and judgement. It is also an essentially interdisciplinary scientific problem requiring input to the modeller from experts in many fields.

In the case of disposal of packaged radioactive waste to the deep sea the general pathway models can be broken down as follows:

(a) Release of radionuclides from the package (container and waste form)

(b) Local mixing with water and adsorption onto sediments

(c) Local biological uptake processes

(d) Physical transport of dissolved or resuspended radionuclides via water column

(e) Biological transport of radionuclides

(f) Sediment transport

(g) Reconcentration of dissolved radionuclides (by fauna and flora) after transport in water column

(1) Exposure of man via ingestion, inhalation and direct irradiation

Many of these pathways may interact and the normal procedure is to construct a block diagram of the system with the interconnections shown.

The appropriate mathematical model for each stage in the pathway will depend on the degree of complexity of the stage, the point of development of such models and the need for refinement of that particular stage in the context of the overall assessment. For example in some simple assessments the release of radioactivity from the package is assumed to be instantaneous as soon as the package reaches the sea bottom. This is clearly not a good representation of the real situation but for some types of assessment may be adequately pessimistic.

When considering transport through the water column a vast range of models of varying degrees of sophistication is available; which of these is selected will depend on many factors, not least of which is the timescale of the assessment.

Even relatively simple models and small numbers of pathways, when combined with the large number of radionuclides which need to be catered for, lead to substantial sets of results, particularly if these are expressed as functions of space or time $/\overline{2.3}/$.

4. RADIOLOGICAL ASSESSMENT

Although it is sometimes assumed that the pathway analysis is the radiological assessment, it is only a component of the assessment. The pathway analysis itself should be conditioned by the form of the radiological assessment. There are, from the simplest viewpoint, two forms of assessment.

(a) The most usual is a pathway analysis using models, parameter values, and assumptions designed to maximize the dose estimate. This maximum dose estimate is then the appropriate value to compare with the ICRP dose limits to ensure compliance with the third condition of the ICRP recommendations. This is the type of calculation carried out in the documents underlying the IAEA definition of high level wastes unsuitable for dumping at sea $/\overline{2.4}/$. It must be made clear, that these type of calculations rightly aim to overestimate the doses to man and therefore they are not appropriate for assessing actual or most likely consequences.

(b) In order to carry out comparisons of different disposal options or optimisation studies it is essential to assess as realistically as possible the most probable distributions of doses from which to calculate the collective dose or collective dose commitment. Although this calculation may utilize the same mathematical models as that for the assessment of maximum doses, it will not in general use the same values for parameters or the same assumptions. Calculation of this type for these purposes have not normally been carried out as part of the radiological assessment of sea disposal of radioactive wastes (or of other disposal options). It is hoped that one outcome of the present seminar will be to focus the attention of researchers on the need for data for use in realistic estimates of dose as well as the extreme values for maximum estimates.

5. INTERACTION OF RADIOLOGICAL ASSESSMENT AND RESEARCH

Once the pathway model has been set up it is possible to tell either directly by inspection or by formal sensitivity studies which parameters and assumptions have the most effect on the calculated doses. It is clearly in these areas that research studies will be of the greatest use in reducing the uncertainty of dose predictions. This interaction is part of a continuing, iterative process:

1st   Construct a model (possibly very crude)

2nd   carry out sensitivity analysis (vary parameters over their range of uncertainty and compare the effect on predicted doses)

3rd   derive recommendations for key research areas

4th   simultaneously carry out research and model development

5th   feed the results of research and model development back into a better overall assessment.

This process is continued until there is sufficient confidence in both the models and the parameter values to rely on the output for making decisions.

Although I have not yet carried out a formal sensitivity study on the disposal of low level waste on the sea bed, nor is there the opportunity in this short paper to go into detail, I felt that a few general comments on research priorities from the viewpoint of a radiological modeller may be of use. These are classified roughly according to the general pathway model given in Section 3.

5.1 Release of radionuclides from the package

Although not of direct concern to this seminar, this is clearly a key area for investigation, particularly with a view to carrying out realistic assessments. The majority of radionuclides are not released immediately even from simple concrete packages such as are currently disposed of but there is little firm information on actual release rates under deep ocean conditions. If this were available there would be a better basis for encouraging the development of more retentive packages, if this were cost-effective in radiological terms, and also for introducing some credit for slower release rates into the framework of the IAEA definition of high-level radioactive wastes unsuitable for dumping at sea [5].

5.2 Local processes

All the radionuclides released from waste canisters will be released directly into the benthic boundary layer (defined for this purpose as the layer of the water column immediately above the sediments extending perhaps 50 to 100m above the bottom together with the layer of sediment immediately below the water column extending downwards about 1 m). For this reason the benthic boundary layer needs more study as a matter of principle. Nonetheless the overall effect of processes in the benthic boundary layer, other than removal processes such as adsorption onto sediment, might be expected to be small. The processes within the benthic boundary layer likely to be of most interest from the radiological assessment viewpoint are :-

(a) Physical transport processes both laterally towards areas of upwelling or rising seabed and vertically into the main water column.

(b) Biological transport processes within and out of the benthic boundary layer. It is important to quantify such processes.

(c) Removal processes, both physical via sediment adsorption and biological via reconcentration and excretion or sheletal deposition.

5.3 Water column

This is the main diluting medium between the released radioactivity and man. There will in reality be a complex interaction of physical and biological processes, some acting to concentrate radioactivity and some to disperse it. Movement of radioactivity both towards and away from man will occur. In practice this system is broken down at the present time for modelling purposes into physical and biological processes.

The study of water movement and dispersion is a major field of work and very many models are available. Measurements made already have identified the major processes which are reasonably well understood. Models either theoretical or semi-empirical, are available or are being developed which are more complex than those used for example in references $\underline{/3,4/}$. It is not clear, however, that the overall results of radiological assessments are very sensitive to either model complexity or input parameters so far as physical transport in the water column is concerned. It also seems that for the long-lived radionuclides such as ^{239}Pu, the spread in time of the dose distribution is so long that it is not worthwhile trying to build detailed models of present day ocean circulation patterns without great confidence that climatic and other changes will not result in major changes in such patterns. This confidence does not yet appear to exist. It is therefore, in my view, better for these radionuclides to rely on simple, robust, oceanographic models which do not require much additional research.

The other aspect of the water column is the biological processes occurring within it. These require more investigation from the viewpoint of quantifying the rates of transport back to man, both for possible "short-circuit" processes from the benthic boundary layer and for general reconcentration processes. More studies should be made for realistic assessments of the biological removal processes from the water column; these have been neglected in all assessments to date even though the net biological transport process is from the surface to the bottom sediments.

5.4 Interactions with man

This topic is outside the precise subject of the seminar but I feel it merits some consideration. The pathway which in all studies leads to the higher doses to man is via ingestion of seafood, including surface and deep water fish, molluscs and crustacea, plankton and seaweeds etc. Preliminary studies $\underline{/3/}$ show maximum doses from prospective pathways such as consumption of plankton or benthic fauna; these results may be a consequence of the conservative assumptions necessitated by an absence of data. Research on the viability of continued harvesting of these seafoods and the likely rates would help to refine the dose estimates.

6. CONCLUSIONS

The proper objectives for the management of the disposal of radioactive wastes in the deep sea are the same as for any other practice involving the exposure of man to radiation; these are to ensure all doses are less than the dose limits and to reduce doses to as low as reasonably achievable, taking economic and social factors into account (ie to optimise protection). The assessment of the attainment of such objectives in advance requires the setting up of a mathematical simulation or "model" of the actual pathways for return of radioactivity to man. Using this model the dose distribution as a function of space and time can be assessed for a given input of radionuclides. Two major outputs are needed for the radiological assessment, maximum individual doses, calculated using pessimistic assumptions for comparison with dose limits; and collective (total) doses, calculating using realistic assumptions, for comparisons of options and for optimisation of protection. In addition, by sensitivity analysis, the model can be used to direct research effort to those topics which have the greatest influence on the dose calculation. The results of such research together with improved models can then be used to provide more reliable and precise radiological assessments to assist in better decision-making.

7. REFERENCES

[1] International Commission on Radiological Protection. "Recommendations of the International Commission on Radiological Protection" ICRP Publication 26. Annals of the ICRP Vol 1, 3, Pergamon Press (1977).

[2] International Atomic Energy Agency. "The Radiological Basis of the IAEA revised definition and recommendations concerning high-level radioactive waste unsuitable for dumping at sea" Report of the Consultants Meeting to review the radiological basis of the provisional definition and recommendations for the convention on the prevention of marine pollution by the dumping of wastes and other matter, organised by IAEA and held at IMCO headquarters, London, 13 - 17 June 1977. Technical Document IAEA-211, Vienna (1978).

[3] Grimwood, P. D. and Webb, G. A. M. "Assessment of the Radiological Protection Aspects of Disposal of High Level Waste on the Ocean Floor" National Radiological Protection Board Report NRPB-R48, Harwell, Didcot, Oxon (1976).

[4] International Atomic Energy Agency "The Oceanographic Basis of the IAEA revised definition and recommendations concerning high-level radioactive waste unsuitable for dumping at sea" Report of the Advisory Group Meeting to review the oceanographic basis of the provisional definition and recommendations for the convention on the prevention of marine pollution by dumping of wastes and other matter organised by the IAEA and held in Vienna 21 - 25 March 1977. Technical Document IAEA-210, Vienna (1978).

[5] International Atomic Energy Agency "Convention on the Prevention of Marine Pollution by Dumping of Wastes and Other Matter : The Defintion and Recommendations". INF CIRC/205/Add.1/Rev 1. IAEA, Vienna (1978).

Session 1

Chairman - Président
Dr. R. ICHIKAWA
(Japan)

Séance 1

RADIOLOGICAL ASPECTS OF SEABED DUMPING IN THE DEEP OCEANS

W. L. TEMPLETON

Ecological Sciences Department
Pacific Northwest Laboratory
Battelle Memorial Institute
Richland, Washington 99352
U.S.A.

ABSTRACT:

Under the Convention of the Prevention of Marine Pollution by Dumping Wastes and Other Matter in the Ocean (The London Dumping Convention), the International Atomic Energy Agency (I.A.E.A.) was charged with the task of defining radioactive wastes unsuitable for dumping at sea and providing recommendations to ensure that dumping of radioactive material into the oceans under special permits involves no unacceptable degree of hazard to humans and their environment. Recently the I.A.E.A. submitted a revised definition and recommendations to the Convention. These were based on certain oceanographic and radiological bases developed by Consultant Groups of the I.A.E.A. This paper presents the radiological basis for the definition and assessment. It describes the steps that were taken, based upon the oceanographic aspects, the pathways to man, and concentration ratios, to arrive at the release rate limits for a wide selection of radionuclides. The acceptance of the concept of applying release rate limits as developed by the I.A.E.A. provides a rational basis for further considering the emplacement of radioactive materials in the seabed as an attractive and acceptable alternative to terrestrial geological repositions.

INTRODUCTION

In order to control coastal discharges or ocean dumping of any kind of material, it is necessary to determine a release rate. This can only come from a knowledge of the composition and chemical form of the source materials, the distribution and bioavailability of these materials in the ocean ecosystem, the degree and rates of bioaccumulation and the actual or potential use of the ocean resources. With this information release rates within acceptable limits for man and the ecosystem can then be determined. Today, probably the only situations which apply this approach are the controlled disposal of radioactive wastes.

In this paper I discuss a recent radiological assessment of the dumping of packaged radioactive wastes on the seabed and describe some environmental aspects of the United States Department of Energy program examining the feasibility of the emplacement of contained radioactive wastes within the deep ocean sediments.

PRESENT PRACTICE

In the 1950's and 1960's many countries used the oceans for dumping of packaged low-level radioactive wastes. In 1972 the United States of America banned the transport, for dumping of high-level radioactive wastes. Although the dumping of medium and low-level radioactive wastes are allowed by law, since that date no permits have been issued for dumping of this type of material.

At the same time there were intensive efforts internationally to reach agreement on ocean dumping of all pollutants. The outcome was the Convention on the Prevention of Marine Pollution by Dumping Wastes and Other Matter in the Oceans (the London Dumping Convention of 1972) [1]. The International Atomic Energy Agency (IAEA) was charged with the task of defining radioactive wastes unsuitable for dumping at sea and providing recommendations to ensure that any dumping of radioactive material into the oceans involves no unacceptable degree of hazard to humans and their environment. In 1974 the IAEA made a provisional definition along with a recommended basis for issuing special permits [2].

The Provisional Definition and Recommendations of 1974 stated that high-level radioactive wastes or other high-level radioactive matter unsuitable for dumping means any material with a concentration in curies per unit-gross mass (in tonnes) exceeding:

(a) 10 Ci/t for α-active waste for half life greater than 50 years. (In the case of ^{226}Ra, not more than 100 Ci/yr may be dumped at any one site);

(b) 10^3 Ci/t for β/γ-active waste (excluding tritium) but the limit for ^{90}Sr plus ^{137}Cs is 10^2 Ci/t; and

(c) 10^6 Ci/t for tritium.

The definition is based on an assumed upper limit to the dumping rate of 100,000 tonnes per year at any one site and averaged over a gross mass not exceeding 100 tonnes.

Since 1967 European dumping operations have been organized and conducted by the Nuclear Energy Agency of the Organization for Economic Cooperation and Development. During the period 1967-1977 a total of about 51,600 tonnes of packaged solid wastes, containing about 5,900 curies of α-active material; about 190,000 curies of β/γ emitters; and about 183,000 curies of tritium have

---

(1) Chairman of the International Atomic Energy Agency Consultants Meeting which developed the Radiological Basis of the IAEA Review Definition and Recommendations Concerning High-Level Radioactive Waste Unsuitable for Dumping at Sea [5].

been dumped in the northeast Atlantic Ocean [3]. The accumulated amounts expressed as fractions of the limiting dumping rates implied in the Provisional Definition are:

(a)   0.1% for α-active materials;

(b)   0.1% β/γ-active materials; and

(c) 10^{-4} for tritium

and only twice approached 10% of the upper limit for mass dumping rate.

BASIS OF THE REVISED IAEA DEFINITION AND RECOMMENDATIONS [3]

The provisional definition and recommendations were actively reviewed by the IAEA during 1976-1978. Three major aspects were reviewed: the oceanographic basis [4], the radiological basis [5], and the implications for the Definition and Recommendations.

The Oceanographic Basis [4]

Assessment of permissible dumping rates of radionuclides to the oceans must include the calculation of the concentration throughout oceanic basins resulting from localized sources. However, our understanding of the processes occurring in the deep oceans is insufficient to permit the construction of a single comprehensive model of the movement of radionuclides. The original oceanographic model used for the provisional definition was inapplicable for long-lived isotopes in finite sized basins. It was considered that the model by Shepherd which includes [6] advection in a finite ocean meets some of the objections raised about the original model used and allows estimates to be made of the entire concentration field, although it only approximates the actual oceanographic processes resulting in the dispersion of radioactive nuclides.

The Shepherd model calculates the equilibrium concentration which would be reached from a continuous release of activity maintained indefinitely into the water near the ocean bottom (>4,000 m). The model ocean is of finite size and has a horizontal (but no vertical) circulation and three dimensional diffusion. Obviously this is an idealization but is adequate for defining large-scale long-term concentrations. However, since little is known about the circulation of the deep ocean water poor vertical mixing cannot be assumed for the isolation of radionuclides, and even if this were so, slow vertical mixing could be short-circuited by direct biological transport. It was concluded then [4] that one should not assume any isolation of the surface waters when estimating the dose to man, and that one should calculate not only the long-term average concentration in the bottom water for the appropriate part of the ocean basin but also the appropriate maximum concentrations arising from short term events.

The model only considers the large-scale average distribution of various oceanographic parameters and does not describe short-term processes, either on the large or small scale such as deep vertical upwelling, effects of large-scale topographic features, or strong convective currents. Since deep vertical upwelling, that is a direct transfer of deep bottom water to the surface, is not explicitly in the model due to our sparse understanding of the rates of vertical diffusion and advection, one cannot assume that disposal of wastes in deep waters provides any isolation from the surface waters when an assessment of the dose to critical groups is being made.

Radiological Assessment [5]

(a)   Oceanographic Aspects

With the oceanographic model as a basis, calculations of the concentrations of radionuclides in water for the dose assessment included both (1) the long-term concentration in the water for the appropriate part of the ocean basin and (2) the appropriate maximum concentrations arising from short-term events. In both cases these are bottom water concentrations which imply that these levels would be acceptable to surface waters and therefore make it unnecessary to distinguish between hypothetical consumption of deep-sea organisms and more realistic consumption of upper-layer organisms.

Since it is difficult to foresee the time scale over which releases of radioactive waste may continue, the calculations have assumed that releases continue for 40,000 years which is approximately the mean lifetime of ^{239}Pu. The release rates limits derived are therefore such that concentrations in the marine environment of long-lived radionuclides will increase very slowly over several thousand years towards their limiting values. This is clearly very conservative; however, it does allow waste dumping operations to be reduced or stopped at any time without exceeding the limiting values. For example, if the dumping of ^{239}Pu is continued at the calculated release rate limit, the concentrations of ^{239}Pu in the ocean will slowly build up approaching the International Commission on Radiological Protection derived concentration after 40,000 years. If the practice ceases after 4,000 years then only 10% of the International Commission on Radiological Protection derived limit will have been reached. For shorter periods of time the oceanographic model suggests the release rate limits might be controlled by short-term processes of advection and upwelling. In order that unrealistic release limits for very short-lived radionuclides are not estimated it was assumed that the containment time on the sea-bed was ten years and that three years decay occurred between the release point and consumption exposure.

Because of a lack of information on the role of sediments in reducing water concentrations, the calculations ignored sorption on sediments. This obviously overestimates water concentrations, and means that release rate limits for pathways that do not involve sediments would be conservative. However, for the radiological assessments of the dose to man or organisms the concentration on the sediment was calculated with the assumption that it is in equilibrium with the bottom water already calculated. This clearly overestimates the concentration on sediments if there is significant partitioning between water and sediment, since it ignores the reduction in overall concentration arising from the sorptive capacity of the sediments themselves.

(b) Assessment of Pathways

The assessment quantified the parameters involved in a number of representative pathways by which man might become exposed to radioactivity after its release on the ocean bottom. The pathways chosen include some of which are known to exist and some which may be important in the future (Table I). For all the possible pathways which were identified the conservative approach was taken. For example, a pathway in the future may include systematic fishing at a depth of 4,000 meters, while the deepest presently known is 2,000 meters. We have no detailed information on the concentration factors for cephalopods or deep-living fish, and for the present calculation it was assumed that these would be sufficiently similar to those for surface organisms.

TABLE I. Pathways, Modes of Exposure, Intake/Occupancy Rates

Pathway	Mode of Exposure	Intake/Occupancy Rate
Fish consumption	Ingestion	600 g/day
Crustacea consumption	Ingestion	100 g/day
Mollusc consumption	Ingestion	100 g/day
Seaweed consumption	Ingestion	300 g/day
Plankton consumption	Ingestion	30 g/day
Desalinated water consumption	Ingestion	2000 g/day
Sea salt consumption	Ingestion	3 g/day
Suspension of sediments	Inhalation	Continuous
Evaporation from seawater	Inhalation	Continuous
Swimming	External irradiation	300 h/yr
Exposure from shore sediments	External irradiation	1000 h/yr
Exposure from fishermen's gear	External irradiation	300 h/yr

The pathways selected are generalized representatives and the consumption parameters selected are sufficiently general to include critical groups in all areas of the world. Where individuals are likely to be members of only a single critical group, the pathways were evaluated independently. Where they might be members of more than one critical group, e.g., shore fishermen and beach dwellers, the limits have been reduced accordingly.

Five individual pathways involving consumption of sea food were considered. These are not intended to represent any particular species but are examples of general pathways. Consumption rates were assumed to be sufficiently large, in a global context, that for each pathway it would be unlikely that members of one critical consumption group would also be members of another critical consumption group.

Four pathways leading to exposure of beach dwellers were considered. Since some individuals would be likely to be exposed to all pathways the derived limits were reduced accordingly. Three miscellaneous pathways were also considered and were combined for convenience.

The IAEA Radiological Assessment conducted the calculations for the pathways for radionuclides which were felt likely to occur in wastes liable to be dumped into the ocean. The list included fission products, actinides, activation products and natural radionuclides.

The uptake, accumulation and depuration of radionuclides by aquatic organisms is a dynamic process, depending upon many variables, such as the physiochemical state of the radionuclide in seawater, on the sediments and in the organisms themselves. Some of the important biological variables which may influence the distribution of trace elements, stable analogs and radionuclides include: (1) differences in the adsorptive capacity and selectivity of external surfaces of marine organisms for multivalent elements, (2) differences in adsorptive capacity of marine organisms for the same element in different oxidation states, (3) variations in the rates at which trace elements are incorporated through limiting membranes of marine organisms, (4) differences in the abilities of individual organisms or species to incorporate trace elements from the soluble, colloidal or particulate form in sea water, (5) retention and excretion rates for given elements by different organisms, (6) differences in the efficiency of conversion of organic and trace element components in food transferred between trophic levels of given food webs, (7) variations in patterns of accumulation of elements by pelagic and benthic organisms comprising the primary producers, (8) variations in feeding habits with respect to particle size selection, (9) differences in patterns of distribution of dominant species of pelagic and benthic marine organisms, (10) variability in the structures of dominant food webs and the distribution patterns of the stable elements within the webs, (11) differences in turnover rates of biomass and associated trace elements in different food webs, (12) the effects of population structures of localized benthic assemblages upon the chemical and physical states of the sediments, and (13) variations in the rates of deposition and incorporation of organic detritus and waste products into the sediments and the local effects of this material upon the chemical state of the sediments.

However, our understanding of how these variables effect the degree of accumulation and retention is not well understood for the majority of the radionuclides under consideration. Therefore, the concept of a concentration ratio, i.e., concentration per unit mass of organism to that concentration in an equivalent mass of seawater has proven useful, particularly in equilibrium situations or one where concentrations change slowly compared with the turnover rate of radionuclides in the organisms comprising the pathway. It ignores, of course, many of the above variables and assumes that the radionuclide in the ecosystem partitions between all segments of system in the same way as the stable analog. Some of the data in Table II are derived from direct measurement of the radionuclide in the marine environment, however, in most cases this is without benefit of information on the physiochemical state in the water or sediments. Additionally, the quoted values are mean values despite our knowledge that many distribution of radionuclides within populations of the same species are not normal but lognormal.

TABLE II. Concentration Factors used for Radiological Assessment

Element	Fish	Crustacea	Molluscs	Seaweed	Plankton	Desal'N	Seasalt	Sediment	Evaporation
H	1.0E 00	1.0E 00	1.0E 00	1.0E 00	1.0E 00	1.0E 00	1.0E 00	1.0E 00	1.0E 00
C	5.0E 04	4.0E 04	5.0E 04	4.0E 03	3.0E 03	(1.0E-04)	3.0E 01	(1.0E 02)	(1.0E-05)
Na	1.0E-01	3.0E-01	2.0E-01	1.0E 00	1.0E 00	(1.0E-04)	3.0E 01	(1.0E 02)	(1.0E-05)
P	2.0E 04	1.0E 04	1.0E 04	1.0E 04	1.0E 04	(1.0E-04)	3.0E 01	(1.0E 02)	(1.0E-05)
S	1.0E 00	1.0E 00	1.0E 00	1.0E 00	1.0E 00	(1.0E-04)	3.0E 01	(1.0E 02)	(1.0E-05)
Cl	1.0E 00	1.0E 00	1.0E 00	1.0E 00	1.0E 00	(1.0E-04)	3.0E 01	(1.0E 02)	(1.0E-05)
Ca	1.0E 00	1.0E 01	1.0E 00	1.0E 00	1.0E 01	(1.0E-04)	3.0E 01	(5.0E 02)	(1.0E-02)
Cr	1.0E 02	5.0E 02	5.0E 02	(3.0E 04)	(3.0E 03)	(1.0E-04)	3.0E 01	(1.0E 04)	(1.0E-02)
Mn	5.0E 02	1.0E 04	1.0E 04	1.0E 04	1.0E 03	(1.0E-04)	3.0E 01	1.0E 04	(1.0E-02)
Fe	1.0E 03	1.0E 03	1.0E 03	1.0E 04	1.0E 04	(1.0E 04)	3.0E 01	1.0E 04	(1.0E-02)
Co	1.0E 02	1.0E 03	1.0E 03	1.0E 03	1.0E 03	(1.0E-04)	3.0E 01	1.0E 04	(1.0E-02)
Ni	5.0E 02	1.0E 02	1.0E 02	5.0E 02	1.0E 03	(1.0E-04)	3.0E 01	1.0E 04	(1.0E-02)
Zn	2.0E 03	4.0E 03	1.0E 05	1.0E 03	1.0E 04	(1.0E-04)	3.0E 01	1.0E 04	(1.0E-02)
Se	1.0E 02	1.0E 03	1.0E 03	1.0E 03	1.0E 04	(1.0E-04)	3.0E 01	1.0E 04	(1.0E-05)
Br	(3.0E 00)	(1.0E 01)	(1.0E 01)	(3.0E 01)	(3.0E 01)	(1.0E-04)	3.0E 01	(1.0E 02)	(1.0E-05)
Sr	1.0E 00	1.0E 01	1.0E 01	1.0E 01	(1.0E 01)	(1.0E-04)	3.0E 01	5.0E 02	(1.0E-05)
Y	1.0E 00	1.0E 03	1.0E 03	1.0E 03	1.0E 02	(1.0E-04)	3.0E 01	1.0E 04	(1.0E-02)
Zr	1.0E 00	1.0E 02	1.0E 03	5.0E 02	(1.0E 04)	(1.0E-04)	3.0E 01	1.0E 04	(1.0E-02)
Nb	1.0E 00	1.0E 02	1.0E 03	5.0E 02	(1.0E 03)	(1.0E-04)	3.0E 01	1.0E 04	(1.0E-02)
Tc	1.0E 01	1.0E 03	1.0E 03	1.0E 05	1.0E 03	(1.0E-04)	3.0E 01	1.0E 04	(1.0E-02)
Ru	1.0E 00	6.0E 02	2.0E 03	2.0E 03	(1.0E 03)	(1.0E-04)	3.0E 01	1.0E 04	(1.0E-02)
Pd	(3.0E 02)	(3.0E 02)	(3.0E 02)	(1.0E 03)	(1.0E 03)	(1.0E-04)	3.0E 01	(1.0E 04)	(1.0E-02)
Ag	1.0E 03	5.0E 03	1.0E 05	1.0E 03	1.0E 03	(1.0E-04)	3.0E 01	1.0E 04	(1.0E-02)
Sn	1.0E 03	3.0E 02	1.0E 02	1.0E 02	1.0E 03	(1.0E-04)	3.0E 01	1.0E 04	(1.0E-02)
Sb	1.0E 03	3.0E 02	1.0E 02	1.0E 02	1.0E 03	(1.0E-04)	3.0E 01	1.0E 04	(1.0E-02)
Te	1.0E 03	1.0E 03	1.0E 03	1.0E 04	1.0E 03	(1.0E-04)	3.0E 01	1.0E 04	(1.0E-05)
I	1.0E 01	1.0E 02	1.0E 02	1.0E 03	1.0E 03	(1.0E-04)	3.0E 01	1.0E 02	(1.0E-05)
Cs	5.0E 01	3.0E 01	1.0E 01	1.0E 01	1.0E 02	(1.0E-04)	3.0E 01	5.0E 02	(1.0E-05)
Ce	(1.0E 01)	1.0E 03	1.0E 03	1.0E 03	1.0E 03	(1.0E-04)	3.0E 01	1.0E 04	(1.0E-02)
Pm	1.0E 02	1.0E 03	1.0E 03	1.0E 03	1.0E 03	(1.0E-04)	3.0E 01	1.0E 04	(1.0E-02)
Sm	(1.0E 02)	(1.0E 03)	(1.0E 03)	(1.0E 03)	(3.0E 03)	(1.0E-04)	3.0E 01	(1.0E 04)	(1.0E-02)
Eu	1.0E 02	1.0E 03	1.0E 03	1.0E 03	1.0E 04	(1.0E-04)	3.0E 01	1.0E 04	(1.0E-02)
Au	1.0E 02	1.0E 03	1.0E 03	1.0E 03	1.0E 04	(1.0E-04)	3.0E 01	1.0E 04	(1.0E-02)
Pb	3.0E 02	1.0E 02	1.0E 02	1.0E 03	1.0E 04	(1.0E-04)	3.0E 01	1.0E 04	(1.0E-02)
Po	2.0E 03	2.0E 04	2.0E 04	1.0E 03	1.0E 04	(1.0E-04)	3.0E 01	1.0E 04	(1.0E-02)
Ra	1.0E 02	1.0E 02	1.0E 02	1.0E 02	1.0E 02	(1.0E-04)	3.0E 01	5.0E 02	(1.0E-05)
Ac	3.0E 01	1.0E 03	1.0E 03	1.0E 03	1.0E 04	(1.0E-04)	3.0E 01	1.0E 04	(1.0E-02)
Th	1.0E 03	1.0E 03	1.0E 03	1.0E 03	1.0E 04	(1.0E-04)	3.0E 01	5.0E 06	(1.0E-02)
Pa	1.0E 01	1.0E 01	1.0E 01	1.0E 02	1.0E 03	(1.0E-04)	3.0E 01	5.0E 03	(1.0E-02)
U	1.0E-01	1.0E 01	1.0E 01	1.0E 01	5.0E 00	(1.0E-04)	3.0E 01	5.0E 02	(1.0E-02)
Np	(1.0E 01)	(1.0E 02)	(1.0E 03)	(1.0E 03)	(2.0E 03)	(1.0E-04)	3.0E 01	(5.0E 04)	(1.0E-02)
Pu	1.0E 01	1.0E 02	1.0E 03	1.0E 03	(2.0E 03)	(1.0E-04)	3.0E 01	5.0E 04	(1.0E-02)
Am	1.0E 01	2.0E 02	2.0E 03	2.0E 03	(2.0E 03)	(1.0E-04)	3.0E 01	5.0E 04	(1.0E-02)
Cm	(1.0E 01)	(2.0E 02)	(2.0E 03)	(2.0E 03)	(2.0E 03)	(1.0E-04)	3.0E 01	(5.0E 04)	(1.0E-02)
Cf	(1.0E 01)	(2.0E 02)	(2.0E 03)	(2.0E 03)	(2.0E 03)	(1.0E-04)	3.0E 01	(5.0E 04)	(1.0E-02)

Concentration Factors in parentheses are estimates.

(c) Release Rate Limits

The output from these calculations for both single site and a finite ocean volume provides the critical pathway for each radionuclide and is that giving rise to the lowest release rate limit. When pathways have been combined under one critical group; i.e., beach dwellers, the critical pathway is that which individually would have the lowest limit. As an example of the output for a finite ocean volume ($10^{17} m^3$) the release rate limits for the forty most restrictive radionuclides are given in Table III.

TABLE III. Release Rate Limits in Ascending Order for a Finite Ocean Volume of $10^{17} m^3$

Limit Curies/Year	Nuclide	Critical Group
2.8×10^3	Thorium-229	Beach dwellers
6.8×10^3	Iodine-129	Seaweed eaters
1.1×10^4	Radium-226	Fish eaters
1.4×10^4	Thorium-232	Fish eaters
1.6×10^4	Thorium-230	Fish eaters
5.7×10^4 (a)	Neptunium-237	Seaweed eaters
5.8×10^4	Tin-126	Beach dwellers
5.9×10^4	Technetium-99	Seaweed eaters
8.6×10^4 (a)	Curium-245	Seaweed eaters
8.7×10^4	Plutonium-242	Seaweed eaters
9.2×10^4	Plutonium-239	Seaweed eaters
1.2×10^5	Americium-243	Seaweed eaters
1.4×10^5 (a)	Curium-246	Seaweed eaters
2.6×10^5	Lead-210	Plankton eaters
3.0×10^5	Plutonium-240	Seaweed eaters
5.9×10^5 (a)	Californium-251	Seaweed eaters
6.1×10^5	Carbon-14	Fish eaters
7.3×10^5	Americium-241	Seaweed eaters
1.1×10^6	Uranium-238	Seaweed eaters
1.5×10^6	Americium-242	Seaweed eaters
3.7×10^6	Nickel-59	Fish eaters
3.9×10^6	Zirconium-93	Beach dwellers
3.9×10^6 (a)	Curium-243	Seaweed eaters
4.4×10^6	Plutonium-238	Seaweed eaters
6.8×10^6	Uranium-235	Seaweed eaters
7.6×10^6 (a)	Curium-244	Seaweed eaters
7.8×10^6	Uranium-234	Seaweed eaters
7.8×10^6	Uranium-233	Seaweed eaters
9.1×10^6	Selenium-79	Seaweed eaters
1.2×10^7	Europium-154	Beach dwellers
1.3×10^7	Cobalt-60	Beach dwellers
1.5×10^7	Europium-152	Beach dwellers
2.0×10^7	Cesium-135	Fish eaters
2.3×10^7	Nickel-63	Fish eaters
2.3×10^7 (a)	Palladium-107	Seaweed eaters
4.0×10^7 (a)	Californium-252	Seaweed eaters
6.6×10^7	Strontium-90	Seaweed eaters
1.4×10^8	Antimony-125	Beach dwellers
1.2×10^8	Silver-110	Mollusc eaters
2.2×10^8	Cesium-137	Fish eaters

(a) Indicates that an estimated concentration factor was used in the most significant pathway.

To meet the present definition under the London Convention the radionuclides were initially grouped according to practical considerations and calculated release rate limits (Table IV). In some cases radionuclides do not appear in the group to which it would seem that they belong. This is because of known factors not included in the calculations or practical considerations such as the very low predicted quantities that will occur. The calculated release rate limits for these groups are given in orders of magnitude based on the more restrictive members of the group.

TABLE IV. Radionuclide Composition of Groups

Group A	Group B	Group C	Group D
Technetium-99	Carbon-14	Sodium-22	Helium-3
Tin-126	Lead-210	Chlorine-36	Phosphorus-32
Iodine-129	Polonium-210	Manganese-54	Sulfur-35
Radium-226	Thorium-229	Iron-55	Calcium-45
	Thorium-230	Cobalt-60	Chromium-51
	Thorium-232	Nickel-59	Iron-59
	Uranium-233	Nickel-63	Cobalt-58
	Uranium-234	Zinc-65	Bromine-82
	Uranium-235	Selenium-79	Strontium-89
	Uranium-238	Strontium-90	Yttrium-90
	Neptunium-237	Zirconium-93	Yttrium-91
	Plutonium-238	Niobium-93m	Zirconium-95
	Plutonium-239	Ruthenium-106	Niobium-95
	Plutonium-240	Palladium-107	Ruthenium-103
	Plutonium-241	Silver-110m	Antimony-124
	Plutonium-242	Antimony-125	Tellurium-125m
	Americium-241	Cesium-134	Iodine-131
	Americium-242	Cesium-135	Barium-140
	Americium-243	Cesium-137	Cerium-141
	Curium-242	Cerium-144	Gold-198
	Curium-243	Promethium-147	Radium-225
	Curium-244	Samarium-151	Actinium-225
	Curium-245	Europium-152	Thorium-234
	Curium-246	Europium-154	Protactinium-233
	Californium-251	Europium-155	Neptunium-239
	Californium-252		

For administrative convenience and analytical simplicity, Groups A and B were combined to give three groupings according to the basic properties of decay type and half-life, as follows:

Group	Release Rate Limits (Ci/yr)	
	Single Site	Finite Ocean Volume ($10^{17} m^3$)
α-emitters, but limited to 10^4 Ci/yr for ^{226}Ra and supported ^{210}Po	10^5	10^5
β/γ-emitters with half-lives of at least 0.5 yr (excluding tritium) and β/γ emitters of unknown half-lives	10^7	10^8
Tritium and β/γ emitters with half-lives of less than 0.5 years	10^{11}	10^{12}

The single-site release rate is more restrictive for short-lived radionuclides, and partitioning of wastes between sites can increase the overall limit for the basin as a whole. For long-lived radionuclides, the long-term finite ocean basin release rate is more restrictive and partitioning of wastes between sites does not affect the limit for the basin as a whole. However, the input of all radionuclides into the basin from all sources, including those from other than dumping of radioactive wastes, must be included in any definitive assessment of a release rate limit.

In all cases the release rate limits derived correspond directly, given the pathways and parameters used, to the ICRP dose limits for individual members of the public. The philosophy underlying this procedure and the use of critical groups is described in publications of ICRP. The annual limit for the effective dose equivalent in individual members of the public applies to the average of this quantity in the "critical group"; namely, the group representing the most exposed individuals. If the critical groups are hypothetical and maximizing assumptions are made in their selection, the ICRP maintains the value of 500 mrem for the annual limit. If, however, real critical groups are identified and realistic models are used to assess the annual effective dose equivalent, the ICRP recommends a limit of 100 mrem in a year for exposures of continuous natures expected year after year. It should be stressed that ICRP dose limits provide a lower boundary of an unacceptable range of values. Values above the ICRP limits are to be avoided while values up to the limit are not automatically permitted, however the values permitted must be justified by assessing the net benefits, considering radiological consequences and alternative procedures. It is anticipated that optimization procedures would usually result in radiation doses lower than the limits [8]. On the other hand the ICRP dose limits are not threshold values above which undesirable effects begin to appear, but represent dose values corresponding to individual risks approaching unacceptable levels. The maximum permissible annual intakes corresponding to those dose limits were taken from the IAEA Basic Safety Standards [9]. Where ingestion is involved following their transport through the water the values for soluble forms have been used. Where the pathway involves inhalation the most restrictive values have been used.

In the provisional Definition and Recommendations of 1974, two explicit safety factors of 10^2 were applied to allow for more than one dumping site and to allow for parameters less favorable than those assumed in the assessment. In the proposed revised Definition and Recommendations explicit account has been taken to account for multiple sites in a finite ocean volume and possible extreme events in ocean areas. It is not appropriate then to apply additional safety factors for the same reasons to the present assessment. The numerical values depend on the particular radionuclide and set of circumstances and can neither be determined precisely nor be guaranteed; however, it is considered that the release rates given are the best possible estimates which can be made for them at the present time.

An assessment of the potential effects on the biota of the marine ecosystem was conducted and it was concluded that radiation doses arising as a result of releases within the limits of the Definition are not expected to lead to significant adverse effects to populations as a whole.

The technical basis for the present radiological assessment is on release rate limits and not on dumping rates. However, to meet the present requirements of the London Convention it is necessary to express the Definition in terms of a concentration for a single site and an assumed upper limit on mass dumping rate at a single site of 100,000 tonnes/year with the added proviso of release rate limits for a finite ocean volume of 10^{17}m [3]. This results in concentration limits for a:

(a) 10 Ci/t for α-emitters but limited to 10^{-1} Ci/tonnes for ^{226}Ra and supported ^{210}Po;

(b) 10^2 Ci/tonnes for β/γ-emitters with half-lives of at least 0.5 years (excluding tritium) and mixtures of β/γ-emitters of unknown half-lives;

(c) 10^6 Ci/tonnes for tritium and β/γ-emitters with half-lives less than 0.5 years.

DEEP SEABED EMPLACEMENT

Since the potential hazards to man and the ecosystem are largely determined by the rates of release of radioactivity to the ocean, the present assessment provides the radiological basis for considering the deep oceans as an alternative ultimate repository for high-level radioactive wastes. If the release rates to the deep ocean waters can be controlled within these limits by suitable containment then there are no radioactive wastes that are intrinsically unsuitable for dumping or sub-seabed emplacement in the deep ocean.

One concept is being explored by the U.S. Department of Energy [10]. It is proposed that sub-seabed geologic formations may be able to contain these high-level wastes in isolation long enough for them to decay to inconsequential levels. The concept is based upon the premise that a set of sequential barriers could balance the rate of decay against the rate of migration to man and his ecosystem. These barriers would be the waste form itself, the containment canister, and the geological medium in which the material is placed. The major task is the selection and definition of the geological formations. These must have tectonic and climatic stability, predictable uniformity over a large area and have a low probability of future resource development. At present the abyssal hill areas appear to be the most promising. These areas are generally covered with 50 to 100 meters of red clay. Where they also occur below the centers of wind-driven surface current gyres they appear to be geologically stable and biologically relatively unproductive. The seabed sediments are considered to be the primary long-term barrier and existing transport models would suggest containment in the sediments for 10^6 to 10^{11} years depending upon physical and chemical characteristics of the sediments. An additional potential barrier is the deep ocean water. The major efforts in this study are an examination of the physical, chemical and mechanical properties of these ocean sediments, assessment of the problems of heat dissipation and the impact upon these properties, deep ocean oceanographic studies and the characterization of deep ocean biological communities. The current assessment of the engineering and environmental feasibility is that nothing has yet been discovered that discredits the concept. However, it will be necessary to intensify the establishment of many oceanographic and ecological parameters in order to develop definitive radiological assessments on a site by site basis.

While the Revised Definition and Recommendations of the IAEA restrict the dumping of radioactive wastes that exceed specified concentration/mass limits, the acceptance of the concept of applying release rate limits as developed by the IAEA provides a rational basis for further considering the emplacement of radioactive wastes in the seabed as an attractive and acceptable alternative to terrestrial geological repositories.

ACKNOWLEDGMENTS

This work was supported by the United States Department of Energy under Contract EY-76-C-06-1830.

REFERENCES

1 International Atomic Energy Agency: Convention on the Prevention of Marine Pollution by Dumping of Wastes and Other Matter, INFCIRC/205, IAEA, Vienna 1974.

2 International Atomic Energy Agency: Convention on the Prevention of Marine Pollution by Dumping of Wastes and Other Matter, The Definition Required by Annex 1, paragraph 6, to the Convention, and the Recommendation Required by Annex II, Section D, INFCIRC/205/Add 1, IAEA, Vienna 1975.

3 International Atomic Energy Agency: Convention on the Prevention of Marine Pollution by Dumping Wastes and Other Matter, The Definition Required by Annex 1, paragraph 6, to the Convention, and the Recommendation Required by Annex II, Section D, INFCIRC/205/Add 1/Rev 1, IAEA, Vienna 1978.

4 International Atomic Energy Agency: The Oceanographic Basis of the IAEA Revised Definition and Recommendations Concerning High-Level Radioactive Waste Unsuitable for Dumping at Sea, Tech. Doc-210, IAEA, Vienna 1978.

5 International Atomic Energy Agency: The Radiological Basis of the IAEA Revised Definition and Recommendations Concerning High-Level Radioactive Waste Unsuitable for Dumping at Sea, Tech. Doc-211, IAEA, Vienna 1978.

6   Shepherd, J. G.: A Simple Model for the Dispersion of Radioactive Wastes Dumped on the Deep Seabed, Fisheries Research Technical Report No. 29, Ministry of Agriculture, Food and Fisheries U.K., H.M.S.O., London 1976.

7   International Commission on Radiological Protection: Principles of Environmental Monitoring Related to the Handling of Radioactive Materials, A Report by Committee 4, ICRP Publication 7, Pergamon Press, 1965.

8   International Commission on Radiological Protection: Recommendations on the International Commission on Radiological Protection, Publication 26, Pergamon Press, 1977.

9   International Atomic Energy Agency: Basic Safety Standards for Radiation Protection, Safety Series No. 9, STI/PUB/147, IAEA, Vienna 1967.

10 Anderson, D. R., W. P. Bishop, V. T. Bowen, J. P. Brannen, W. N. Caudle, R. J. Detry, T. E. Ewart, E. E. Haynes, G. R. Health, R. R. Hessler, C. D. Hollister, K. Keill, J. A. McGowan, R. W. Rohde, W. P. Schimmel, C. L. Schuster, A. J. Silva, W. H. Smyrl, B. A. Taft, D. M. Talbert: Release Pathways for Deep Seabed Disposal of Radioactive Wastes", Proc. Symp. on Impacts of Nuclear Releases into the Aquatic Environment, pp. 483-502, IAEA, Vienna, 1971.

Discussion

Y. MIYAKE, Japan

In order to consider the effects of dumping radioactive wastes into the sea, we also have to take into account the radioactive contamination originating from nuclear weapons testing.

W.L. TEMPLETON, United States

The IAEA model assumes that the calculated release rates represent the total environmental capacity of the ocean volume ($10^{17} m^3$). Therefore any definitive assessment of the release rate limit must include all inputs of all radionuclides into the ocean basin from all sources. These include weapon test fallout, controlled coastal discharges of radioactive wastes, and those inputs other than the controlled dumping of radioactive wastes.

M. ISHIKAWA, Japan

What are the recent trends of countries participating in your Organization (OECD) with regard to seabed dumping ?

What are the opinions of the countries which are not bordered by any sea coast ?

What are the views of the under-developing countries ; and also of the communist countries ; especially the USSR ?

W.L. TEMPLETON, United States

The revised definition of defining radioactive wastes unsuitable for dumping at sea, and the provision of recommendations to ensure that any dumping of radioactive wastes into the oceans does not involve any unacceptable degree of risk to man and the marine resources has been submitted for ratification to the signatories of the London Convention. These include member countries covered by your four categories.

My personal opinion is that those involved at the scientific and technical level must submit their unbiased findings, assessments and recommendations to these international bodies. Whether the political decision makers accept or reject these is an entirely different matter, which I am not qualified to address.

At the international level I have always been impressed that despite the policy considerations of their own country, individuals always give their views based only on the scientific merits of the problem, rather than based upon their particular domestic problems.

BIOLOGICAL STUDIES OF THE U.S. SUBSEABED DISPOSAL PROGRAM

Subseabed Biology Team: (L.S. Gomez[1], R.R. Hessler[2], D.W. Jackson[3], M.G. Marietta[1], K.L. Smith, Jr.[2], D.M. Talbert, A.A. Yayanos[2].)

[1]Sandia Laboratories
Albuquerque, N.M. 87185

[2]University of California
Scripps Institution of Oceanography
La Jolla, CA 92093

[3]Falcon Research and Development
2350 Alamo Avenue S.E.
Albuquerque, N.M. 87106

ABSTRACT

The Subseabed Disposal Program (SDP) of the U.S. is assessing the feasibility of emplacing high level radioactive wastes (HLW) within deep-sea sediments and is developing the means for assessing the feasibility of the disposal practices of other nations. This paper discusses the role and status of biological research in the SDP. Studies of the disposal methods and of the conceived barriers (canister, waste form and sediment) suggest that biological knowledge will be principally needed to address the impact of accidental releases of radionuclides. Current experimental work is focusing on the deep-sea ecosystem to determine: (1) the structure of benthic communities, including their microbial component; (2) the faunal composition of deep midwater nekton; (3) the biology of deep-sea amphipods; (4) benthic community metabolism; (5) the rates of bacterial processes; (6) the metabolism of deep-sea animals, and (7) the radiation sensitivity of deep-sea organisms. A multi-compartment model is being developed to assess quantitatively, the impact (on the environment and on man) of releases of radionuclides into the sea.

1. INTRODUCTION

On a world-wide scale the nuclear fuel cycle is producing an ever increasing amount of waste. A recent review (McElroy and Burns, [1]) describes concisely the several options being investigated in the U.S. for the disposal of high level radioactive wastes (HLW). The deep sea is among the loci being considered as potentially suitable. Anderson [2] has written a current report on the U.S. Subseabed Disposal Program (SDP) which has two overall objectives: (1) to determine the feasibility of emplacing HLW within selected deep-sea sediments; and (2) to develop a capability for assessing the feasibility of the practices of other nations disposing radioactive wastes into the marine environment. The research and development goals at the operational level of the SDP are to achieve or analyze: site selection, multibarrier description, emplacement processes, transportation considerations, societal-political aspects, risk, economics and environmental impact. The biological aspects of the research, the subject of this paper, are relevant to most of these goals.

The disposal of appropriately packaged and solidified HLW into the seabed would differ from many of the other geological options being evaluated in that the marine environment could be affected. In addition, other geologic disposal methods might also require transportation by ship if one nation were to use the land disposal site or reprocessing facility of another nation. A ship carrying HLW destined for disposal on land or into the seabed could accidently be lost at sea. Another kind of accident could occur if there were a failure to emplace packaged HLW into the sediments. Once the waste is successfully buried, the canister and waste form should contain the radionuclides for hundreds of years. During this period the ecosystem should be minimally affected. After the canister and waste form corrode and disintegrate, then radionuclides will interact with sediments. These interactions should result in virtually all of the radionuclides being sorbed onto the sediments (Duursma and Gross,[3]; Heath, et al., [4]). But we do not yet know if all radionuclides and their chemical forms will be sorbed and processes yet to be fully evaluated may result in some release to the benthic boundary layer where interaction with deep-sea life would occur. Clearly, the evaluation of accidental releases make biological research necessary. Finally, biological research may allow for the design of sensitive monitoring programs (Schaefer, [5].)

Radioactivity of man-made origin has been entering the marine environment by several mechanisms for decades. The continuing study of the pathways of these radionuclides provides much of the necessary information needed for assessing the impact of a deep-sea disposal plan. Low level wastes have been disposed by the U.S. (Ocean Dumping, [6]) and are being disposed by a consortium of nations (European Nuclear Energy Agency, [7]) onto the sea floor. Some of the low level disposal sites in the deep sea have been studied (e.g., Noshkin, et al., [8]). Nevertheless, the bulk of our knowledge of the pathways to man and of the impact of radioactivity in the marine environment is derived mainly from shallow water situations such as the Bikini tests affecting the Pacific Ocean (e.g., Japan Society for the Promotion of Science, [9]), the Windscale effluents into the Irish Sea (e.g., Preston, [10]), and the Hanford discharges into the Pacific Ocean via the Columbia River (e.g., Pearcy, et al., [11]). The point is that the radioecological data base is most deficient in the deep sea. Indeed, biological knowledge in general about the deep sea is lagging behind that of shallow waters. In part this situation is due to the virtual virgin nature of many deep-sea resources.

Research to achieve management of them has thus been unnecessary. The burgeoning world's population will result in closer scrutiny of the food resources of the deep sea; the need for minerals is already causing examination of the deep-sea floor; waste disposal (some radioactive) into the marine environment is already a reality-- even if not yet for HLW. There is a desire to use a given resource of the deep sea without affecting the other ones and consequently research in deep-sea biology is beginning to be stimulated world wide.

On account of this dearth of basic biological knowledge of the deep sea, much of the research in the SDP to date has had mainly to do with deep-sea ecology. The radioecology of the deep sea will be more directly addressed this coming year because of the successes in basic deep-sea ecology over the past few years. What follows is a description of the highlights of the biological aspects of the ongoing U.S. SDP program and an indication of the directions it will be taking over the next year.

Workshops internal to the Subseabed Disposal Program as well as international workshops help to identify the important problems of HLW in the deep sea. There is a Biology Task Group which is one of seven Task Groups of the Seabed Working Group of the NEA-Radioactive Waste Management Committee. The Seabed Working Group meetings provide an international forum on all aspects of HLW in the deep sea by identifying cooperative research efforts, providing information exchange and evaluating progress.

Finally, it should be noted that the SDP has a schedule aiming to provide a demonstration of the capability to emplace canisters by approximately 1990-95. A more immediate milestone is to show the technical and environmental feasibility of the SDP concept by the end of 1983.

2. DEVELOPMENT OF AN ECOSYSTEM MODEL

There are a large number of biological and physical events whose interaction may transport radionuclides released inadvertently into the environment. Food for the organisms of the sea is synthesized by phytoplankton in surface waters. This is the beginning of the nourishment of marine organisms ranging in size from bacteria to the largest of animals, blue whales. Feeding and excretion transfer material through the fabric of the seas. Migrations of the animals themselves are also important. Certain marine mammals and birds exhibit annual horizontal migrations of thousands of miles. Salmon, shad and other fishes have dramatic anadromous migrations. Many commercial fishes reproduce with pelagic eggs transported horizontally for tens of miles. (Russell, [12]). Vertical migrations in the water column are no less profound, ranging from a diel pattern (Longhurst, [13]) over hundreds of meters to the ontogenetic migration of rattails (Merrett, [14]) that can extend over a few thousand meters. Some benthic invertebrates have pelagic eggs or larvae that may disperse horizontally. These and other biological interactions are influenced by an assortment of physical processes such as currents, eddies, upwellings, and thermal structure.

A mathematical model which describes the migration of radionuclides by the various transport mechanisms (physical as well as biological) of the sea is being developed in order to quantify the impact due to any accidental release of waste material. This model is directed towards answering two questions. What is the effect upon the marine environment, and what is the effect upon man?

The goal is to estimate the levels of radioactivity in marine organisms and the dose to man.

The model is made up of the following modules: near field transport in sediments with heat effects; far-field isothermal transport in the sediments; physical and biological transport; human dosimetry; affected human populations; and, finally human health effects. The resultant human health effects will answer the question concerning the effect on man. The concentrations of radioactivity predicted in the physical and biological transport module will provide part of the answer for the question concerning ecosystem effects. Knowledge of how a given concentration of radioactivity will affect particular organisms will be gained through an investigation of the radiosensitivity of deep-sea organisms. Predictions of effects on the ecosystem inferred from the studies on individuals should be confirmed experimentally..

The oceanic ecosystem has been compartmentalized in a manner begun by Wishner ([15]) by grouping organisms according to their function. The species within a compartment are assumed to behave similarly both in their relationships with the environment and in the way that they pass along radionuclides. Group bounds on the standing crop and metabolic rates may then be assigned to each compartment by observing the constituents of the group. Further, the module for the water column and biological transport has been separated into sub-models reflecting the different mixing layers of the water column. The benthic boundary layer part of the model extends from the lower limit of biological activity in the sediments and extends into the bottom 20-50 meters of the water column. Above the boundary layer is a multi-layered model.

The mass transports between biological compartments and physical compartments are most complex. Besides predator-prey relationships, the radionuclides may be transported by excretion, mortality, adsorption, migration (feeding, breeding and ontogenetic) and molting. In addition, both organic and inorganic sediments and/or particulate materials may be consumed. Radionuclides may also be taken up directly from the water column or from the interstitial water of the sediments. There are various physical transports to be considered such as association-disassociation in the sediments and water column for both organic and inorganic matter, erosion-sedimentation, vertical and horizontal water dispersion, bioturbation, horizontal migration of benthic megafauna, vertical migration of mid-water animals, and the dispersion of host particulate matter. Each of these radionuclide transports between the compartments are being examined.

Modelers and experimentalists of the SDP meet bimonthly to develop the proper mathematical expressions for the biological transports in the model. While a deep-sea data base is being obtained, it is being supplemented with shallow water data in order to exercise the model. In addition, some algorithms for computing the transfer rates have been adopted.

3. BIOLOGY OF NEAR-FIELD AND FAR-FIELD REGIONS

In the SDP concept, HLW would be buried in sediments 30 to 100 meters beneath the sediment water interface. The near-field region is that immediately surrounding the canister; the far-field region extends to a spherical surface whose radius comes up to the zone of active animals in the sediments. The biological questions in these regions are mainly microbiological. These along with chemical and physical processes (see articles in Krumbein, [16] ; Gieskes, [17])

are involved in the diagenetic events in sediments. The need for
a greater understanding of the microbiological processes is widely
recognized. Are there metabolically active or potentially
metabolically active microbes at these sediment depths? Can
microbiological activities affect (adversely or favorably) the
canister, the waste form, or the sediment-radionuclide interactions
at 30 to 100 meter sediment depths? We would not expect
organisms to survive in the near-field region or in the adjoining
parts of the far-field regions because of thermal and radiation
effects.

About three years ago a core about 24.5 m in length was
taken (Hollister, [18]) with a giant piston corer and was sampled
for microbes (Yayanos and Van Boxtel, [19]). The samples were
preserved by several methods (freezing; freeze drying and chemical
fixation). Organic carbon content ranged from 0.06% to 0.15%
(w/w). ATP (adenosine triphosphate--a component of living but not
of dead cells) was present at depths shallower than ten meters
and in a single sample at about a 24 m depth. This could have
been due to a contaminant. The use of ATP is a tenuous method
for determining the presence of cells. In using it to detect
low cellular concentrations, for example, contamination
from gear and the experimenter can be a difficulty. These
things could be minimized. Yet the chief source of uncertainty
would remain due to the length of time it takes to
retrieve and dismantle a large corer to allow sampling. ATP content
of cells would most likely change during such delays. We plan
to analyze some of the other samples (frozen and freeze dried)
using microscopy this coming year. In order to achieve an analysis
of many microscopic fields of view we will use computer assisted
image analysis. The variety of methods of sample preparation to
be investigated includes fluorescent labelling for microbes.

Cultivation of microbes is one of the most sensitive ways
to detect their presence. Because of the well-known uncertainty
in being able to cultivate microbes, the above indirect methods
of microbial detection are being explored. We have endeavored,
however, to find cultivation conditions for deep-sea microbes.
Rather than tackle the cultivation problem for the microbes of
deep sediments first -- even though these might be of most
interest during the first few hundred years of emplacement -- we
have chosen to solve cultivation difficulties using microbes
whose existence is quite apparent -- those that participate in
putrefaction. Our success has been encouraging (Dietz and Yayanos,
[20]; Yayanos, Dietz and Van Boxtel, [21]) with the isolation of
genuine deep-sea microorganisms. We have many isolates but have
only studied one in detail. This isolate (designated CNPT-3,
and originating in the central North Pacific Ocean) grows rapidly
(a generation time of about 9 hours at deep-sea pressures and
temperatures). We are studying its growth requirements to find
out which of the many cultivation variables are significant.
CNPT-3 dies when decompressed at increased temperatures (above 15°C).
We thus imagine that cultivation of microbes from core samples
will necessitate that sediments be kept cold during retrieval.
Decompression during recovery of samples from the deep sea does
not seem to harm them. Microbes sustain decompressions
(growth is often stopped or inhibited) but warming is intolerable
(they die at about 15 to 20°C). Once the cold sample is on board
ship, we attempt the cultivations both at low temperatures and at
high pressures. The brief decompression of cultures to achieve
the transfer of inocula has been found to have no detectable adverse
effects. Cultivation at atmospheric pressure has been avoided to
eliminate or reduce the chance of contamination of cultures with
shallow water microbes which continuously enter the deep sea. We
hope to use this procedure in the near future to isolate deep-
sediment bacteria.

Spores of thermophilic (that is prefering high temperatures) bacteria have been found in sediments (Bartholemew and Paik, [22]). The ability of these spores to germinate in the presence of heat from HLW needs to be determined. Evidence to date with the bacterium CNPT-3 and a few other isolates suggests that the deep-sea microbial population will stop functioning above 15°C at 580 bars. Thus, in the 100° C regions of the waste, only spores of thermophilic bacteria could have an effect.

4. BENTHIC BOUNDARY LAYER STUDIES

If and when the emplaced canister and waste form disintegrate, radionuclides may interact with sediments so that some substantial fraction or possibly all could be sorbed (Duursma and Gross, [3] ; Heath, et al., [4]). An accident, however, might place radionuclides into the upper portion of the overlying sediment where microorganisms and macrobiota live. We will need to understand the composition, structure, function and dynamic processes of this benthic community and how it interacts with radionuclides. These kinds of data would help determine if recoverability, which is possible, would be necessary.

4.1 Community Structure

A box corer technique for retrieving a reasonably undisturbed portion (0.25 m^2) of sediment and its community is described by Hessler and Jumars [23]. Analyses of the contents of such cores inform us of the composition and three dimensional distribution of the organisms living in the sediments. Hessler and Jumars [23,24] reported the analyses of twelve box cores from the central North Pacific Ocean -- a mid-plate, mid-gyre region (Hollister, [18]) of the kind where HLW might be emplaced. Only a few of the many results from this type of study will be noted here.

The analyses show the area to be oligotrophic (that is, with an apparently small food supply). There were only 84 to 160 individual macrofauna per m^2 of sediment. The species present form a diverse set but any given species is rare. Many of the possible taxa are not represented in the samples and of the 108 species, 67 were found only once. Polychaetes and tanaidaceans together accounted for 75% of the individuals with the polychaetes representing 55%. A substantial number of polychaetes were found to have sediment in their guts. Although foraminifera and nematodes were not sampled adequately, there was evidence of a substantial population of these meiofauna (animals which pass through a 1 mm mesh size screen). Subsequent work by Jumars [25] shows, among other things, that sediment-ingesting polychaetes stay mainly in the upper 1 cm of the sediment but occasionally a few individuals will extend ten to twenty centimeters into the sediment. Bernstein et al.[26] and Bernstein and Meador [27] have observed that there is an order of magnitude more foraminifera than macrobiota and that foraminifera have a patchy distribution. Because of their abundance in the samples, the foraminiferan species composition and distributional patterns reveal a patch structure and a persistence of this structure over more than one generation. Details such as this one of community structure may allow for the detection of changes in deep-sea community disturbed by an abnormal level of radionuclides. Burnett [28] found that nematodes do not penetrate much beyond 5 to 10 mm into the sediment and that they too are abundant as suggested by Hessler and Jumars [23] . The determination of the food web of all of the infauna, of their elemental composition, and of their metabolic rate should greatly augment our capability to predict important radionuclide transport pathways. Of course the locomotory and feeding activities of the infauna move both sediments and pore water -- a process refered to as bioturbation.

The above discussion has dealt with those organisms whose existence is tied to the sediments in an intimate manner. The visible effects of their activities can be seen in the pictures in Heezen and Hollister [29]. Photography with baited cameras (Isaacs, [30]; Hessler, et al. [31]) has revealed the presence of large mobile apparently epibenthic animals. Liparids, rattails, shrimp, amphipods and other animals have been photographed at abyssal depths in the vicinity of bait. There appears to be some interest in the study of rattails (macrourids) as evidenced by the study of McLellan [32] on their feeding and the study of Merrett [33] on the distribution of their larval and juvenile stages. A sufficient data base for understanding the role of rattails in mobilizing radionuclides to upper portions of the water column would include a knowledge of the size of the population, its age structure, fecundity, their migratory behavior, their elemental composition, and their radiosensitivity.

We have chosen to pursue at this time the study of one component of this mobile epibenthic group of animals -- amphipods. It is practically an axiom of deep-sea biology that these animals will always come to bait on the sea floor (Hessler, et al., [34]). Since they can be caught, many aspects of their population structure can be learned. The age structure of a catch, for example, is mirrored by a plot of the frequency of a given size class versus the size class of the animals. This is a consequence of their growth by steps necessitated by molting. An abnormal event in the population can lead to a distinctive cohort. If this could be followed by sampling a population over time, then a good approximation of age would result.

Studies are also in progress to determine the size of the populations using tag-recapture data. The methodology for tagging amphipods has been developed. We are now dealing with the fielding of tag-recapture experiments at 5700 m depths and have learned that labeled animals can be recovered.

Among the several species of amphipods (Hessler and Ingram, [34]), Eurythenes gryllus appears to be the most ubiquitous and to have the greatest range in its distribution having been trapped at least 500 m above the sea floor. Their mobility and ingestion of sediments suggests that they could disperse radionuclides.

Another method for dispersion of matter vertically may exist because of the feeding manner of amphipods and because many lipids are less dense than water. Amphipods are scavengers and do not consume necessarily their entire food source. It is likely that lipid particles are often released and float to the sea surface. This may not be a major pathway in a normal ocean. A poor epi-seabed disposal practice may result in an abundance of dead animals on the sea floor. The resultant attraction of mobile scavengers may result in the movement of lipid-containing particles (and any bound radionuclides) rapidly and directly to the sea surface (Yayanos and Nevenzel, [35]).

4.2 Community and organismal processes

One way to gauge the function of the deep-sea benthic community is to determine its metabolism in a fashion which disturbes it as little as possible. An instrument, a free vehicle grab respirometer (FVGR), has been designed and constructed (Smith, et al., [36]) to achieve the capability of measuring metabolism *in situ*. The instrument is deployed with ballast. On the sea floor, it will respond to commands from the ship. It can gently place 4 boxes (grabs) over the sediments trapping water and undisturbed sediment. Electrodes detect the utilization of oxygen by the entrapped organisms. Syringes can sample or inject substances into the entrapped water. On command, the grabs are closed retaining the sediment and water on which measurements were made. A final command releases the ballast and the FVGR is recovered. The water samples in the syringes are analyzed chemically. The sediments from the grabs are analyzed for biomass and faunal composition.

Measurements with this instrument (Smith, [37]) revealed that benthic community respiration diminished by three orders of magnitude with depth along the Gay Head Bermuda transect of the Atlantic Ocean. Two stations vertically spaced by 2600 m in the Pacific Ocean, however, showed little difference in community respiration. This similarity was hypothesized to be due to the enriching effects of the California Current system.

The FVGR is a useful tool for doing in situ experiments with radionuclides. Already a tracer experiment using a neutron-activatable isotope has been fielded.

We are also studying methods for capturing and retrieving deep-sea animals in a live condition (Yayanos, [38]). If these can be maintained in the laboratory, their behavior, nutrition, and physiology can be studied in detail. To date, we have been successful in capturing animals (amphipods) at a 5900 m depth and keeping them at 590 atm and 2°C for nearly 17 days. If maintenance of these animals can be extended indefinitely, then they should be valuable for studies of radiosensitivity.

5. ABYSSOPELAGIC FAUNA

The depths considered for the disposal of radioactive wastes are about 5,000 m and are known as the abyssal depths of the sea. The extensive amount of water defined by about ten to a few hundred meters above this sea floor and below the 4°C isotherm is known as the abyssopelagic zone. This is a poorly understood volume of the oceans because of the difficulty in sampling mobile animals that might live there. In an effort to determine what lives there, new methods of sampling and catching organisms are being developed.

Fisheries that are commercially important at the present time are above this zone. Radionuclides released from a hypothetical deep-sea disposal site would have to penetrate this zone in order to reach man's sea food. One of the mechanisms by which this could happen is by the trophic relationships among the animals in this zone which could link the benthic ecosystem to the shallow water one. But we really do not yet know what lives in this abyssopelagic zone.

We have sampled this zone with a few new approaches. A free vehicle midwater net system consisting of gill nets, baited traps and baited hooks has been developed. The gill nets are 30.5 m by 3.3 m with a 7.6cm mesh. The system of nets, hooks and traps has been deployed extensively. Examples of species taken with it are: the fish Coryphaenoides armatus (up to 685 m off of the sea floor); the fish Nomoctes alvifrons (730 m above the sea floor ; and the amphipod Eurythenes gryllus (400 m above the sea floor) (Smith et al. [39]. These are widely regarded as benthic animals. We do not yet know to what extent these animals penetrate the abyssopelagic realm or for what reason they would leave the sea floor and enter the presumably relatively food-poor deep midwater regions of the sea. This evidence for life in this zone reinforces the notion that there exists a mechanism for actively transporting matter from the sea floor towards the sea surface. Whether the transport is all of the way to the upper few hundred meters of the sea remains to be shown.

Now that we have evidence that there are, indeed, large mobile organisms living in the abyssopelagic region, we are developing another type of net to capture them. This net has a 100m diameter and 200m depth. It is deployed in a folded configuration with ballast. Upon reaching the sea floor, the ballast is released, the net rises slowly towards the sea surface, and samples a 100m diameter column of the ocean for nearly the full depth of the ocean. The slow ascent, large diameter and vertical motion should all combine to obviate the net avoidance problem that is attendant with towed nets.

6. RADIOSENSITIVITY OF DEEP-SEA ORGANISMS AND ECOSYSTEMS.

As far as we know, the radiosensitivity of a deep-sea organism has not yet been determined. But there is a body of research pertaining to the radiosensitivity of shallow water organisms and ecosystems (Templeton, et al., [40]; Blaylock and Trabalka, [41]). There are many problems in dealing with radiosensitivity: determining the dose to the biota; defining and measuring a meaningful biological endpoint -- this is difficult in experiments both with single organisms as well as with ecosystems; assessing genetic effects; dealing with low-level doses; and, assessing many other biological, radiation and synergistic aspects.

In an effort to gain data on the radiosensitivity of deep-sea organisms we will do the following: (1) determine the radiation sensitivity of a few bacteria since these can be cultivated at deep-sea conditions in the laboratory; (2) determine aspects of the radiation sensitivity of deep-sea amphipods since these can be recovered in a living state and maintained in the laboratory for a few weeks (hopefully this can be extended); (3) attempt to identify situations peculiar to the deep sea that might exhibit an enhanced radiosensitivity. For example the pelagic eggs of rattail fish may be radiosensitive because of the vertical ontogenetic migration they undergo. This stresses them with decompressions of several hundred atmospheres and possibly with temperature changes if the eggs penetrate the thermocline. The adult rattails may be radiosensitive if their swimbladder tissues contain oxygen which is known to greatly enhance radiosensitivity.

This kind of information on radiosensitivity in the deep sea will allow an estimate of the appropriateness of using the data on shallow water organisms in assessing the impact of accidents on individual organisms in the deep sea. A determination of the effects on a deep-sea community may ultimately require an in situ experiment.

REFERENCES

[1] McElroy, J.L. and Burns, R.E.: "Nuclear Waste Management Status and Recent Accomplishments", Report NP-1087, Electric Power Research Institute, Palo Alto, California (1979).

[2] Anderson, D.R.: "Nuclear Waste Disposal in Subseabed Geologic Formations: the Seabed Disposal Program" SAND78-2211, National Technical Information Service, Springfield, Virginia, (1979).

[3] Duursma, E.K. and Gross, M.G.: "Marine Sediments and Radioactivity", U.S. National Academy Sciences, Washington, D.C. (1971)

[4] Heath, R., Epstein, G.P., Leinen, M. and Prince, R.A.: "Geotechnical and Sedimentological Assessment of Deep Sea Sediments" in SAND78-1359, National Technical Information Service, Washington, D.C. (1979). p. 33-55.

[5] Schaefer, M.D.: "New Research Required in Support of Radioactive Waste Disposal" in Disposal of Radioactive Wastes, IAEA, Vienna (1960)

[6] "Ocean Dumping", Council on Environmental Quality, Superintendent of Documents, U.S. Government Printing Office, Washington, D.C. (1970)

[7] European Nuclear Energy Agency: "Radioactive Waste Disposal Operation into the Atlantic, 1967" ENEA/OECD, Paris (1968) 74pp.

[8] Noshkin, V.E., Wong, K.M., Jokela, T.A., Eagle, R.J. and Brunk, J.L.: "Radionuclides in the Marine Environment Near the Farallon Islands", Report UCRL-52381, University of California, Livermore, California (1978) 17 pp.

[9] Japan Society for the Promotion of Science: "Research in the Effects and Influences of the Nuclear Bomb Test Explosions" Vol. I and II., Ueno, Tokyo (1956) 1824 pp.

[10] Preston, A.: "The radiological consequences of release from nuclear facilities to the aquatic environment " in Impacts of Nuclear Releases into the Aquatic Environment, IAEA, Vienna (1975), pp3-23.

[11] Pearcy, W.G., Krygier, E.E. and Cutshall, N.H.: "Biological transport of zinc-65 into the deep sea" Limnology and Oceanography 22, 846-855 (1977)

[12] Russell, F.S.: "The Eggs and Planktonic Stages of British Marine Fishes" Academic Press, N.Y. (1976) 524 pp.

[13] Longhurst, A.R.: "Vertical Migration" in The Ecology of the Seas (Cushing and Walsh, Editors) W.B. Saunders Co., Philadelphia (1976) p 116-140.

[14] Merrett, N.R.: "On the identity and pelagic occurrence of larval and juvenile stages of rattail fishes (Family Macrouridae) from 60°N, 20°W and 53°N, 20°W", Deep-Sea Research 25, 147-160 (1978)

[15] Wishner, K.: "An Ocean Ecosystem Model" in SAND77-1270, National Technical Information Service, Springfield, Virginia (1977) pp.356-364.

[16] Krumbein, W.E. (Editor): "Environmental Biogeochemistry and Geomicrobiology" Vol. 1, 2, 3, Ann Arbor Science, Ann Arbor, Michigan (1978)

[17] Gieskes, J.M.: "Chemistry of interstitial waters of marine sediments" Annual Review of Earth and Planetary Sciences 3, 433-453 (1975)

[18] Hollister, C.D.: "Annual Report for Sandia Seabed Program" in Report SAND77-1270, National Technical Information Service, Springfield, Virginia (1977) p25-38.

[19] Yayanos, A.A. and Van Boxtel, R.: "Progress Report: Determination of the Microbial Content of Ocean Sediments" in ibid. p. 281-288.

[20] Dietz, A.S. and Yayanos, A.A.: "Silica gel method for isolating and studying bacteria under hydrostatic pressure", Applied and Environmental Microbiology 36 , 966-68 (1978)

[21] Yayanos, A.A., Dietz, A.S. and Van Boxtel, R.: "Isolation of a deep-sea barophilic bacterium and some of its growth characteristics", Science 205 808-810 (1979)

[22] Bartholemew, J.W. and Paik, G.: "Isolation and identification of obligate thermophilic spore-forming bacilli from ocean basin cores" Journal of Bacteriology 92 635-638 (1966)

[23] Hessler, R.R., and Jumars, P.A.: "Abyssal community analysis from replicate box cores in the central North Pacific", Deep-Sea Research 21, 185-209 (1974)

[24] Hessler, R.R., and Jumars,P.A. :"Abyssal Communities and Radioactive Waste Disposal", Oceanus 20, 41-46 (1977)

[25] Jumars, P.A. : "Spatial autocorrelation with RUM (Remote Underwater Manipulator): vertical and horizontal structure of a bathyal benthic community", Deep-Sea Research 25, 589-604 (1978)

[26] Bernstein, B.B., Smith, R., Jumars, P.A. : "Spatial dispersion of benthic Foraminifera in the abyssal central North Pacific", Limnology and Oceanography 23, 401-416 (1978)

[27] Bernstein, B.B. and Meador, J.P. : "Temporal Persistence of Biological Patch Structure in an Abyssal Benthic Community", Marine Biology 51, 179-183 (1979)

[28] Burnett, B.R.: "Benthic Biological Studies: Microbiota and Meiofauna" in Seabed Disposal Program Annual Report, 1978. National Technical Information Service, Springfield, Virginia (1979)

[29] Heezen, B.C. and Hollister, C.D.: "The Face of the Deep", Oxford University Press, New York, 659 pp. (1971)

[30] Isaacs, J.D.: "The nature of oceanic life", Scientific American, 221, 146-162 (1969)

[31] Hessler, R.R., Isaacs, J.D, and Mills, E.L.: "Giant amphipod from the abyssal Pacific Ocean", Science 175, 636-637 (1972)

[32]   McLellan, T.: "Feeding strategies of the macrourids",
       Deep-Sea Research 24, 1019-1036 (1977)

[33]   Hessler, R.R., Ingram, C.L., Yayanos, A.A., and Burnett,B.R.:
       "Scavenging amphipods from the floor of the Philippine Trench"
       Deep-Sea Research 25, 1029-1047 (1978)

[34]   Hessler, R.R. and Ingram, C.L. :"Benthic Biological Studies"
       in Seabed Disposal Program Annual Report, 1978, National
       Technical Information Service, Springfield, Virginia (1979)

[35]   Yayanos, A.A. and Nevenzel, J.C. : "Rising-Particle Hypothesis:
       Rapid Ascent of Matter from the Deep Ocean",
       Naturwissenschaften 65   (1978)

[36]   Smith, K.L., White, G.A., and Laver, M.B.: "Oxygen uptake
       and nutrient exchange of sediments measured in situ
       using a free vehicle grab respirometer", Deep-Sea Research, 26,
       337-346 (1979)

[37]   Smith, K.L., Jr.: "Benthic Community Respiration in the
       N.W. Atlantic Ocean: in situ Measurements from 40 to 5200 m",
       Marine Biology 47, 337-347 (1978)

[38]   Yayanos, A.A.: "Recovery and Maintenance of Live Amphipods
       at a Pressure of 580 Bars from an Ocean Depth of 5700 Meters",
       Science 200, 1056-1059 (1978)

[39]   Smith, K.L. Jr., White, G.A., Laver, M.B., McConnaughey, R.R.,
       and Meador, J.P. : "Free vehicle capture of abyssopelagic
       animals", Deep-Sea Research 26, 57-64 (1979)

[40]   Templeton, W.L., Nakatani, R.E., Held, E.E.: "Radiation
       Effects" in Radioactivity in the Marine Environment
       National Academy of Sciences, Washington, D.C.
       (1971)

[41]   Blaylock, B.G. and Trabalka, J.R.: "Evaluating the effects
       of ionizing radiation on aquatic organisms", Advances in
       Radiation Biology 7, 103-152 (1978)

CONTRIBUTION AU CONTRÔLE RADIOLOGIQUE DU MILIEU MARIN

Ortins de Bettencourt, A., Vaz Carreiro, M.C. et Sequeira, M.M.
L.N.E.T.I., Département de Protection et Sureté Radiologique
Sacavém, Portugal

RÉSUMÉ

On discute la nécéssité et la validité d'effectuer un contrôle radiologique du milieu marin en rapport avec les opérations d'immersion des déchets radioactifs, ayant essentiellement en vue l'établissement du fond radiologique et l'observation des tendances d'évolution de la radioactivité.

On présente quelques résultats de l'analyse d'échantillons d'eau, de sédiments et de poissons prélevés surtout au long des côtes portugaises, dont le poisson sabre de Madère (Aphanopus carbo) qui vit à 1800-2200 m de profondeur.

ABSTRACT

We discuss the need and usefulness of a radiological surveillance program of the marine environment in relation with radioactive waste dumping, essentially with the aim of establishing a radiological baseline and observe the radioactivity trends.

We present some results of water, sediments and fish samples collected namely along portuguese coasts, in particular of the scabbard fish (Aphanopus carbo), living at depths about 1800-2200 m, near Madeira island.

1. INTRODUCTION

En rapport avec les opérations d'immersion de déchets radioactifs en mer, il est recommandé de "contrôler, pour autant qu'il est possible et utile de le faire les conditions de la mer ... au voisinage des lieux d'immersion". /1_/.

Le sens de ces mots "possible" et "utile" a déjà été longuement discuté dans différentes réunions sur ce sujet. Cette difficulté d'interprétation est en rapport avec celle soulevée par la recommandation de l'ICRP "aussi bas que raisonnablement possible, ayant en compte les aspects économiques et sociaux". Il nous semble, en effet, que selon certains points de vue on envisage surtout les avantages économiques de l'immersion, tandis que sous d'autres, on y voit surtout les inconvénients, puisqu'aucun intérêt direct, des programmes nucléaires qui sont à la source des déchets immergés, n'est ressenti.

Laissant donc de côté le mot "possible" qui a, à notre avis, surtout un trait économique, nous aborderons brièvement l'utilité des programmes de surveillance du milieu marin en rapport avec les opérations d'immersion des déchets radioactifs en mer et nous réfèrerons ensuite les travaux effectués au Portugal dans ce domaine.

2. OBJET ET VALIDITÉ DU CONTROLE RADIOLOGIQUE DU MILIEU LIÉ À L'IMMERSION DES DÉCHETS RADIOACTIFS EN MER

Un contrôle radiologique simplifié sur des échantillons de surface a très peu de chances, à l'heure actuelle, de permettre de détecter la présence de radioéléments éventuellement en provenance de l'immersion des déchets et, surtout, de les distinguer de ceux provenant des retombées dûes aux essais nucléaires.

Nous pensons, cependant, qu'il faut tenir compte des possibilités d'un court-circuit dans les chaînes alimentaires, soit par la pêche à de plus grandes profondeurs, soit par le recours croissant dans l'alimentation à des espèces à grandes migrations verticales comme les céphalopodes, soit encore par l'alimentation des poissons en des zones d'upwelling accentué.

En effet, il n'est pas possible de garantir que l'eau du fond des bassins océanographiques soit maintenue isolée de l'homme et des chaînes alimentaires. Il est connu que dans des régions voisines du site d'immersion dans l'Océan Atlantique nordeste des courants marins du type anticyclonique ont été observés, qui sont susceptibles de correspondre à des courants d'upwelling. Il est également connu que les courants de surface dans cette région de l'Atlantique ont une direction SE-S se rapprochant de la côte portugaise et de la Madère /2_/.

Nous considérons, donc, qu'outre les indispensables recherches sur la libération des radioéléments par les fûts de déchets, sur l'interaction eau-sédiments sur les facteurs de concentration des organismes vivant dans les eaux profondes, etc., il est nécessaire et utile de commencer un programme de contrôle aux environs des lieux d'immersion utilisés ou prévus.

De telles mesures doivent avoir comme objet l'établissement du fond radiologique et l'analyse des tendances d'évolution de l'activité dans le milieu marin. Les données ainsi obtenues permettront, en conjugation avec des recherches scientifiques dans ce domaine, de prouver les hypothèses actuelles quant au comportement des radioéléments. En outre, un contrôle radiologique pourra permettre d'interrompre suffisamment tôt les opérations d'immersion si l'analyse des tendances d'évolution montre que celà est nécessaire.. En effet, une fois qu'une augmentation significative de la radioactivité dans un milieu aussi vaste que l'Océan commencerait à se faire sentir, elle serait irréversible avant plusieurs années, même si on arrêtait immédiatement les immersions.

Il est, par ailleurs, indispensable de connaître les tendances d'évolution de la radioactivité dans le milieu marin, de façon à pouvoir approfondir les connaissances sur le comportement des différents radioéléments dans ce milieu. Ainsi, par exemple, les facteurs de transfert des radioéléments dans la chaîne alimentaire des espèces marines de fond, comme l'Aphanopus carbo, ne sera probablement possible qu'avec des données obtenues in situ, en rapport avec les opérations d'immersion actuelles / 3 /.

Il serait également du plus grand intérêt d'effectuer des mesures de radioactivité sur des sédiments et des organismes benthoniques au voisinage des sites actuels d'immersion. Bien qu'un tel contrôle soit surtout valable sur des échantillons de profondeur, nous pensons qu'il sera également utile de poursuivre, en parallèle, des mesures en surface pour permettre les indispensables comparaisons et détecter un éventuel, bien que peu probable, transport vers la surface.

Pour qu'un contrôle radiologique du milieu marin soit significatif il est nécessaire de distinguer la contribution des différentes sources de pollution. De simples mesures de radioactivité dans l'eau, surtout en ce qui concerne la concentration des produits de fission dans les couches superficielles, permettront difficilement de distinguer une éventuelle contamination dûe à l'immersion des déchets radioactifs en mer, de celle dûe aux retombées atmosphériques. Une analyse systématique de la variation de cette activité avec la localisation et avec la profondeur et l'étude des rapports de certains radioéléments pourront cependant fournir, très probablement, des résultats utiles. Ainsi, par exemple, l'observation des variations avec la profondeur des rapports ^{137}Cs/^{239}Pu et ^{239}Pu/^{241}Am dans l'eau, aux environs du site d'immersion utilisé depuis dix ans, pourrait vraissemblablement permettre la détection d'une éventuelle contribution des déchets immergés pour la radioactivité des eaux profondes. Ceci nous semble un projet qui mériterait un effort de coopération internationale dans ce domaine.

D'autre part, étant donné les facteurs de concentration élevés des sédiments et des organismes marins pour certains radioéléments, leur contrôle pourrait fournir des indications précieuses. Ainsi, il nous semble utile de prélever autant d'échantillons d'organismes et de sédiments que possible dans les régions profondes de l'Océan Atlantique en vue d'établir le fond radiologique et d'observer les tendances d'évolution de la radioactivité dans ce milieu.

3. ACTIVITÉS DE CONTRÔLE RADIOLOGIQUE DU MILIEU MARIN MENÉES PAR LE PORTUGAL

Il est évident que l'ampleur d'un contrôle radiologique du milieu marin suppose une coopération internationale étroite sans laquelle l'effort d'un pays isolé ne sera utile qu'en tant que contribution à des études plus élargies et comme une tentative de dépistage de voies de transport d'intérêt. C'est dans ce sens et pour permettre d'obtenir des résultats qui puissent contribuer à l'établissement du fond radiologique de l'Océan Atlantique nordeste, que nous réalisons au Portugal, depuis plusieurs années, des mesures sur les échantillons marins prélevés surtout au long de la côte occidentale portugaise et dans les régions des Açores et de Madère.

Ainsi, en octobre 1968 (au cours de la croisière NR 15 du navire océanographique Meteor), en oct.-nov. 1970, jan.-fev., avr.-mai, juil.-août et oct.-nov. 1971, des échantillons d'eau ont été prélevés dans une région définie par les coordonnées 41°31' N, 36°19' N et 10°30' W, 7°22' W et analysés pour la recherche du strontium-90 et du cesium-137 / 4 / / 5 /.

En 1973 nous avons effectué une campagne de pêche de chinchard (Trachurus trachurus L.) au long de la côte, aux environs de Nazaré et au sud de Lisbonne. En fev.-mars 1976 et en mars-avr. 1978, des échantillons d'eau de surface et de poissons ont également été prélevés au long de la côte occidentale portugaise et aux Açores et Madère.

Étant donné le plus grand intérêt dans l'obtention de données sur des échantillons d'eaux profondes, en 1979 nous avons commencé une étude du poisson sabre noir de Madère, Aphanopus carbo, qui est une espèce benthonique vivant à des profondeurs de 1.500 à 2.200 m dans des régions limitées de l'Océan Atlantique nord. Il se nourrit essentiellement de céphalopodes et de crevettes géantes et a une taille moyenne d'environ 1,2 m avec à peu-près 14 cm de largeur et 2 à 3 kg de poids. La population aux environs de Madère est constante depuis plusieurs années et estimée

TABLEAU I

Eau – 1976 et 1978 (en pCi/l)

Région	Coord. appr.		Césium-137			Strontium-90	
	N	W	1976 (Mars)	1978 (Mars)	1978 (Avril)	1978 (Mars)	1978 (Avril)
Leixões (Porto)	41°00' 41°20'	8°50' 9°20'	0,28 ± 0,07	0,27 ± 0,03	0,27 ± 0,04	0,18 ± 0,02	0,22 ± 0,02
Aveiro	40°30' 40°50'	8°55' 9°25'	0,32 ± 0,08	–	–	–	–
Nazaré	39°25' 39°45'	9°15' 9°45'	0,26 ± 0,07	0,30 ± 0,03	0,25 ± 0,06	0,19 ± 0,02	0,23 ± 0,02
L.Albufeira	38°20' 38°40'	9°15' 9°45'	0,26 ± 0,07	0,23 ± 0,02	0,24 ± 0,03	0,16 ± 0,02	0,20 ± 0,02
Arrifana	37°10' 37°30'	9°00' 9°30'	0,26 ± 0,06	0,31 ± 0,03	0,27 ± 0,04	0,24 ± 0,02	0,20 ± 0,02
Madeira	32°20' 32°40'	16°45' 17°15'	0,30 ± 0,07	–	–	–	–

– 50 –

à environ 300.000 exemplaires [6].

Nous avons également commencé l'analyse de certains échantillons de sédiments prélevés pour des études de pollution non radioactive à des profondeurs entre près de 2.000 et 5.000 m.

4. RÉSULTATS

Les échantillons d'eau (50 l) ont été soumis à des séparations radiochimiques pour le césium-137 et le strontium-90 et ceux de poissons (1 à 2 kg) et de sédiments (50 à 100 g) ont été analysés par spectrometrie gamma au Ge(Li). Les échantillons prélevés en 1979 sont gardés en vue d'une analyse ultérieure pour les transuraniens.

Les résultats des analyses d'eau de surface en 1976 et 1978 sont groupés sur le tableau I, ayant déjà été publiés ceux relatifs à 1968, 1970 et 1971 [4] [5]. On n'y remarque aucune variation sensible ni avec la date ni avec les coordonnées du point de prélèvement (dont les écarts sont d'ailleurs faibles). Le rapport césium-137/strontium-90 pour les échantillons d'eau prélevés en 1978 donne une valeur moyenne de 1,33 ± 0,17, avec des écarts entre 1,58 et 1,09, résultat qui est de l'ordre de ceux observés par d'autres auteurs pour l'Océan Atlantique [7].

Sur le tableau II nous avons groupé les moyennes des mesures de césium-137 et de strontium-90 dans les échantillons d'eau de surface pendant les campagnes ci-dessus mentionnées. Ce tableau laisse apercevoir une certaine homogeneité pour tous les résultats.

Sur le tableau III sont présentés les résultats de la concentration du césium-137 dans les sédiments prélevés en début d'année et on y observe une très faible activité en césium-137, en dessous des limites de détection pour la plupart des sédiments et, en tout cas, inférieure à celles des sédiments de la plateforme continentale.

En ce qui concerne le poisson sabre, on procéde à des prélèvements réguliers de 15 exemplaires qui sont également déstinés à des études de pollution non radioactive. Nous prenons environ 150 g de muscle de chacun et nous analysons l'ensemble par spectrometrie gamma, après incinération. Les résultats sont présentés sur le tableau IV et on y observe que les concentrations du césium-137 sur les trois échantillons de poissons sont du même ordre de grandeur et toutes assez faibles.

L'ensemble des résultats obtenus sur des poissons en 1973, 1976, 1978 et 1979 est finalement rassemblé sur le tableau V [1]

Bien que le nombre de résultats soit encore peu expressif, on observe une tendance pour le plus faibles concentrations de césium-137 dans l'Aphanopus carbo que dans d'autres poissons de couches superficielles, ce qui nous semble être en accord avec la diminution des concentrations des produits de fission avec la profondeur.

5. CONCLUSIONS

D'après les résultats obtenus, nous pouvons conclure, en premier lieu, qu'étant donné les très faibles activités des produits de fission dans les échantillons de sédiments et de poissons des eaux profondes, il serait très probablement possible d'y distinguer une augmentation d'activité éventuellement dûe aux déchets radioactifs immergés.

---

(1) Pour les sédiments et les poissons analysés en 1979 les résultats présentent une meilleure précision et une meilleure sensibilité, car nous avons utilisé un nouveau détecteur, de plus gros volume, et un temps de comptage plus long.

TABLEAU II

Eau - comparaison des concentrations moyennes
obtenues pour les différentes campagnes
(en pCi/l)

Année	Césium-137	Strontium-90
1968	0,29 ± 0,08	0,20 ± 0,17
1970/71	0,23 ± 0,07	0,19 ± 0,08
1976	0,28 ± 0,02	—
1978	0,27 ± 0,03	0,20 ± 0,03

TABLEAU III

Sédiments - 1979

Profondeur	Coordonnées		Césium-137
m	N	W	pCi/g sec
1750	37°04'	13°49'	≤ 0,08
2112	41°15'	10°41'	0,051 ± 0,036
2170	34°40'	24°24'	≤ 0,08
2270	36°58'	10°43'	0,054 ± 0,035
3625	32°52'	16°10'	0,084 ± 0,047
4850	46°55'	08°10'	≤ 0,06

TABLEAU IV

Poisson sabre de Madère - 1979

Date de prélèvement	Césium-137		
	pCi/g cendres	pCi/kg frais	pCi/gK
Fevrier 79	0,56 ± 0,10	6,8 ± 1,2	2,0 ± 0,3
Avril 79	0,81 ± 0,14	10,5 ± 1,8	2,7 ± 0,4
Mai 79	0,74 ± 0,14	10,7 ± 2,0	2,7 ± 0,5

La régularité dans les concentrations du césium-137 pour chaque type d'échantillons permet de commencer à établir des données de référence qui sont, cependant, encore en nombre très réduit et qui devront être suivies au long du temps.

Nous considérons donc utile et valable de faire un contrôle radiologique du milieu marin dans le but d'établir des niveaux de référence et d'étudier les tendances d'évolution de la radioactivité dans ce milieu. En particulier, une campagne de prélèvement d'échantillons de sédiments, de matériel biologique et d'eau, sous coordination internationale, aux environs du site d'immersion dans l'Océan Atlantique nordeste pourrait apporter des résultats du plus grand intérêt.

REMERCIEMENTS

Les auteurs tiennent à remercier le Comité Executif du Poligone d'Accoustique sous-marine des Açores, en la personne du Ct. Ataíde, qui leur a fourni la plupart des échantillons prélevés et, encore, leurs collègues du Service de Radioactivité Ambiante qui ont contribué à l'analyse des échantillons.

RÉFÉRENCES

[1] Définitions et Recommandations revisées de l'AIEA aux fins de la Convention de Londres. INFCIRC/205/Add 1/Rev.1, 1978

[2] Ataíde, J.C. et als: Introduction à l'étude de l'occurrence des métaux lourds dans les eaux adjacentes aux Iles de S.Maria et S. Miguel. Lisboa, 1979

[3] The Oceanographic basis of the AIEA revised definition and recommenda - tions concerning high-level radioactive waste unsuitable for dumping at sea, IAEA-210, 1978

[4] Marques, B.E. et al.: A poluição radioactiva no Oceano Atlântico Norte. 2ª Com., Centro de Estudos de Radioquímica, IAC, Lisboa, 1974

[5] Marques, B.E. et als: Sur la contamination radioactive dans l'Océan Atlantique Nord. Centro de Estudos de Radioquímica, VI - SFRP/12, Lisboa,1972

[6] Barros, M.C. Bioaccumulation of organochlorine compounds in scabbard (Aphanopus carbo) from North Atlantic-Madeira Island (Preliminary Report) - Funchal Meeting of the working group on degrability - Oslo Convention, May 1975

[7] Bowen, V.T. et als: Cesium-137 to strontium 90 ratios in the Atlantic Ocean 1966 through 1972. Limnology and Oceanography, 19:4, pp 670-681, 1974.

TABLEAU V

Poissons - comparaison des concentrations en césium-137 (en pCi/g cendres)

Région	Coordonnées		Espèce	1973	1976	1978	1979
	N	W					
Leixões	41°00' 41°20'	8°50' 9°20'	Trachurus trachurus (chinchard)			2,5 ± 0,2[b]	
Peniche	39°00' 39°40'	9°30' 10°00'	Trachurus trachurus	4,7 ± 0,8[a]		1,6 ± 0,1[b]	
			Scomber scombrus (maquereau)		0,8 ± 0,1[b]		
			Sparus aurata (dorade)		1,4 ± 0,2[b]		
L.Albufeira	38°20' 38°40'	9°15' 9°45'	Trachurus trachurus	4,6 ± 1,0[a]			
St^a.Maria (Açores)	36°50' 37°10'	25°00' 25°15'	Trachurus trachurus		2,0 ± 0,6[a]		
			Sparus pagrus (pagre)		1,7 ± 0,2[b]		
Madeira (Madère)	32°20' 32°40'	16°45' 17°15'	Aphanopus carbo (sabre noir)		1,0 ± 0,5[a]		0,70 ± 0,13[a]

(a) déviation standard
(b) erreur statistique

Session 2

Chairman - Président
Mr. A.W. VAN WEERS
(The Netherlands)

Séance 2

HYDROGRAPHIC SURVEYS IN THE CENTRAL WESTERN NORTH PACIFIC
IN RELATION TO DEEP-SEA DISPOSAL OF RADIOACTIVE WASTES

H. Sudo

Tokai Regional Fisheries Research Laboratory, Tokyo, Japan

ABSTRACT

The deep water of the North Pacific is characterized by very uniform properties, but their distributions are influenced to some extent by bottom topography. There is evidence to show that the deep flow along the Izu-Ogasawara Trench is southward and northward east and west of the trench axis, respectively. It is hardly probable that the Pacific Bottom Water southeast of Japan steadily spreads north of 30°N or west of 145°E. From deep water properties observed in the central western North Pacific, it is suggested that neither the northward bottom current nor the deep western boundary current is well defined at least at middle latitudes.

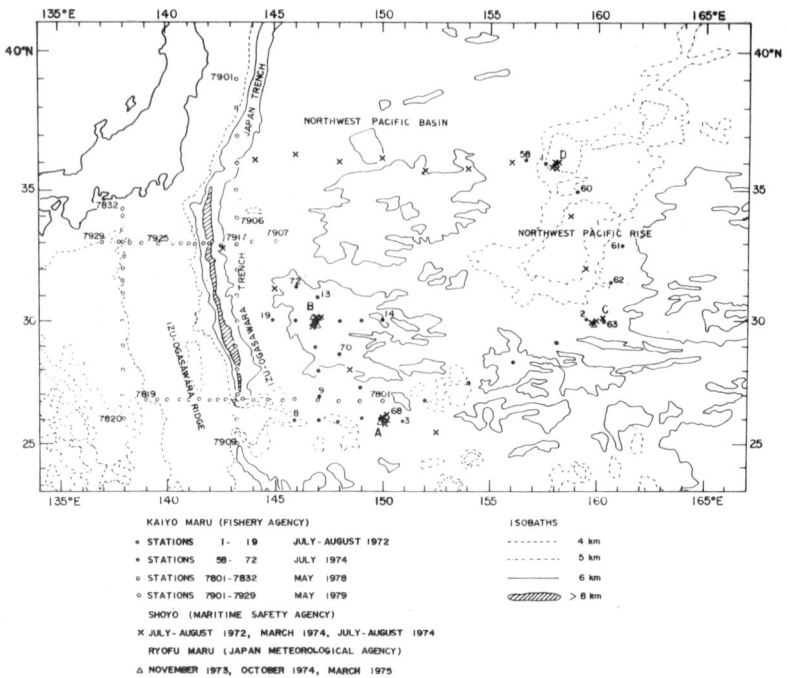

Figure 1. Location of hydrographic stations.

Introduction

A three-year oceangraphic survey on four selected areas (centered at A, B, C and D; each of them is about 110 km in diameter) for a tentative disposal of radioactive wastes was made in 1972 to 1974 by three governmental agencies of Maritime Safety Agency, Meteorological Agency (including Meteorological Research Institute) and Fishery Agency under the auspices of Science & Technology Agency. During the survey deep hydrographic casts down to near bottom were made on 68 stations in the southwestern part of the Northwest Pacific Basin including above four areas (Figure 1).

Further scientific investigations concerning deep-sea disposal of radioactive wastes have been carried out since 1977. One of the most important problems in our country related to sea-bottom dumping southeast of Japan main island is the influence of the Izu-Ogasawara Ridge on the circulation, centered on near-bottom currents, boundary currents and upwelling. For the purpose of a systematic survey of the subject the Fishery Agency took four hydrographic sections, two zonal of 26°50'N and 33°00'N and two meridional of 138°00'E and 143°15'E, in 1978 and 1979 (Figure 1). Station spacing was generally 110 km or less in the deep basin, but was reduced to about 30 km over the Izu-Ogasawara Trench and the eastern slope of the ridge. Vertical spacing did not exceed 250 m and reduced to 50 m or 100 m for the deepest several Nansen bottles.

Property distribution and water mass

Variations of temperature, salinity and dissolved oxygen of the deep water in each of the four selected areas are within measurement errors with a few exceptions (Table I).

In the northwestern Pacific potential temperature decreases and both salinity and dissolved oxygen concentration increase monotonically with depth below the main thermocline. Their gradients diminish exponentially with depth, but they are still discernible down to the bottom except for in trenches.

A few slight thermoclines are often detected in the deep water in the southwestern part of the Northwest Pacific Basin. At Sta. 63 (Site C), for example, rapid decreases of 0.06°C and 0.04°C in potential temperature were observed in vertical spacings of 3901 to 4163 m and 4913 to 5163 m, respectively. Immediately above these thermoclines there were relatively rapid increases with depth in dissolved oxygen concentration (3658 to 3901 m and 4663 to 4913 m). Deeper ones are both discernible through Figures 4 and 5. This relatively sharp gradient found at a depth of about 5 km implies the vertical discontinuity separating the underlying Pacific Bottom Water, which is colder, more saline and higher oxygen content, from the Pacific Deep Water. Discontinuities detected at shallower depths may be generated by layering of locally formed water masses. The water mass structure on smaller scales is not well defined and it may be worthy of future investigation.

Though the deep water of the North Pacific is characterized by very uniform properties, their distributions are influenced to some extent by bottom topography. At Sta. 59 (Site D) on the western edge of the Northwest Pacific Rise, the deepest water is slightly lower in temperature (Figure 2) and higher in dissolved oxygen and the salinity distribution is influenced up to about 2700 m (Figure 3). At Sta. 62 taken on the easternmost portion of a small branch of the rise, the deep water below 3000 m is relatively cold and rich in oxygen, but less saline. This suggest that the deep water lying north of Site C may have passed through various sea areas mixing with neighbored water masses rather than have flowed directly from the south. Figures 4 and 5 show that on the vertical section between Sites C, A and B the deep water at depths greater than 4000 m is colder, more saline and higher in dissolved oxygen toward the east. In general these isopleths increase in steepness with depth, but actual configurations are rather complicated. The extreme values observed during *Kaiyo-Maru* Cruises were 0.99°C at 5663 m on Sta. 63 and at 5765 m on Sta. 7801 for potential temperature, 34.706‰ at the former for salinity and 4.13 ml/l at 5868 m on Sta. 68 for dissolved oxygen. It is supposed that the bottom water splits into two or three water masses between Sites A and C.

On the Izu-Ogasawara Trench isopleths are lowered, especially in salinity distribution (Figures 6 and 7). These remarkable horizontal gradients of properties must reflect the geostrophic motion (Figures 10 and 11). From rise or fall of isopleth of property distribution in accordance with bottom configurations, it is conjectured that bottom currents have some components rectangular to isobaths or crawling up and down on the bottom as well as geostrophically balanced components along depth contours.

Geostrophic flow and deep circulation

The discontinuity at a depth of about 5 km means that it is vertically the most stable in the deep water. On the assumption that this stability maximum represents the depth of minimum flow, the 5000-decibar depth was fundamentally chosen as a reference for geostrophic flow computation. At depths of 4 km or more, however, not only the geostrophic velocity but also the vertical geostrophic shear is slight for the most part. Therefore, some shifts of reference level would not seriously affect the following results.

Except for the western boundary of the basin, the calculated geostrophic speeds in the deep water are generally less than 1 cm/sec and even directions are indefinite (Figures 9 to 11). It should be noted that the deep currents at the western boundary of the basin can be divided into three parts. Both sections of 26°50'N (Figure 10) and 33°00'N (Figure 11) are consistent in geostrophic structure. The first is the westernmost part of the Northwest Pacific Basin with weak, rather northward current. On the trench there are distinct counter currents adjoining at the trench axis. Both transports are nearly equivalent below 2 km (Table II). Whether these currents are the western boundary currents suggested by general circulation theory or not is a matter for argument. The reason is that these currents may form meso-scale eddies.

It is not possible to construct available current charts because of lack of data, simultaneously taken data in particular (Figures 14 and 15). Geostrophic flow at 5500-decibar relative to 4500-decibar in Area B seems to be counter-clockwise. The oxygen distribution at 5750 m shows an indication of northwestward flow in the area (Figure 13). From the potential temperature, however, it seems likely that there are a southeastward intrusion of warm water near the site and a northwestward spreading of cold water southwest of the area (Figure 12), but on the other hand this feature is naturally associated with the geostrophic flow. Though above results are not inconsistent in possible circulation, there needs accounting for them on the basis of direct current measurements.

The Pacific Bottom Water, entering the Northwest Pacific Basin around 15°N, 170°E, partly flows northward east of the Northwest Pacific Rise and partly spreads westward west of the rise. If the water of less than 1.10°C for potential temperature, more than 34.690‰ for salinity and more than 3.80 ml/l for dissolved oxygen may be referred to as the Pacific Bottom Water, its spreading is shown in Figure 16. Of these properties the oxygen is the most attributive in the area under consideration. It is hardly probable that the Pacific Bottom Water southeast of Japan steadily spreads north of 30°N or west of 145°E.

From deep water properties observed in the central western North Pacific, it is suggested that neither the northward bottom current nor the deep western boundary current is well defined at least at middle latitudes.

Table I. Means and standard deviations of potential temperature, salinity and dissolved oxygen observed in four selected areas during 1972 to 1975.

	Depth (m)	Potential temperature (C°)		Salinity (‰)		Dissolved oxygen (ml/l)	
Area A centered at 26°N, 150°E 6 stations	3000	1.37	0.014	34.662	0.005	3.05	0.05
	4000	1.16	0.017	34.684	0.002	3.48	0.04
	5000	1.07	0.008	34.692	0.003	3.77	0.13
	6000	1.02	0.012	34.694	0.003	3.91	0.06
Area B centered at 30°N, 147°E 9 stations	3000	1.38	0.017	34.665	0.009	3.09	0.11
	4000	1.18	0.009	34.683	0.007	3.48	0.08
	5000	1.08	0.004	34.692	0.007	3.72	0.11
	6000	1.02	0.012	34.698	0.005	3.87	0.06
Area C centered at 30°N, 160°E 5 stations	3000	1.34	0.013	34.665	0.006	2.99	0.09
	4000	1.14	0.005	34.686	0.005	3.49	0.08
	5000	1.02	0.017	34.694	0.005	3.83	0.07
	5500	0.99	0.012	34.699	0.006	3.99	0.05
Area D centered at 36°N, 158°E 5 stations	3000	1.29	0.040	34.667	0.006	3.03	0.12
	3500	1.14	0.014	34.678	0.001	3.40	0.02

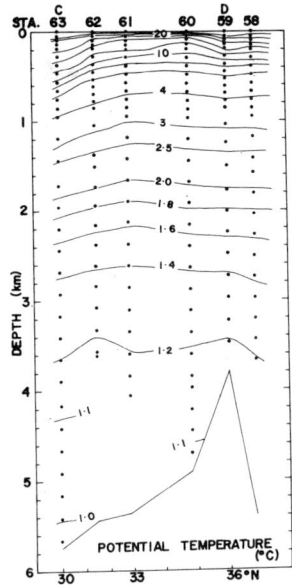

Figure 2. Potential temperature on the vertical section between Sites D and C.

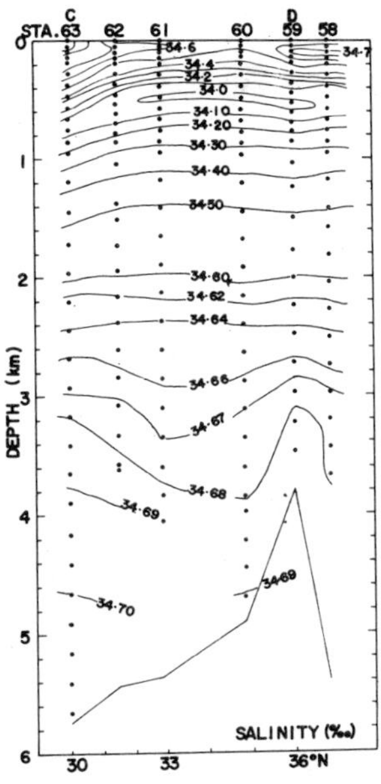

Figure 3. Salinity on the vertical section between Sites D and C.

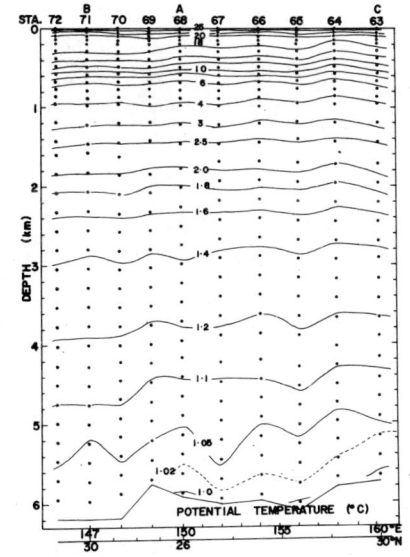

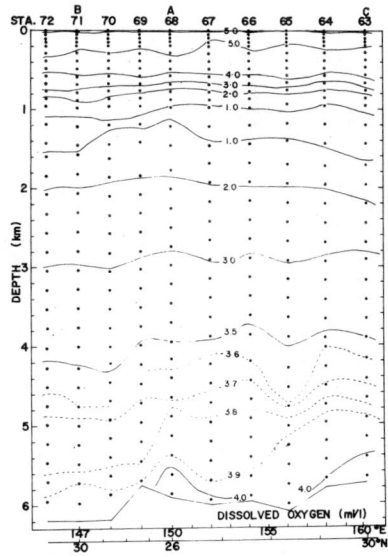

Figure 4. Potential temperature on the vertical section between Sites C, A and B.

Figure 5. Dissolved oxygen on the vertical section between Sites C, A and B.

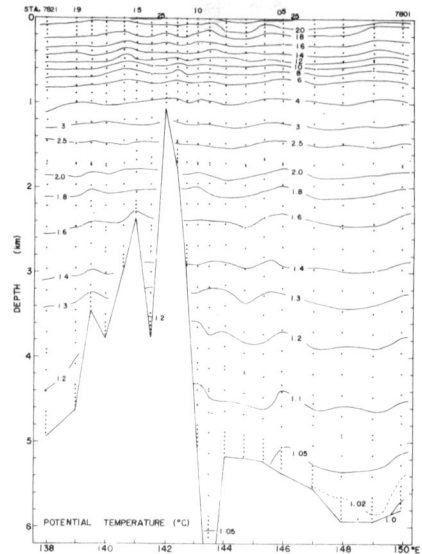

Figure 6. Potential temperature on the vertical section along 26°50'N.

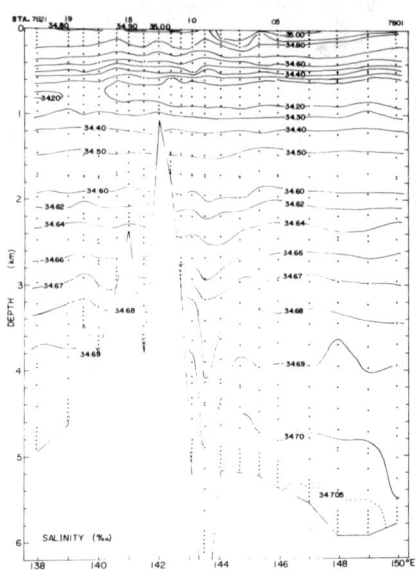

Figure 7. Salinity on the vertical section along 26°50'N.

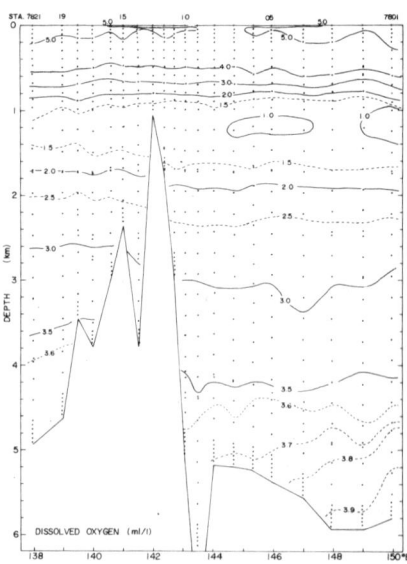

Figure 8. Dissolved oxygen on the vertical section along 26°50'N.

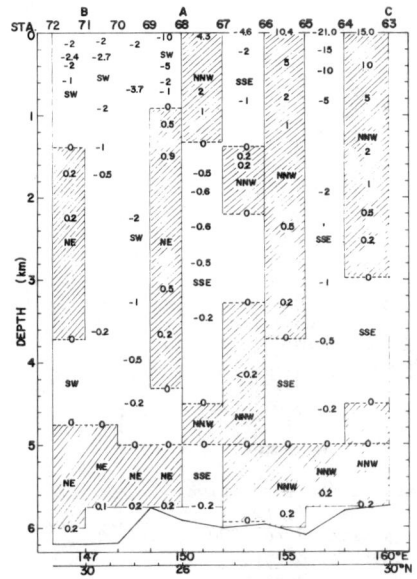

Figure 9. Geostrophic speed (cm/sec) relative to 5000-decibar surface (east of Sta. 70) and 4750-decibar surface (west of it) on the vertical section between Stations 63 and 72. Shaded areas indicate flow to the north-northwest or northeast.

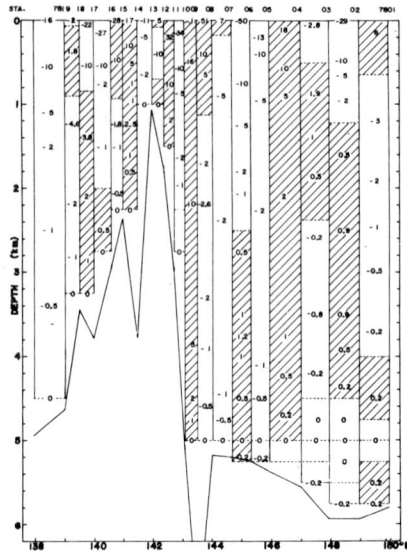

Figure 10. Geostrophic speed (cm/sec) relative to 5000-decibar surface (between adjacent stations sampled to 5000 m or more) or to depth of deepest samples on the vertical section along 26°50'N. Shaded areas indicate northward flow.

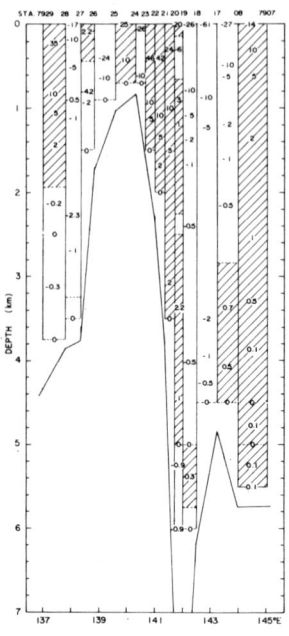

Figure 11. Same as Figure 10 except for along 33°00'N.

Table II. Geostrophic transport on the Izu-Ogasawara Trench and east of it. (10^6 m^3/sec, positive means northward)

Depth (km)	On the eastern slope of the ridge or the western slope of the trench		On the eastern side of the trench		On the western-most part of the Northwest Pacific Basin		Total
33°00'N	Sta. 7924	7919		7908		7907	Total
0~2	24.7		-40.7		9.3		-6.7
2~4	2.1		- 3.0		1.3		0.4
4~6	0.1		- 0.1		0.1		0
Total	26.8		-43.9		10.8		-6.3
26°50'N	Sta. 7813	7809		7805		7801	Total
0~2	11.3		-18.2		4.9		-2.0
2~4	6.7		- 6.1		2.5		3.2
4~6	0.9		- 0.9		0.4		0.5
Total	19.0		-25.2		7.9		1.7

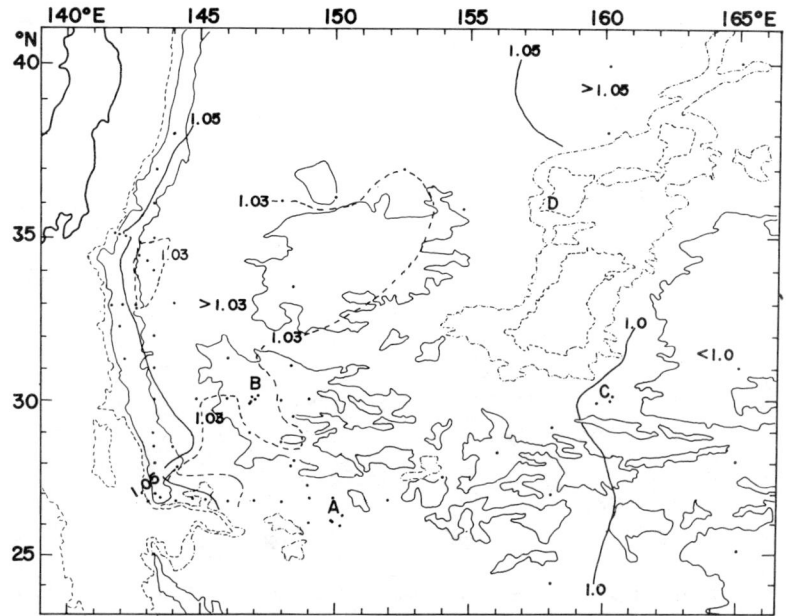

Figure 12. Potential temperature (C°) at a depth of 5750 m or at a depth greater than 5000 m and within 150 m above the bottom. Depths as in Figure 1.

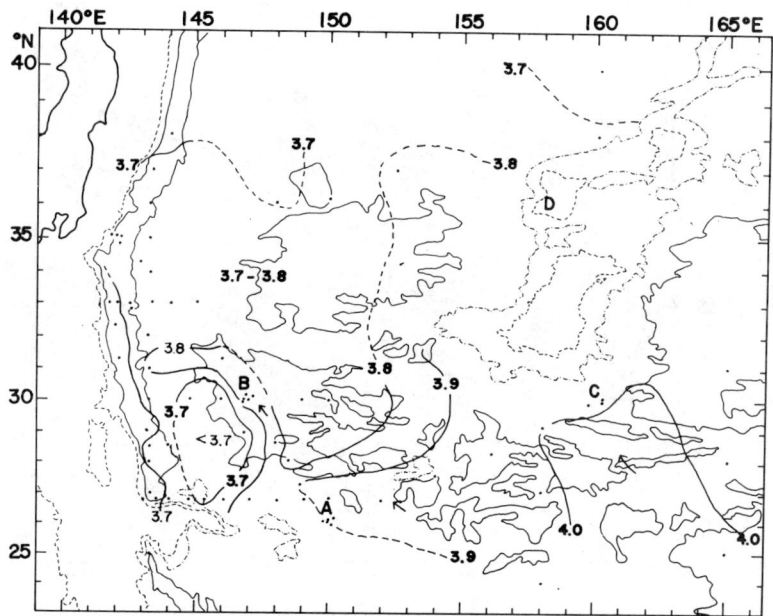

Figure 13. Dissolved oxygen (ml/l) at a depth of 5750 m or at a depth greater than 5000 m and within 150 m above the bottom. Depths as in Figure 1.

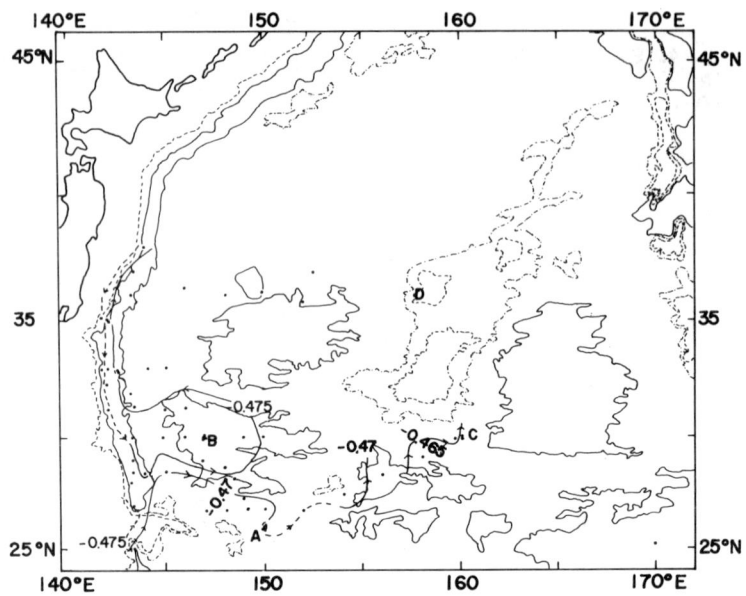

Figure 14. Geopotential anomaly (dynamic meters) at 5500-decibar surface relative to 4500-decibar surface. Depths as in Figure 1.

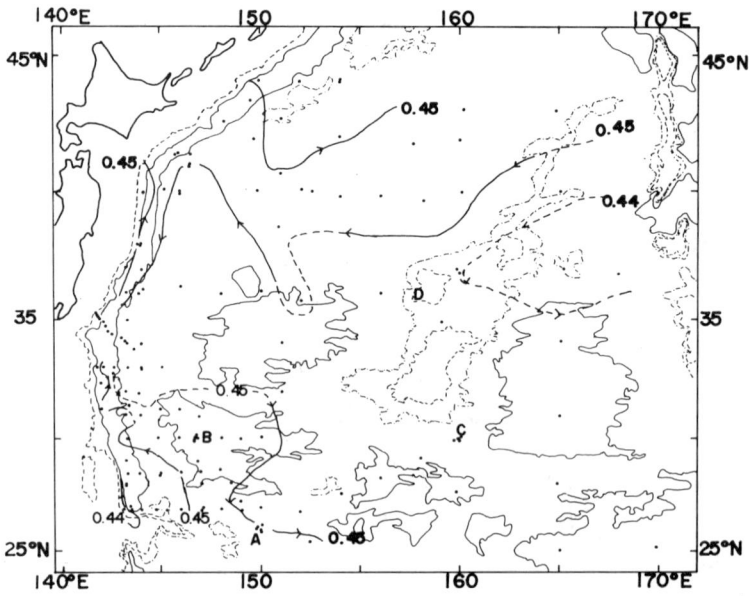

Figure 15. Geopotential anomaly (dynamic meters) at 3500-decibar surface relative to 4500-decibar surface. Depths as in Figure 1.

NORMALISATION INTERNATIONALE DES FRUITS ET LÉGUMES
INTERNATIONAL STANDARDISATION OF FRUIT AND VEGETABLES

PÊCHES - PEACHES
(Révision)

CORRIGENDUM

Texte français seulement, page 30, lire :

A. *Tolérances de qualité*

Catégorie «Extra»: 5 % en nombre ou en poids de pêches ne correspondant pas aux caractéristiques de la catégorie, mais conformes à celles de la catégorie I ou exceptionnellement admises dans les tolérances de cette catégorie.

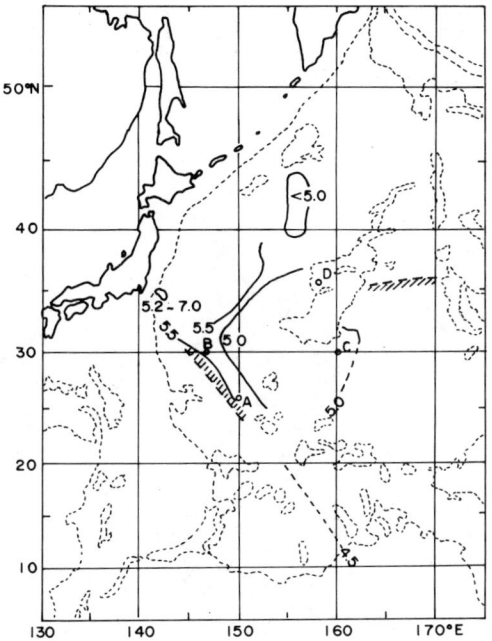

Figure 16. The spreading area of the Pacific Bottom Water and its upper boundary depth (km). It is probable that the Bottom Water always reaches to the shaded sides of thick broken lines. Thin broken lines denote 5000 m depth contours.

Discussion

K. TAIRA, Japan

Sudo shows that geostrophic flows are to the north at 142°E, to the south at 144-145°E, and to the north at 146-147°E. Mean speeds measured by Taira and Imawaki are found to be in agreement with Sudo's results, I think.

DEEP AND BOTTOM CURRENT MEASUREMENTS IN AN AREA ADJACENT TO THE PROPOSED
WASTE DUMP SITE

K. Taira [*], S. Imawaki [**] and T. Teramoto[*]
[*] Ocean Research Institute, University of Tokyo, Tokyo 164 Japan
[**] Geophysical Institute, Kyoto University, Kyoto 606 Japan

ABSTRACT

Deep and bottom currents in a layer from 25m to 2750m high above the bottom were measured at Station TA(30°N,145°45'E, 5845m deep), adjacent to the proposed waste dump site located at 30°N,147°E. Observational periods are from December 4, 1977 to February 27,1978 and from September 30, 1978 to March 16,1979. We found an intense current flowing steadily to the south and uniformly in a layer from 25m to 400m high above the bottom. The maximum values of daily mean speeds are recorded to be about 20cm/s in the uniform layer. In a layer from 800m to 2750m, the intensity of the southward flow is decreasing exponentially with an e-folding height of 1500m. Some characteristics of this observed decreasing flow are found to be in agreement with those of bottom-trapped topographic Rossby waves reported by Rhines(1970) and Thompson and Luyten(1976).

1. Introduction

Our present knowledge on deep and bottom currents in deep ocean is poor due to limit of methods to detect them. Surface currents are sometimes detected by deflection of ship's course and by trajectories of drifters of various kinds. We can survey rapidly with the GEK. We can also estimate subsurface currents through dynamic computations from distribution of water density assuming a reference level on a appropriate layer. Tracking of a surface marker buoy attached to a subsurface drogue is effective to detect the subsurface currents (Taira, et al. 1978). These methods can not be applied to detect the deep and bottom currents.

When we utilize the vast space in the deep and bottom layers of the oceans to store away industrial waste of any kind, it is required to fully understand the motions in these layers. One of the most effective methods to measure the currents is direct measurements with moored current meters of self-recording type. By making good use of this method, international cooperative studies on the ocean currents have been carried out successively in the Atlantic Ocean in the 1970's. An idea that water motions are extremely small especially in the deep layers of the central portion of the oceans has been changed largely by observations of intense currents associated with meso-scale eddies of a time scale of several tens days and horizontal scale of several hundreds kilometers.

Our research group has started in 1975 to solve technical problems in deep-sea mooring of current meters. Since 1977, we have been engaged in deep and bottom currents measurements in an area adjacent to the proposed waste dump site located to the east of the Izu-Ogasawara Trench. We observed an intense current flowing steadily to the south and uniformly in a layer from 25m to 400m high above the bottom. The intensity of the flow is decreasing exponentially with an e-folding height of about 1500m in the layer from 800m to 2750m. Velocity variations with a period of 50-60 days and an amplitude of about 3cm/s are outstanding in current-records for all the layers. Since there is no significant phase lag among the layers, these variations are considered to be barotropic motions of meso-scale prevailing in the area.

2. Observations

On December 4, 1977, the first mooring line of five current meters was deployed at Station TA located at 30°N,145°45'E where the water depth is 5845m. The current meters are moored at heights of 25m, 50m, 100m, 845m and 1845m high above the bottom. The mooring line was retrieved on February 27,1978. The second mooring line was set at the same station on September 30, 1978. Seven current meters are moored at heights of 50m, 100m, 200m, 400m, 800m. 1850m and 2750 m,respectively. The second line was retrieved on March 16, 1979. The third was set on March 20, 1979 and retrieved on July 23, 1979. Successive cruises of research vessels are scheduled to the end of the 1980 fiscal year to keep mooring lines in the area including the Station TA.

Bottom topography along 30°N, surveyed during the cruise of the R/V Hakuho-Maru in March, 1979, is schematically shown in Figure 1. Both the Izu-Ogasawara Ridge and the Izu-Ogasawara Trench are extending in the north-south direction. Station TA is located on the midway of gentle slope from the east edge of the Trench to a basin at 150°E. Scale of the slope is 300km in the north-south direction and 600km in the east-west direction, approximately. The proposed waste dump site is located at 30°N, 147°E.

We are adopting recording current meters of the Model 5, manufactured by Aanderaa Instruments. Although this model is of highest confidence among current meters in the market, we have been suffered from several kinds of troubles. One is the leakage due to pressure sensors. Three current meters were broken in the first observation.

Figure 1. Bottom topography along 30°N.

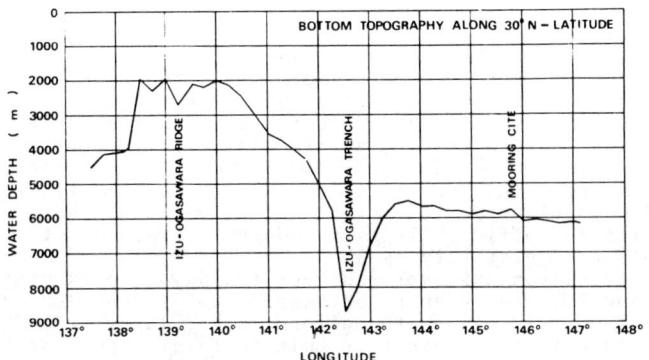

Figure 2. Observed profile of the southward velocity V (cm/s).

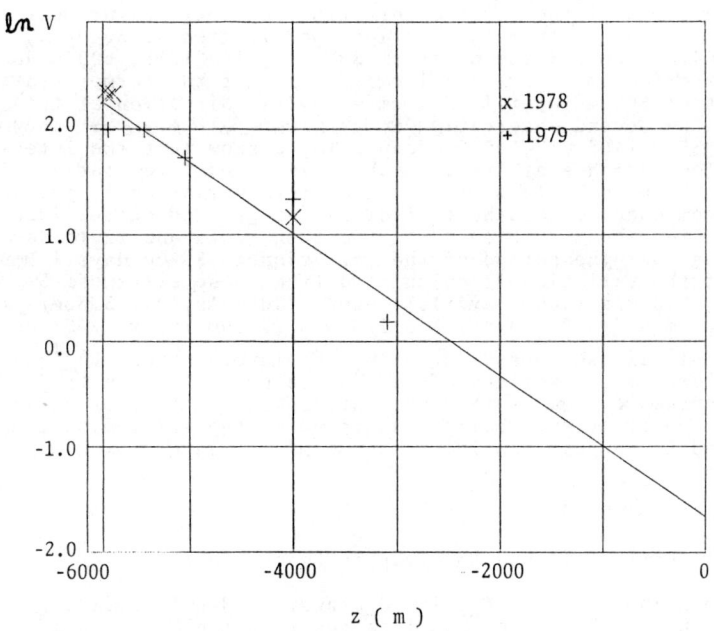

Shortening of recording periods was sometimes caused by deficiency of batteries. Deformation of pressure case was caused when moored near to 6000m, capability-depth of the model. The current meter moored at 5745m depth was crashed in the second observation.

3. Obseravational results

Speed sensor of an Aanderaa current meter is of rotor type. Minimum speed detected by the sensor is limited to be 1.5 cm/s. Revolution of the rotor is counted for a selected sampling period and the number is recorded on a magnetic tape. Mean speed is calculated from recorded numbers for each current meter in the first mooring line at Station TA. The results are tabulated in Table I. Sampling period is selected to be 20 min. Frequency of zeros of rotor counter is also tabulated. The results of the second mooring line are shown in Table II. Sampling period is 30 min. These show that quality of the records is sufficiently high and that errors due to motions of mooring lines are relatively small.

Velocity components averaged over ten days are tabulated in Tables III and IV. An intense southward flow is recorded in the lower layer. The velocity is quite uniform in the layer from 25m to 400m high above the bottom. Above this uniform layer, speed of the southward flow is decreasing. Logarithms of southward component V (cm/s) are plotted against water depth z (m) in Figure 2. Velocity profile is found to be

$$\ln V = -z/1500 - 5/3$$

or $V = V_0 \exp(-z/1500)$

where velocity at the sea surface V_0=0.19 cm/s. This shows that velocity is decreasing exponentially above the uniform layer and that an e-folding height is approximately 1500m.

In the perid of the first observation, current measurements were carried out at another two stations one located at 30°02'N, 146°13'E, 6047m deep and the other at 30°15'N, 146°15'E, 6097m deep. Mean current of 4000m-layer is 4.1 cm/s in 14° at the former station. Mean current of 5000m-layer is 3.4 cm/s in 311° direction at the latter. In the second observation, it is 1.5 cm/s, 311°degree at 4000m-layer at 29°35'N, 146°40'E, 6040m deep. These show that the intense southward flow vanishes at stations about 45 km east to Station TA.

It is shown in Table IV that a long-term variation is predominant in v-components. Southward flow is intensified on the 40th and 90th day. Intensity is decreased on the 20th, 70th and 130th day. This suggests that the period of the variation is 50-60 days. Double-amplitude of the variation is calculated from these extremes: 5.8cm/s (50m-layer), 5.8cm/s (200m-layer), 4.0cm/s (400m-layer), 5.3cm/s (800 m-layer), 4.2cm/s (1850m-layer) and 5.3cm/s (2750m-layer). Since there is no significant phase lag among the layers, these variations seem to be barotropic motions of meso-scale prevailing in the area.

The southward flow is the most intense in the earlier period of the first observation. Daily-mean speed is the maximum on the third day and it ranges from 19cm/s to 20cm/s at layers of 25m, 50m and 100m.

4. Bottom-trapped topographic Rossby waves

As shown in Figure 1, the Izu-Ogasawara Ridge is deviding the deep waters in the Phillipine Sea from those in the North Pacific Ocean. The station TA is located at the western boundary of the Ocean. We can antipate that there may exist a western boundary current of the deep circulation in the area. Horizontal scale of the

Kuroshio or the Gulf Stream, the western boudary currents in the surface layer, is of order of 100km. It suggests that southward flow may vanish only 45 km east to a station for a boundary current. In July 1979, three mooring lines are deployed along 30°N at 144°E, 145°E and 145°45'E to examine whether the southward flow is existing at these stations.

Station TA is located on a gentle slope of 500m/300km starting from the east edge of the Trench. Rhines(1970) shows theoretically that topographic Rossby waves are trapped on a slope. Thompson and Luyten(1976) shows that these waves are trapped on the slope extending in the north-south direction of the Site-D. In their case, period of the waves is suggested to be 1-2 weeks.

According to Rhines(1970), characteristics of the bottom-trapped topographic Rossby waves are examined for a situation shown in Figure 3. We take x-axis to the east and y-axis to the north. The origin is on the sea surface and z is upward positive. Vertical walls are assumed at $x = 0$ and $x = L$. The bottom is sloping to the east: $z = -H - \alpha(x - L/2)$. Let u, v and w be the velocity components, boundary conditions are $w = 0$ at $z = 0$; $w/u = \alpha$ at $z = -H$, and $u = 0$ at $x = 0$ and L. The Brunt-Vaisala frequency N is assumed to be constant. Horizontal velocity components associated with low frequency waves are described by

$$u = -A' \ell \sin(\ell y - \sigma t) \sin k_n x \cosh(NH\sqrt{k_n^2 + \ell^2}\, z/f)$$

$$v = A' k_n \cos(\ell y - \sigma t) \cos k_n x \cosh(NH\sqrt{k_n^2 + \ell^2}\, z/f)$$

where f is Coliolis parameter, σ angular frequency and k_n and ℓ are components of wave number. Horizontal boundary condition requires $k_n L = \pi, 2\pi, \ldots$ Dispersion relation is

$$\sigma = -(\ell/\sqrt{k_n^2 - \ell^2})\, N \coth(NH\sqrt{k_n^2 + \ell^2}/f)$$

A condition that $\sigma > 0$ requires $\ell < 0$, showing that phase speed is in the y-negative direction, or to the south. Frequency vanishes when $\ell/\sqrt{k_n^2 + \ell^2}$ or α vanishes.

When we take $\ell = 0$, then both σ and u vanish. We obtain the velocity component v:

$$v = A \cos k x \, \cosh(kNz/f)$$

Profile of v is shown in Figure 4 by taking $kNH/f = 10$. When kNH/f is larger than unity, the profile becomes

$$v \sim (A/2) \cos kx \, \exp(-kNz/f)$$

The velocity is decreasing exponentially upward from the bottom. Observational e-folding height is about 1500m as described in the previous section. We have a relation: $kN/f = 1/1500$ (/m). This relation gives horizontal wave length of 100km for $N \sim 0.001$ (/s).

5. Summary

Since 1977, we have carried out current measurements at Station TA (30°N, 145°45'E, 5845m deep). We obtain the following results:
1) In a layer from 25m to 400m high above the bottom, an intense southward flow is existing. The maximum daily mean speed is about 20 cm/s. 2) Speed of the southward flow is decreasing exponentially with height in a layer from 800m to 2750m. An e-folding height is

Figure 3. Definition graph: Bottom is sloping uniformly to the east.

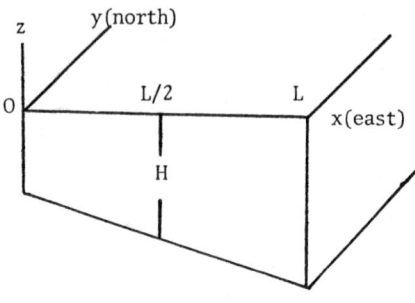

Figure 4. Profile of $V = V_0 \cos kx \cosh(kNz/f)$ for a case $KNH/f=10.0$.

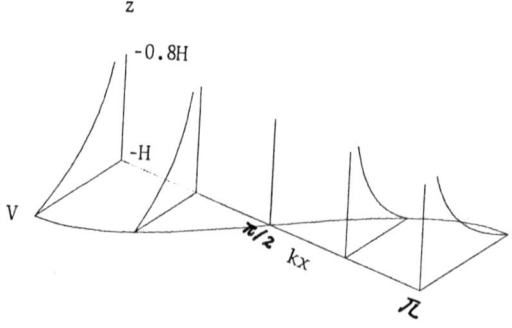

found to be about 1500m. 3) Long-term variation is prevailing with a period of 50-60 days and an amplitude of about 3cm/s. 4) Northward flow is observed at three stations (30°02'N,146°13'E; 30°15'N,146°15'E; and 29°35'N,146°40'E) east to Station TA. 5) Some characteristics of the observed southward flow are in agreement with those of bottom-trapped topographic Rossby waves.

It is important to investigate the intense flow in the area adjacent to the proposed waste dump site since it may carry materials to a wide area in a relatively short time. Measurements of the deep and bottom currents at the site is also important. Depth capability of current meters and acoustic releases in the present market is limited to be less than 6000m. The depth of the site exceeds 6000m. It is required to develop new instruments of higher confidence to meet the depth. A Lagrangean method of current measurements is much effective to follow the movement of deep and bottom waters. Tracking of SOFAR buoys has been carried out to elucidate the deep circulation in the Atlantic Ocean. Investigation of physical process in the benthic boundary layer just below the observed uniform flow is also important and it reqires new instruments.

This study has been sponsored by the Ministry of Education, Science and Culture since 1975. We are much indebted to Professor Y. Horibe for his excellent leadership. We thank to the crew of the R/V Hakuho-Maru, her onboard scientists, K. Takeuchi and S. Kitagawa for thier help.

References

Rhines, P. : Edge,bottom-, and Rossby waves in a rotating stratified fluid.   Geophysical Fluid Dynamics, Vol.1,pp.273-302.(1970)
Taira, K., T. Teramoto, N. Shikama and K. Takeuchi : Currents measurements with surface and subsurface drifters. Journal of the Oceanographical Society of Japan, Vol. 34,pp.73-77. (1978)
Thompson, R.R.Y. and J.R. Luyten : Evidence for bottom-trapped topographic Rossby waves from single moorings. Deep-Sea Research, Vol. 23,pp.629-635. (1976)

Table I   Mean speeds calculated from the current meters in the first mooring line at Station TA.

Current meter height	25m	50m	100m	1850m
Mean speed	11.7 ± 4.3 (cm s^{-1})	14.9 ± 3.4 (cm s^{-1})	11.22± 4.8 (cm s^{-1})	5.7 ± 3.1 (cm s^{-1})
Zeros of rotor revolution	0.2%	0.0%	1.9%	14.4%

Table II. Mean speeds calculated from the current meters in the second mooring line at Station TA.

Current meter height	50m	200m	400m	800m	1850m	2750m
Mean speed	9.0 ± 3.0 (cm s^{-1})	8.9 ± 3.1 (cm s^{-1})	8.4 ± 2.7 (cm s^{-1})	7.2 ± 2.5 (cm s^{-1})	5.7 ± 2.2 (cm s^{-1})	4.3 ± 2.4 (cm s^{-1})
Zoros of rotor revolution	0.4%	0.4%	1.5%	1.9%	9.8%	18.8%

Table III Velocity components averaged over ten days for the first observation.

	Period	25m		50m		100m		1850m	
		u	v	u	v	u	v	u	v
10	Dec.4, 1977 ～	-5.8	-16.7	-5.7	-16.5	-5.1	-17.4	-2.3	-11.0
20	Dec.14,1977 ～	-2.5	-13.1	-2.4	-12.6	-2.3	-13.0	-1.4	-5.9
30	Dec.24,1977 ～	1.5	-13.0	—	—	2.0	-12.7	1.9	-5.5
40	Jan.3, 1978 ～	6.3	-12.2	—	—	5.9	-12.0	4.4	-4.4
50	Jan.13,1978 ～	4.5	-6.8	—	—	4.2	-6.6	2.9	0.0
60	Jan.23,1978 ～	1.5	-6.9	—	—	1.2	-6.2	0.7	1.1
70	Feb.2, 1978 ～	2.6	-7.0	—	—	2.0	-6.4	1.6	0.3
80	Feb.12,1978 ～	2.8	-7.7	—	—	2.2	-7.1	1.9	-0.1
	Mean	1.7	-10.6	-2.9	-14.1	1.5	-10.3	1.3	-3.2

Table IV Velocity components averaged over ten days for the second observation.

	Current meter height	50m		200m		400m		800m		1850m		2750m	
	Period	u	v	u	v	u	v	u	v	u	v	u	v
10	Sept.30,1978~	5.4	-8.5	2.7	-9.3	2.7	-9.3	4.6	-6.6	3.8	-4.2	3.5	-2.6
20	Oct.10, 1978~	5.7	-7.9	3.3	-9.1	3.4	-9.4	5.0	-6.1	4.0	-4.1	4.3	-2.8
30	Oct.20, 1978~	5.0	-8.5	2.3	-9.5	2.1	-9.2	5.1	-7.2	4.6	-5.7	4.5	-4.5
40	Oct.30, 1978~	1.9	-11.6	-3.1	-11.1	-2.7	-9.6	1.0	-8.6	-0.1	-6.7	-0.2	-5.9
50	Nov. 9, 1978~	2.1	-9.5	-1.8	-9.7	-1.8	-9.4	1.1	-8.1	0.5	-4.7	-0.2	-3.3
60	Nov.19, 1978~	-0.3	-5.9	-2.9	-5.7	-2.7	-5.4	-1.1	-4.5	-1.6	-1.0	-1.2	-0.6
70	Nov.29, 1978~	0.7	-5.0	-1.1	-4.1	-1.3	-4.2	-0.6	-2.8	−	−	-0.5	2.1
80	Dec. 9, 1978~	4.1	-8.3	0.7	-9.0	0.7	-8.7	2.1	-7.2	−	−	2.8	-1.3
90	Dec.19, 1978~	2.8	-10.9	-1.8	-11.3	-1.7	-9.5	-0.5	-9.4	−	−	1.6	-3.5
100	Dec.29, 1978~	3.0	-8.4	0.1	-8.1	0.2	-8.7	0.3	-7.2	−	−	-1.1	-1.3
110	Jan. 8, 1979~	3.8	-6.7	1.4	-7.4	1.3	-6.9	1.8	-4.3	−	−	-0.8	0.9
120	Jan.18, 1979~	2.9	-5.0	0.9	-5.8	0.9	-5.5	1.8	-3.5	−	−	-0.5	1.3
130	Jan.28, 1979~	6.4	-3.9	6.0	-4.4	6.0	-4.3	5.7	-3.0	−	−	2.7	1.0
140	Feb. 7, 1979~	6.2	-4.6	5.8	-5.1	6.4	-5.6	6.2	-4.3	−	−	3.0	0.4
150	Feb.17, 1979~	6.2	-5.8	4.6	-4.2	5.5	-5.2	4.8	-3.3	−	−	1.8	1.1
160	Feb.27, 1979~	6.2	-6.6	4.6	-4.6	6.1	-6.5	5.5	-4.3	−	−	2.0	-0.7
	Mean	4.0	-7.3	1.6	-7.4	1.8	-7.3	2.8	-5.6	1.9	-4.5	1.3	-1.2

A DEEP CURRENT MEASUREMENT AT A PROPOSED WASTE DUMP SITE

S. Imawaki
Geophysical Institute, Kyoto University, Kyoto, 606 Japan

K. Takano
Rikagaku Kenkyusho, Wako, Saitama, 351 Japan

ABSTRACT

A deep current measurement was made at a proposed radioactive waste dump site (30°N,147°E) in 1978 to 1979. Current velocity and temperature data were obtained at depths of 4000 m and 5000 m at three stations for about 170 days. There are two pronounced peaks in the power spectra of the zonal and meridional components of velocities, whose periods are 12.5 hours and 22 to 24 hours. The former is due to the semidiurnal tidal currents and the latter may be the composite of the diurnal tidal currents and the inertial currents. Mesoscale eddies are outstanding in the daily mean velocity records, whose periods are estimated to be 50 to 80 days by maximum entropy spectral method.

RESUME

Une mesure de courants profonds est effectuée par des courantomètres mouillés dans une zone destinée aux décharges des déchets radioactifs (30°N,147°E) en 1978 à 1979. Les données de la vitesse et de la température à 4000 m et à 5000 m pour 170 jours mettent en relief deux périodes prédominantes : 12,5 heures et 22 à 24 heures, la première relative aux marées semi-diurnes et la seconde relative aux marées diurnes aussi bien qu'aux courants d'intertie. Par ailleurs, les tourbillons à échelle intermédiaire se montrent saillants. Leur périodes sont évaluées aux environs de 50 à 80 jours.

1.   INTRODUCTION

    The ocean environment could not be kept clean without correctly estimating the possible extent of spreading of the disposed materials prior to the dumping of radioactive wastes at the ocean bottom. One of the Japanese proposed dump sites of radioactive waste, called Site B, is located at $30°N, 147°E$, about 500 km south to the Kuroshio Extension and about 500 km east to the Izu-Ogasawara Ridge, which is far from any intense, steady currents. Its location should be in a typical mid-ocean. It was not believed until 1960 that there might be considerable motions of deep water in the mid-ocean. The recent progress in physical oceanography revealed, however, the presence of intense low-frequency current fluctuations or mesoscale eddies almost everywhere in the mid-ocean, whose swirl speeds are 5 to 50 cm/sec and space and time scales are tens to hundreds of kilometers and weeks to months, respectively (see the MODE Group [1]).

    Previously a direct deep current measurement was made 500 km southwest to the site by Takano [2]. He observed velocities at intermediate and deep layers for about eight days and pointed out the predominance of the diurnal fluctuation. Nagasaka [3] observed the velocity 110 m above the bottom at the site for about ten days and showed, though not so convincingly, predominant diurnal and semidiurnal fluctuations. Durations of these two measurements, however, are too short to examine the mesoscale variability. Teramoto et al. [4] made a long-term current measurement near the site and obtained good quality velocity data for about 90 days at deep and bottom layers at three stations. They suggested that there might exist mesoscale eddies of about 100-day period near the site as well as a strong current above the bottom. But the lengths of records are not long enough to estimate precisely their time scales.

    A long-term current measurement program was started in 1978 to observe the fluctuations of deep currents and to estimate the dispersion of materials in the deep sea. The present paper is concerned with a preliminary result of the program, which is still under way.

2.   OBSERVATION

    An array of three mooring lines was deployed in a triangular pattern at the proposed dump site B during the R.V. Hakuho Maru Cruise KH-78-4 of Tokyo University in October 1978. The locations of mooring stations and the bottom topography are shown in Fig. 1. The water depth is about 6000 m and the bottom topography is fairly flat in the observation area. The distance between Stations RA8A and RB8A is about 45 km and that between Stations RA8A and RC8A is about 90 km. Two Aanderaa current meters were set at nominal depths of 4000 m and 5000 m on each mooring line. They were recovered successfully during the R.V. Hakuho Maru Cruise KH-79-1 in March 1979. The mooring locations, water depths, deployment and recovery dates and durations are summarized in Table I.

    Some fundamental properties of the records are tabulated in Table II. Five of the six current meters deployed yield good quality velocity and temperature data throughout the mooring durations. Only a small portion (7%) of the record RA8A2 is unusable, becouse of a battery trouble. Sampling intervals were set to 30 minutes. Current speeds were recorded with a resolution of 0.09 cm/sec. Their nominal accuracy is ± 1 cm/sec. Temperatures were recorded in an optional, high resolution range, "Arctic Range" of Aanderaa current meter, which gives a resolution of 0.008 °C. The nominal accuracy of temperature is ± 0.15 °C.

    Table II shows that mean velocities are rather small as is expected. Standard deviations of the zonal and meridional components of velocities are three to four cm/sec, which include short-term and long-term fluctuations as is mentioned later. Horizontal excursions of instruments due to the mooring line motions are estimated numerically to be several tens of meters in the case of a current velocity of 10 cm/sec, which might contaminate horizontal velocity data for fluctuating currents. For example, the extent of contaminations is estimated to be about 0.5 cm/sec for the semi-diurnal tidal current. It is not serious in the present study, however.

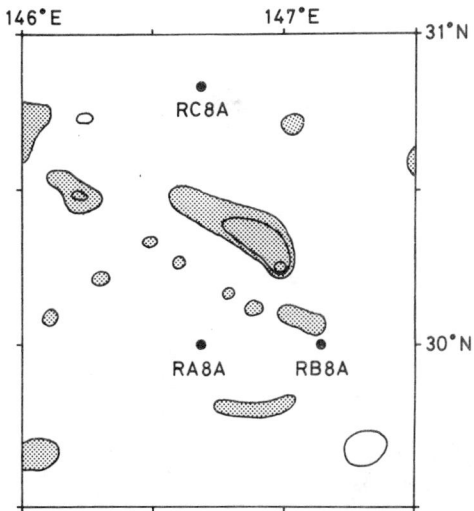

Figure 1. Locations of the mooring stations and a sketch of the bottom topography. The shaded area indicates water depth less than 6000 m. The contour interval is 500 m.

Table I. Description of the mooring stations

Mooring Station No.	Location	Water Depth (m)	Deployment Date 1978	Recovery Date 1979	Duration (days)
RA8A	29°59.2'N,146°40.7'E	6210	October 1	March 17	167
RB8A	30°00.1'N,147°08.6'E	6240	October 2	March 19	168
RC8A	30°49.8'N,146°41.1'E	6180	October 2	March 17	166

Table II. Description of records and basic properties. $\bar{U}$ and $\bar{V}$ are the time averages of the eastward and northward components of velocities, σU and σV their standard deviations, $\bar{T}$ the average of temperature and σT its standard deviation. The first four symbols of the record number refer to the mooring station number and the last digit is the instrument number.

Record No.	Nominal Depth(m)	Duration (days)	$\bar{U}$ (cm/sec)	σU (cm/sec)	$\bar{V}$ (cm/sec)	σV (cm/sec)	$\bar{T}$ (°C)	σT
RA8A1	4000	166	1.1	3.4	0.2	3.6	1.42	0.007
RA8A2	5000	154	1.2	3.2	-0.3	3.5	1.41	0.005
RB8A1	4000	168	0.1	3.8	-0.4	3.7	1.41	0.007
RB8A2	5000	168	-0.7	3.4	-0.4	3.4	1.44	0.004
RC8A1	4000	165	-1.0	3.9	2.3	4.3	1.41	0.009
RC8A2	5000	165	-1.6	3.7	2.2	3.7	1.41	0.005

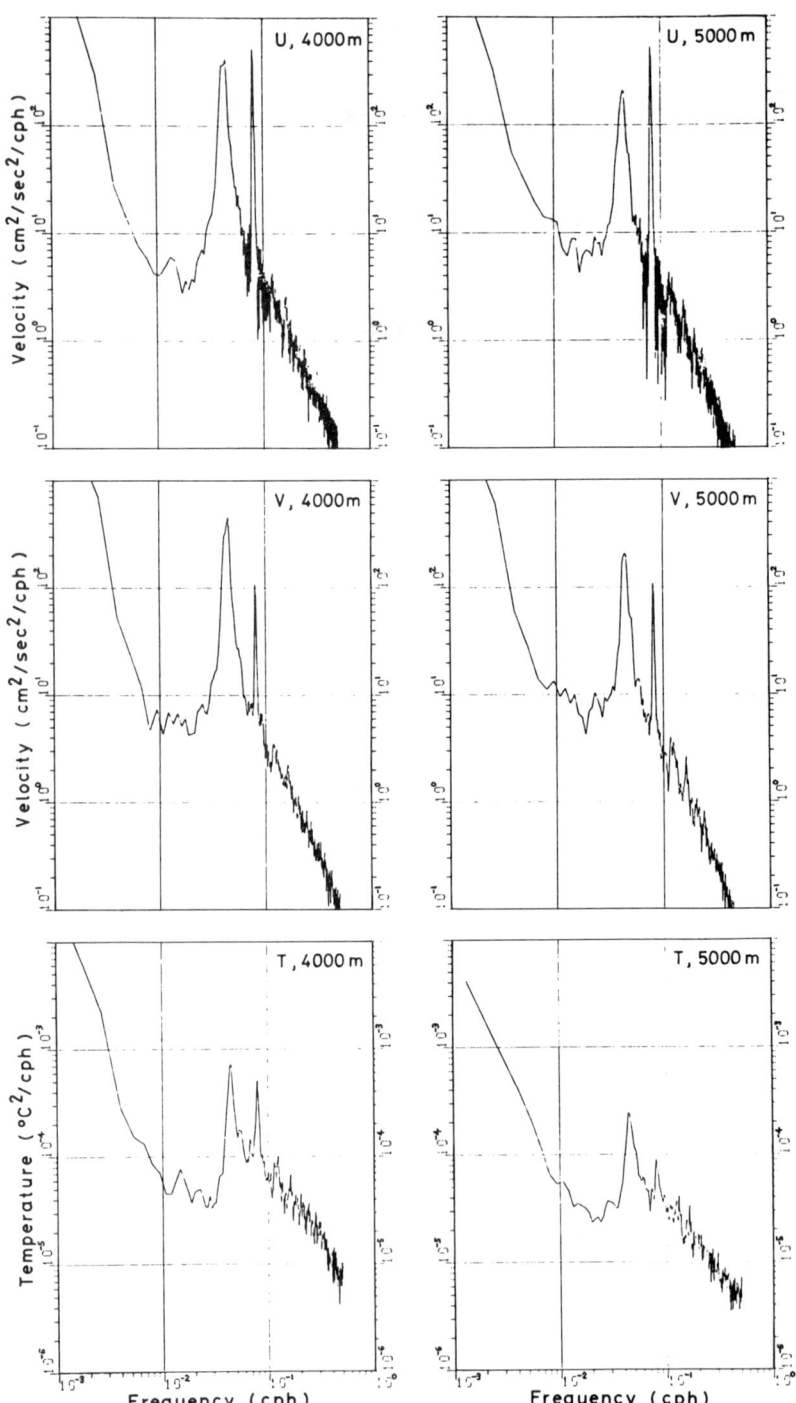

Figure 2. Spatial mean spectra of the zonal (upper) and meridional (middle) components of velocity and temperature (lower) at depths of 4000 m (left) and 5000 m (right)

Values of mean temperatures are a little smaller than those
obtained by the serial observations carried out at the stations,
which were about $1.50°C$ at a depth of 4000 m and about $1.53°C$ at a
depth of 5000 m. Although the absolute values of temperatures are
not accurate enough, their variations may be meaningful in the
present study. A part of the changes in observed temperature may be
attributed to the vertical displacements of instrument due to the
mooring line motions. They should be sufficiently small, less than
$0.0007°C$ at a depth of 4000 m and $0.0003°C$ at a depth of 5000 m,
under the same circumstance as mentioned above.

3.   SHORT-TERM FLUCTUATIONS

Spectra of the zonal and meridional components of velocities and temperatures are obtained by Blackman-Tukey method. They
are one-sided spectra, whose numbers of degrees of freedum are about
20. Periods and spectral densities of the pronounced peaks near the
diurnal and semidiurnal periods are summarized in Table III. Here
the symbol "mean" denotes the spatial mean spectra obtained by
averaging the individual spectra at the three stations. The mean
spectra are illustrated in Fig. 2. Periods of dominant peaks of
velocity spectra are 22 to 24 hours near the diurnal period and 12.5
hours near the semidiurnal period. The latter is common to all
velocity records and comes from the semidiurnal tidal currents. The
former differs a little among different records. The period of
local inertial oscillations at Stations of RA8A and RB8A is 23.93
hours and that at Station RC8A is 23.35 hours, which are close to
the periods of diurnal tidal currents. Hence it is difficult to
distinguish the fluctuations due to the inertial oscillations from
those due to the diurnal tidal currents. Observed periods of
dominant diurnal fluctuations are slightly shorter than both the
nominal diurnal tidal periods and the local inertial oscillation
period, as is studied by Munk and Phillips [5] for the inertial
oscillation.

Spectral densities of the zonal and meridional velocity
components at a depth of 4000 m are almost the same as those at a
depth of 5000 m for the semidiurnal fluctuations, while those at a
depth of 4000 m are about two times as large as those at a depth of
5000 m for the diurnal fluctuations. Effective amplitudes of the
predominant fluctuations are estimated as roots of the spectral
densities multiplied by frequencies and are listed in parentheses in
Table III. Amplitudes of the diurnal and semidiurnal fluctuations
are several cm/sec.

There are also dominant peaks near the diurnal and semidiurnal periods in the spectra of temperatures, whose periods are
about 22 hours and 12.5 hours. Spectral densities at a depth of
4000 m are two to three times larger than those at a depth of 5000 m
for the diurnal fluctuations and several times larger for the semidiurnal ones. Effective amplitudes of these fluctuations should be
about $0.006°C$ at a depth of 4000 m and about $0.003°C$ at a depth of
5000 m. Diurnal and semidiurnal spectral peaks of temperature are
also shown in the MODE area at a depth of 4000 m as well as at a
depth of 515 m by Richman et al. [6], whose spectral densities are
one order of magnitude greater than the present ones.

4.   LONG-TERM FLUCTUATIONS

The short-term fluctuations described in the last
section are eliminated by taking the daily mean to investigate the
long-term fluctuations (with periods longer than one day) of
velocities and temperatures. The results are shown in Fig. 3. At
first sight the time series of daily mean current vectors at depths
of 4000 m and 5000 m surprisingly resemble each other. Root mean
squares of differences between the velocity components at depths of
4000 m and 5000 m are estimated at the three ststions to be 0.6 to

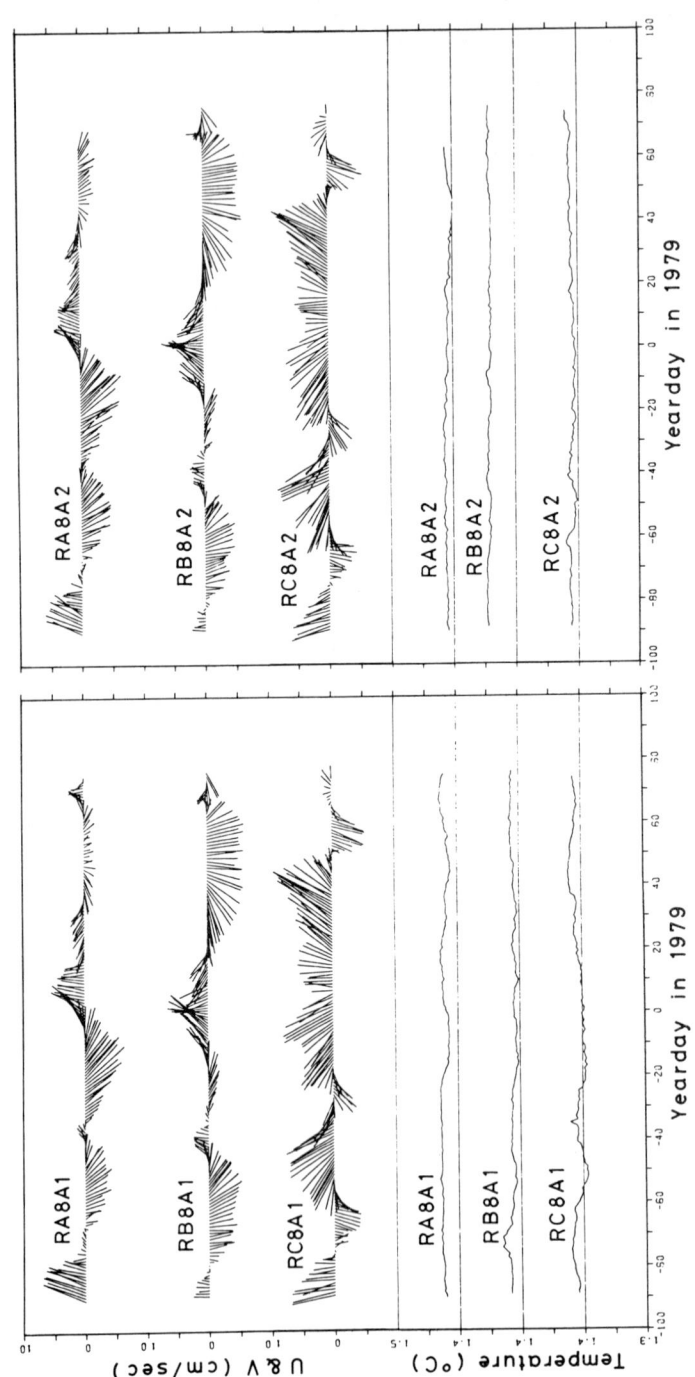

Figure 3. Time series of velocities (upper) and temperature (lower) at depths of 4000 m (left) and 5000 m (right). In the upper panels each stick represents a daily mean velocity vector. Its length is proportional to the current speed and its direction is the current direction (upward north). The abscissae are the year-days of 1979.

1.1 cm/sec. It suggests that the long-term fluctuations in this area are almost the same at depths of 4000 m and 5000 m or in the deep water layer.

Maximum entropy spectra are calculated to examine predominant periods in the long-term fluctuations of the zonal and meridional velocity components and temperatures. The results are listed in Table IV. Dominant periods are 50 to 80 days in the fluctuations of the zonal and meridional components and about 40 days in those of temperatures. The time scales and intensities of these fluctuating currents are similar to those observed in the MODE area (see the MODE Group [1]) and these low-frequency fluctuations are considered to be the mesoscale eddies. The dynamics of the mesoscale eddies has not yet been well understood. Analyses of their horizontal scales, horizontal propagation rates and dynamics are now under way.

Table III. Spectral peaks of the zonal and meridional components of velocities and temperatures. In each column the left-hand side figures are the periods in hour of pronounced peaks and the right-hand side ones are their spectral densities in $cm^2/sec^2/cph$ for the velocity components and in $10^{-3}(°C)^2/cph$ for the temperature. Figures in parentheses are the effective amplitudes of fluctuations in cm/sec for the velocity components and in °C for the temperature.

Station No.	4000m Diurnal Period		4000m Semidiurnal Period		5000m Diurnal Period		5000m Semidiurnal Period	
	\multicolumn{8}{c}{Zonal Component of Velocity}							
RA8A	24.2	571.	12.5	683.	22.7	299.	12.5	798.
RB8A	22.1	458.	12.5	461.	23.4	202.	12.5	390.
RC8A	22.1	517.	12.5	424.	22.1	179.	12.5	421.
Mean	22.1	417.	12.5	523.	22.7	210.	12.5	537.
		(4.3)		(6.5)		(3.0)		(6.6)
	\multicolumn{8}{c}{Meridional Component of Velocity}							
RA8A	24.2	493.	12.5	185.	22.7	299.	12.5	198.
RB8A	22.1	541.	12.5	87.	23.4	258.	12.5	80.
RC8A	22.1	546.	12.5	49.	22.1	135.	12.5	47.
Mean	22.1	455.	12.5	107.	22.7	206.	12.5	109.
		(4.5)		(2.9)		(3.0)		(3.0)
	\multicolumn{8}{c}{Temperature}							
RA8A	22.1	1.03	12.5	0.60	22.7	0.36	12.5	0.12
RB8A	22.1	0.81	12.5	0.45	21.4	0.28	12.1	0.09
RC8A	22.7	0.36	12.5	0.48	22.1	0.23	12.3	0.09
Mean	22.1	0.72	12.5	0.51	22.1	0.23	12.5	0.09
		(0.006)		(0.006)		(0.003)		(0.003)

Table IV. Periods of predominant peaks in maximum entropy spectra of the zonal and meridional velocity components and temperatures (units:days)

Station No.	Zonal Component		Meridional Component		Temperature	
	4000m	5000m	4000m	5000m	4000m	5000m
RA8A	64	88	67	82	42	—
RB8A	61	58	84	64	42	—
RC8A	59	—	47	47	43	27

5.   SUMMARY

A deep current measurement was made at a proposed radioactive waste dump site in the mid-ocean in 1978 to 1979. Current velocity and temperature data were obtained at depths of 4000 m and 5000 m at three stations for about 170 days.

Diurnal and semidiurnal fluctuations are dominant in the short-term fluctuations of the zonal and meridional components of velocities, whose periods are 22 to 24 hours and 12.5 hours, respectively. The former may be the composite of the diurnal tidal currents and the local inertial currents. The latter is due to the semidiurnal tidal currents. There are also diurnal and semidiurnal fluctuations in the short-term fluctuations of temperatures, whose periods are about 22 hours and 12.5 hours, respectively.

Mesoscale eddies are dominant in the daily mean velocity fluctuations, whose periods should be 50 to 80 days by maximum entropy spectral method. There are also dominant fluctuations in the long-term temperature fluctuations, whose periods are estimated to be about 40 days by the same method. These fluctuations as well as the general circulation of the ocean may have possibly important effects on the spreading and dispersion of materials in the deep sea, as shown, for example, by a numerical study (Takano and Matsuyama [7]).

ACKNOWLEDGMENTS.

The authors are indebed to Captain I. Tadama, officers and crews of the R.V. Hakuho Maru of Tokyo University, and Chief Scientist T. Teramoto and the other participated scientists for their kind cooperations and skillful assistances in the deployment and recovery of mooring lines.

REFERENCES

[1] The MODE Group : "The Mid-Ocean Dynamics Experiment", Deep-Sea Research 25, 859-910 (1978).
[2] Takano, K. : "A Current Measurement by Moored Savonious Meters", Journal of Oceanographical Society of Japan 30, 91-95 (1974).
[3] Nagasaka, K. : "Direct Measurement of Deep Sea Current by Richardson-Type Current Meters in the Western-North Pacific", The Oceanographical Magazine 27 (1), 25-36 (1976).
[4] Teramoto, T. et al. : "Current Measurement near the Proposed Dump Site of Low-Level Radioactive Waste", in Environmental Marine Sciences, Vol. 3, edited by Y. Horibe, Tokyo University Press (1979).
[5] Munk, W. and N. Phillips : "Coherence and Band Structure of Inertial Motion in the Sea", Reviews of Geophysics 6 (4), 447-472 (1968).
[6] Richman, J. G. et al. : "Space and Time Scales of Mesoscale Motion in the Western North Atlantic", Reviews of Geophysics and Space Physics 15 (4), 385-420 (1977).
[7] Takano, K. et M. S. Matsuyama : "Dispersion de Polluants d'une Source au Fond Profond par des Tourbillons à Echelle Intermédiaire", This Proceedings (1979).

EVALUATION ON THE DISPOSAL OF RADIOACTIVE WASTES INTO THE
NORTH PACIFIC -   THE EFFECT OF STEADY FLOW AND UP-WELLING

Y. Sugiura*, K. Saruhashi** and Y. Miyake**
* institute of chemistry, college of liberal arts,
kagoshima university, korimoto, kagoshima 890, japan
** meteorological research institute,
koenji-kita, suginami, tokyo 166, japan

ABSTRACT

Regarding the evaluation on the disposal of radioactive wastes into the North Pacific, the effect of steady flow and up-welling was studied by solving the three dimensional diffusion eqation assuming that the ocean is a closed system. Resultantly, it was concluded that the estimation of the concentration at the depth of 1 km above the point of release, taking only the influence of diffusion into consideration was appropriate for the safety assessment on radioactive waste disposal on the deep sea floor, regardless the horizontal advection and up-welling are taken into account or not.

1. INTRODUCTION

In relation to the dumping of wastes at sea, the International Atomic Energy Agency [1] adopted a definition of high-level radioactive waste or other high-level radioactive matter unsuitable for dumping at sea. This definition was based on the Webb-Morley report [2] in which the ocean was regarded as an infinite medium. In view of the great importance of definition for the future developments in peaceful uses of atomic energy and also for the health and safety of human being, as well as, for the conservation of marine environment, Miyake and Saruhashi[3] made a criticism on the Webb-Morley report. Following this critical study, the present authors [4] estimated by using a simple model, the concentrations at the depth of 1 km above the dumpsite which was assumed to be located at the center of the bottom of the North Pacific. In their calculation, they considered in a closed system only the Fickian turbulent diffusion without introducing any term of advection. The final concentration after the lapse of infinite length of time was estimated during which the dumping of an equal amount of radioactive waste was done every year. As a result, they arrived at the conclusion that the concentration of nuclides with a longer half-life such as ^{239}Pu are much higher than postulated by Webb and Morley. Consequently, they proposed a more conservative definition of the high-level radioactive waste as compared with the definition given by IAEA in 1975. Recently, IAEA[5] revised the definition which are lower than the former definition on the basis of the ocean model of closed system as pointed out by Miyake and Saruhashi. Generally, the large scale mixing of water is incessantly taking place in the ocean under the additive action of diffusion and advection. As in the previous paper, only the effect of diffusion was discussed, the effect of advection i.e. horizontal current and up-welling was considered in this paper.

2. A MODEL AND EQUATIONS USED

In general, in a semi-infinite medium the concentration of a radioactive material at any point (x, y, z) at time t after the release of one curie at one time can be expressed in a following way:

$$C(\text{ci cm}^{-3}) = \{4^{-1}(\pi t)^{-3/2}(k_x k_y k_z)^{-1/2}\} [\exp\{-(x-ut)^2(4k_x t)^{-1}\}$$
$$\exp\{-(y-vt)^2(4k_y t)^{-1}\} \exp\{-(z-wt)^2(4k_z t)^{-1}\}] \exp(-\lambda t)$$
$$= A \cdot B \exp(-\lambda t) \qquad (1)$$

where k_x, k_y and k_z are the turbulent diffusion coefficients in direction of x, y and z, and u, v and w are the x, y and z components of the current velocity, respectively. In Equation (1), 'A' expresses the concentration at the cloud center. 'B' expresses the relative concentration through out the cloud. It is a product of three exponential functions as shown in Equation (1), in the case of an infinite or semi-infinite medium (B), but it somewhat differs depending on the location of the point to be considered (ΣB) in the case of the closed system. In the closed system, a material is supposed to be reflected at the boundary without receiving any absorption or lost of the material as given in the images in the electrostatistics. A simple model of the North Pacific used in this paper is a rectangular prism, 18,000 km long (L), 5,000 km wide (r) and 5 km high (z), as shown in Figure 1, which is filled with ocean water. A pycnocline is assumed to exist at the depth of 1 km. For the sake of simplicity, a point of release is placed at the center of the bottom which is the origin of the coordinate system.

Some examples of formulea of ΣB, which is equivalent to B in the case of infinite medium are shown as follows:

$$\Sigma B(0, 0, 0) = [1 + 2\sum_{1}^{\infty} \exp\{-(nL)^2(4k_x t)^{-1}\}]$$

$$\times [1 + 2\sum_{l=0}^{\infty}\exp\{-(nr)^2(4k_y t)^{-1}\}]$$

$$\times [1 + 2\sum_{l=0}^{\infty}\exp\{-(2nz_0)^2(4k_z t)^{-1}\}] \quad (2)$$

$$\Sigma B(0,0,z_0) = 2[1 + 2\sum_{l}^{\infty}\exp\{-(nL)^2(4k_x t)^{-1}\}]$$

$$\times [1 + 2\sum_{l}^{\infty}\exp\{-(nr)^2(4k_y t)^{-1}\}]$$

$$\times \sum_{0}^{\infty}\exp\{-(1+2n)^2 z_0^2 (4k_z t)^{-1}\} \quad (3)$$

$$\Sigma B(L/2,0,z_0) = 4[\sum_{0}^{\infty}\exp\{-(0.5+2n)^2 L^2 (4k_x t)^{-1}\}$$

$$+ \sum_{0}^{\infty}\exp\{-(1.5+2n)^2 L^2 (4k_x t)^{-1}\}] \times [1 + 2\sum_{l}^{\infty}\exp\{-(nr)^2(4k_y t)^{-1}\}]$$

$$\times \sum_{0}^{\infty}\exp\{-(1+2n)^2 z_0^2 (4k_z t)^{-1}\} \quad (4)$$

and

$$\Sigma B(L/2, r/2, z_0) = 8[\sum_{0}^{\infty}\exp\{-(0.5+2n)^2 L^2 (4k_x t)^{-1}\}$$

$$+ \sum_{0}^{\infty}\exp\{-(1.5+2n)^2 L^2 (4k_x t)^{-1}\}]$$

$$\times [\sum_{0}^{\infty}\exp\{-(0.5+2n)^2 r^2 (4k_y t)^{-1}\} + \sum_{0}^{\infty}\exp\{-(1.5+2n)^2 r^2 (4k_y t)^{-1}\}]$$

$$\times \sum_{0}^{\infty}\exp\{-(1+2n)^2 z_0^2 (4k_z t)^{-1}\} \quad (5)$$

The formulea shown above are those in the case of diffusion alone with no advection. z_0 in Equations (3), (4) and (5) denotes the vertical distance from the origin of co-ordinate system to the pycnocline. In considering the horizontal advection, simply a constant velocity (U cm s^{-1}) was assumed throughout the whole region. Accordingly, a patch of the material cloud released from the source always moves in a certain direction, which is chosen as an x-direction, leading to the right hand side in Figure 1. It is assumed that as soon as the patch has reached the boundary at the right hand side, it appears at the left hand side immediately. In this way, the model can be used to represent a circulation of material cloud [6]. When we put U cm s^{-1} as a rate of horizontal steady flow, $\Sigma B(0,0,z_0)$ becomes as follows:

$$\Sigma B(0,0,z_0) = 2[\sum_{0}^{\infty}\exp\{-(\Delta + nL)^2 (4k_x t)^{-1}\}$$

$$+ \sum_{l}^{\infty}\exp[-(nL - \Delta)^2 (4k_x t)^{-1}\}]$$

$$\times [1 + 2\sum_{l}^{\infty}\exp\{-(nr)^2(4k_y t)^{-1}\}] \times \sum_{0}^{\infty}\exp\{-(1+2n)^2 z_0^2 (4k_z t)^{-1}\} \quad (6)$$

where Δ = Ut when $0 < Ut < L/2$, Δ = L-Ut when $L/2 < Ut < L$, Δ = Ut - L when $L < Ut < (3/2)L$, Δ = 2L - Ut when $(3/2)L < Ut < 2L$, Δ = Ut - 2L when $2L < Ut < (5/2)L$, Δ = 3L - Ut when $(5/2)L < Ut < 3L$, etc.

In the next place, when an up-welling motion with the velocity of w cm s^{-1} is considered, $\Sigma B(0,0,z_0)$ becomes as follows:

$$\Sigma B(0,0,z_0) = [1 + 2\sum_{l}^{\infty}\exp\{-(nL)^2 (4k_x t)^{-1}\}]$$

$$\times [1 + 2\sum_{l}^{\infty}\exp -(nr)^2 (4k_y t)^{-1}] \times [\sum_{0}^{\infty}\exp\{-(z_0 - wt + 2nz_0)^2 (4k_z t)^{-1}\}$$

$$+ \sum_{0}^{\infty}\exp\{-(z_0 + wt + 2nz_0)^2 (4k_z t)^{-1}\}] \quad (7)$$

3. RESULTS AND DISCUSSIONS

Let us consider a role of pycnocline in the diffusion process. If the pycnocline interferes the penetration of the material effectively, it works as if there is a boundary wall on which the material is reflected. On the contrary, if the pycnocline does not

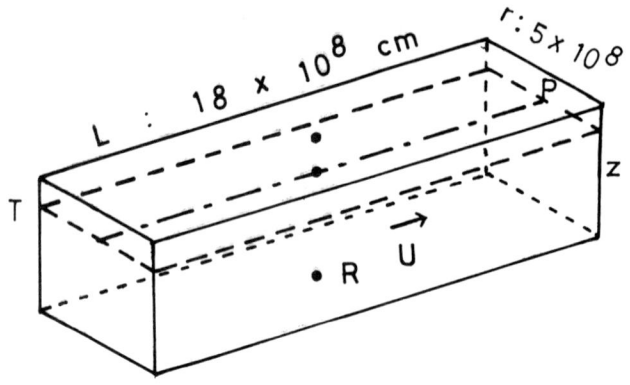

Figure 1. A model of the North Pacific. U, R and T indicate current velocity, the point of release and transition layer, respectively.

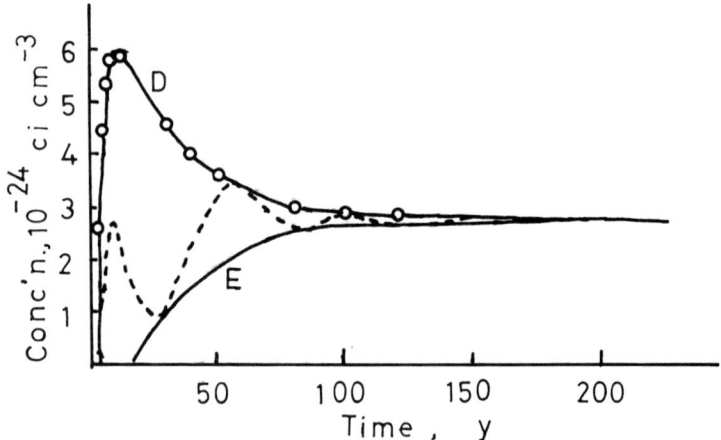

Figure 2. The variation in the activity concentration with time after the release of radioactive material. Solid lines 'D' and 'E', a dotted line and open circles indicate the case of no horizontal advection at the locations $(0,0,z_o)$ and $(L/2,0,z_o)$, that of horizontal flow rate of 1 cm s^{-1} and that of up-welling velocity of 1 x 10^{-4} cm s^{-1}, respectively.

work as a boundary, the material is reflected at the ocean-atmosphere boundary passing through the pycnocline. Needless to say that the resulting concentration of the material is higher in the former case than in the latter. However, an actual state seems to be somewhere between two cases. Therefore, the further calculation was done assuming the former case which leads to the more conservative results. Equations (3), (4) and (5) mentioned previously show those for the former case.

Results of calculation are shown in Figure 2. In this Figure, a curve 'D' shows the secular variation of concentration at the depth of pycnocline above the point of release, that is, the point $(0,0,z_o)$ after one instantaneous release of one curie of the radioactive material at the dump site under the condition of 1×10^8 cm^2 s^{-1} of k_x and k_y, and 1×10^2 cm^2 s^{-1} of k_z but no advection. A curve 'E' shows the secular variation of concentration at the point $(L/2,0,z_o)$, that is the point P in Figure 1 under the same condition as above. A broken line meandering between curves 'D' and 'E' shows the secular variation of concentration at the point $(0,0,z_o)$ under the influence of diffusion as above with a horizontal advection of $U = 1$ cm s^{-1}. Small open circles on the curve 'D' show concentrations at respective time which was elapsed after release. Concentration was obtained by the calculation under influence of both diffusion as above with an up-welling of $w = 1 \times 10^{-4}$ cm s^{-1}. From the results shown in Figure 2, it can be concluded that under the condition of $k_x=k_y=1 \times 10^8$ and $k_z = 1 \times 10^2$ cm^2 s^{-1} for diffusion and 1 cm s^{-1} for horizontal advection, the concentration at the depth of pycnocline above the dump site is always below the value obtained by considering the effect of diffusion alone. The concentration calculated under the same condition as above for diffusion and that of 1×10^{-4} cm s^{-1} for up-welling, the concentration is equal to the value obtained under the inflence of diffusion alone. The same trend as above was confirmed also in the case of $k_x = k_y = 1 \times 10^7$ and $k_z = 1 \times 10^2$ cm^2 s^{-1}.

According to the study by Miyake and Saruhashi [7], [8] on the distribution of man-made radioactive materials in the open ocean, k_x and k_y were of the order of 1×10^7 to 1×10^8 cm^2 s^{-1} and k_z was of the order of 1×10^2 cm^2 s^{-1}. Defant [9] obtained the horizontal diffusion coefficient of 5.5×10^7 cm^2 s^{-1} in the Atlantic on the basis of the spreading of Mediterranean water flowing out through the Straits of Gibraltar. In conclusion, it may be said that, as explained in the previous paper, the estimation of concentration at the point $(0,0,z_o)$ obtained by taking only the influence of diffusion into consideration is appropriate for the safety assessment on radioactive waste disposal on the deep sea floor, regardless the horizontal advection and up-welling are taken into account or not.

Summing up each value in every year in the secular variation of concentration for infinite length of time, the accumulated concentration ΣC can be obtained. Figure 3 shows the relation between the accumulated concentration and a half-life of each nuclide with respect to three locations i.e. $(0,0,0)$, $(0,0,z_o)$ and $(L/2,r/2,z_o)$ in both cases of $k_x = k_y = 1 \times 10^8$ and $k_z = 1 \times 10^2$ cm^2 s^{-1} (shown by thick lines) and $k_x = k_y = 1 \times 10^7$ and $k_z = 1 \times 10^2$ cm^2 s^{-1} (shown by broken lines). Each line shown in Figure 3 was obtained at a point whose location is shown in Table 1 under the condition of diffusion shown in the same table. The accumulated concentration at any location in the North Pacific is expressed by that at a point found somewhere between both curves 'a' and 'c' in the case of $k_x = k_y = 1 \times 10^8$ and $k_z = 1 \times 10^2$ cm^2 s^{-1} and between both curves 'd' and 'f' in the case of $k_x = k_y = 1 \times 10^7$ and $k_z = 1 \times 10^2$ cm^2 s^{-1}. Figure 3 shows that as the horizontal diffusion coefficient increases, the highest concentration decreases, on the contrary, the low-

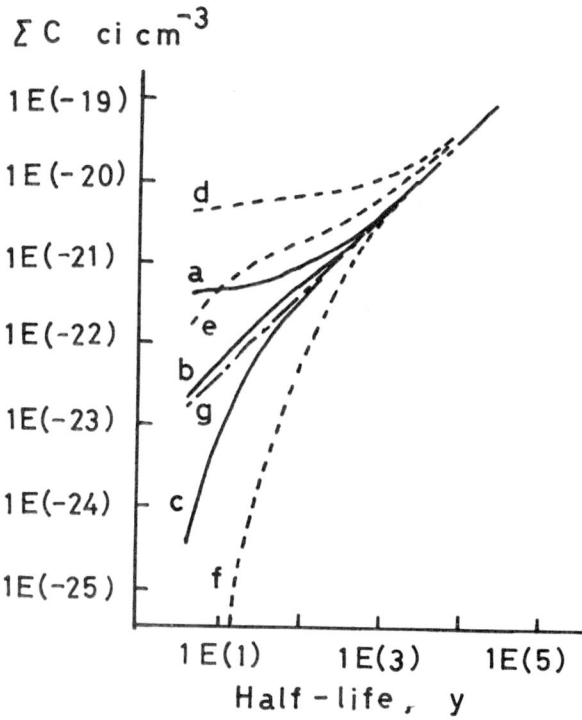

Figure 3. Relation between the accumulated concentration of a radioactive nuclide and its half-life.

Table I Diffusion coefficients and location of the point for calculation of the concentration of radioactive material in sea water in connection with Figure 3.

Curve in Fig. 3	Diffusion coefficient, $cm^2 \ s^{-1}$ Horizontal	Vertical	Location of point cm
a	1×10^8	1×10^2	(0,0,0)
b	do.	do.	$(0,0, \ 4 \times 10^5)$
c	do.	do.	$(9 \times 10^8, \ 2.5 \times 10^8, 4 \times 10^5)$
d	1×10^7	1×10^2	(0,0,0)
e	do.	do.	$(0,0, \ 4 \times 10^5)$
f	do.	do.	$(9 \times 10^8, 2.5 \times 10^8, 4 \times 10^5)$

est concentration increases and the concentration at any part of the system approaches the homogeneous concentration with the increase in the diffusion coefficient.

A straight line 'g' shows the relation between the accumulated concentration and half-life under the extreme condition of the instantaneous, complete mixing which is equivalent to the diffusion coefficient large enough. At the instantaneous, complete mixing, the relation between the accumulated concentration and half-life can be expressed by the next equation

$$\Sigma C = (v \cdot \lambda)^{-1} \qquad (8)$$

where V, the volume of the North Pacific is 3.6×10^{23} cm^3. Then, the following equation is obtained:

$$\log(\Sigma C) = \log(4 \times 10^{-24}) + \log T \qquad (9)$$

where T is half-life. Equation (9) shows that a plot of the common logarithm of the accumulated concentration against the common logarithm of half-life makes a straight line with a slope of 45° and gives a y-intercept of 4×10^{-24} ci cm^{-3}.

REFERENCES
[1] IAEA : "Convention on the Prevention of Marine Pollution by Dumping of Wastes and Other Matter." Information Circular INFCIRC/205/Add. 1, 10 Jan. 1975.
[2] Webb, G.A.M. and F. Morley : "A Model for the Evaluation of the Deep Ocean Disposal of Radioactive Waste." NRPE-R14, National Radiological Protection Board, UK (1973).
[3] Miyake, Y. and K. Saruhashi : "A Critical Study on the IAEA Definition of High Level Radioactive Waste Unsuitable for Dumping at Sea." Pap. Met. Geophys., 27, 75-80 (1976).
[4] Sugiura, Y., K. Saruhashi and Y. Miyake : "Evaluation on the Disposal of Radioactive Wastes into the North Pacific." Pap. Met. Geophys., 27, 81-87 (1976).
[5] IAEA : "The IAEA Revised Definition and Recommendations of 1978 Concerning Radioactive Wastes and Other Radioactive Matter Referred to Annexes I and II of the Convention on the Prevention of Marine Pollution by Dumping of Wastes and Other Matter." INFCIRC/205/Add. 1/Rev. 1, Aug., 1978.
[6] Shepherd, J.G. : "A Simple Model for the Dispersion of Radioactive Wastes Dumped on the Deep Sea Bed." MAFF. Fish. Tech. Rep. No. 29, 1976.
[7] Miyake, Y. and K. Saruhashi : "Distribution of Man-Made Radioactivity in the North Pacific through Summer 1955." J. Mar. Res., 17, 383-389(1958).
[8] Miyake, Y. and K. Saruhashi : "Vertical and Horizontal Mixing Rates of Radioactive Material in the Ocean." Disposal of Radioactive Waste, pp. 167-173, 1960.
[9] Defant, A. : "Die Ausbreitung des Mittelmeer wassers im Nordatlantischen Ozean." Pap. in Marine Biology and Oceanography, Suppl. to Vol. 3 of Deep-Sea Res., 465-470(1956).

A PRELIMINARY ASSESSMENT OF BIOLOGICAL TRANSPORT OF RADIONUCLIDES
DUMPED AT DEEP SEA BOTTOM

T. Doi*, T. Kidachi*, K. Honjo*, Y. Matsushita**, T. Nemoto***,
M. Shimizu****, H. Sudo* and H. Tsuruga*

* Tokai Regional Fisheries Research Laboratory, Tokyo (Japan)
** Rad Waste Management Center, Tokyo (Japan)
*** Ocean Research Institute, University of Tokyo, Tokyo (Japan)
**** Faculty of Agriculture, University of Tokyo, Tokyo (Japan)

ABSTRACT

In hazard evaluation of deep sea disposal of solid radioactive wastes, biological transport through food chain has not so far been fully considered. In the present paper we examined how to include in the computation of nuclide concentration in predator species the transport of nuclide through prey organisms and obtained two equations. Then a model network was constructed to describe food chain from detritus upto main commercial species in northwest Pacific, supposed disposal area. Biological transport through this model network was then calculated using two equations for nuclides released at sea bottom of 5 km deep. Influence of changes in various conditions and values of parameters was examined.

Introduction
    Deep sea disposal of solid radioactive wastes has been carried out by European countries and United States and hazard evaluation concerning this has been performed already. In Japan also, deep sea disposal is now under consideration and necessary surveys of proposed sites on oceangraphy, ecology and fisheries are now underway together with dose assessment for this disposal. [1], [2]
    The evaluation so far has been based on the concentration of a nuclide in marine organisms which are estimated using the concentration factor and the concentration of the nuclide in sea water which was assumed to distribute from the dumping site to the surface by physical transport. In this process biological transport. In this process biological transport is not usually considered. If the distribution of a nuclide is uniform, i.e. the concentration of a nuclide is same from the deep to the surface, biological transport, migration of a nuclide through marine food chain, need not be taken into consideration. However, if there exists any gradient in the concentration of a nuclide, transport through food chain should be included in the calculation. When a fish living in the surface layer, for instance, feeds not only on preys living in the same surface layer but also living in the deeper layer where the concentration of a nuclide is higher than in the surface layer, the estimation of the concentration in that fish by multiplying the concentration in sea water of the surface layer with the concentration factor will make an underestimates. Therefore the evaluation so far usually used the concentration in the deeper layers than the fish living layer to avoid this underestimation.
    The case is not the none where the magnitude of biological transport was estimated. According to Ketchum and Bowen (1958)[3], Feldt (1967)[4] and Lowman et al. (1971)[5], biological transport is smaller than physical transport in either down to up or up to down in open ocean with its small biomass unless the concentration factor is sufficiently large.
    In the present paper it is aimed to make quantitatively clear the difference between the case where migration of a nuclide through foodchain is taken into consideration and that where without any consideration to this pathway.

Calculating method of nuclide concentration in marine organism in prey-predator system
1. Concept
    First a simple system is taken. A predator P lives in a layer I and feeds on a prey F_I living in the same layer and also another prey F_{II} living in a deeper layer II.
    The concentration of nuclide A in P, P_A, is the sum of a portion taken up directly from water, and another derived from preys. It is assumed here that the portion from water and that from food are mutually independent and the former is proportional to the concentration in water and the latter to that in prey. Namely P_A is written as

$$P_A = K_W \cdot W_{IA} + K_F \cdot U[\alpha F_{IA} + (1-\alpha) F_{IIA}] \quad (1)$$

where K_W and K_F are constants of transport from water and from prey respectively, U ration and α the ratio of F_I consumption to the total. K_W and K_F are constants which are determined for each species-nuclide combination and not to be varied with changes in concentration neither in water nor in prey.
    Further it is assumed that the ratio of the part derived from water in the whole P_A is a constant in the case where the concentration of nuclide in water is the same through all layers. If we put it w, the ratio derived from prey is 1-w.

2. K_W and K_F
    If we use the ratio w, K_W can be written as $[CF]P_A \cdot w$. K_F on the otherhand, is a ratio to the concentration of prey of that in predator at equilibrium which is attained by continuous consumption of prey having some concentration of the nuclide.
    If we put F_A the concentration of nuclide A in prey, ration U, absorption rate from alimentary canal ε and turnover rate λ_b, the concentration of nuclide A in the predator derived from food at an equilibrium $[P_A]_F$ is

$$[P_A]_F = \frac{U \cdot F_A \cdot \varepsilon}{\lambda_b} \quad (2).$$

Therefore K_F is $[P_A]_F/F_A = U \cdot \varepsilon/\lambda_b$. Strictly speaking, we should use effective decay constant λ_{eff} instead of biological decay constant λ_b but if physical decay constant λ is quite small, we can approximately use λ_b. Thus K_F is a constant determined by the metabolic parameters of the predator such as ration, absorption rate, turnover rate or biological decay constant.

K_F can be also derived in another way as the following. Since K_F is a ratio of the concentration in the predator derived from prey to that in prey, K_F is written as follows if the predator P feeds on F_I only,

$$K_F = \frac{[P_A]_{FI}}{F_{IA}} = \frac{P_A(1-w)}{F_{IA}} = \frac{[CF]P_A(1-w)}{[CF]F_{IA}} \quad (3).$$

However, in case the predator P feeds also another prey (here F_{II}), the equation changes into

$$K_F = \frac{[CF]P_A(1-w)}{\alpha[CF]F_{IA} + (1-\alpha)[CF]F_{IIA}} \quad (4).$$

Thus we have two ways to obtain K_F, namely one is based on the dynamics of feeding and metabolism and another on the ratio of concentration factors of predator and prey modified by contribution rate of food. These two methods should give the same result if the metabolism of a nuclide in an organism is fully understood. At the moment, however, the informations on these parameters are very scarce and therefore the two methods will give different results.

3. General formulae

Up to here, we only examined a simple case where one predator with two prey species living in different layers. As a next step, we need a general formulae which can describe more complicated system. We consider the marine ecosystem as a network shown in Fig. 1. (shown later) In this figure, each group designated by D, P, MN, F etc. is called biobox. Here it is necessary to describe the migration of nuclide into a biobox from prior bioboxes in general form.

As is the same with the above mentioned simple system, transport of a nuclide into a biobox can be devided into two, one is directly from water and another is through food. Namely,

$$R_j^{Total} = R_j^W + R_j^F \quad (5),$$

where R_j^{Total}: concentration of nuclide in biobox j

R_j^W : concentration of nuclide in biobox j derived directly from environmental water

R_j^F : concentration of nuclide in biobox j derived through food chain.

Here again R_j^W and R_j^F are assumed to be mutually independent.

The portion derived directly from water can be written as

$$R_j^W = K^W \cdot R^W \quad (6),$$

where R^W is the concentration of nuclide in the water layer biobox j inhabiting and K^W a constant, namely $[CF]_j \cdot w$.

The another portion transported through food chain can be written as

$$R_j^F = K^F \cdot (\sum_k \alpha_k \cdot R_{ik}^{Total}) \quad (7).$$

Here, i means biobox prior to biobox j in network, namely prey for predator j, k the number of prey species and α_k a ratio of consumption of kth prey biobox to total consumption. This may be called feeding ratio and $\sum_k \alpha_k = 1$. K^F is a constant and defined as

$$K^F = \frac{\varphi \cdot \varepsilon}{\lambda_b} \quad (8),$$

or

$$K^F = \frac{[CF]_j \cdot (1-w_j)}{\sum_k \varepsilon_k \cdot [CF]_{ik}} \quad (9).$$

In the equation (8) ration is expressed as annual feeding rate φ, namely annual

Table I. Distribution model of radionuclides in waters characterized with selected locality on the North-West Pacific.

Locality*	Nuclides	Types of nuclide's distribution**	Concentrations of nuclides allocated to layers divided by depth***, pCi/m^3				
			0-0.2 km	0.2-1.0 km	1.0-3.0 km	3.0-5.0 km	4.5-5.0 km
A	^{60}Co	I	1.18 E4	1.18 E4	4.03 E4	1.68 E3	2.28 E3
		II	1.58 E3	1.58 E3	1.58 E3	1.58 E3	1.58 E3
		III	---	---	---	1.58 E2	1.58 E2
	^{90}Sr, ^{137}Cs	I	8.32 E4	8.32 E4	1.35 E3	3.14 E3	3.93 E3
		II	2.00 E3	2.00 E3	2.00 E3	2.00 E3	2.00 E3
		III	---	---	---	2.01 E2	2.01 E2
	^{60}Co	I	9.88 E5	9.88 E5	3.00 E4	1.10 E3	1.46 E3
		II	5.98 E4	5.98 E4	5.98 E4	5.98 E4	5.98 E4
		III	---	---	---	5.98 E3	5.98 E3
B	^{90}Sr, ^{137}Cs	I	7.75 E4	7.75 E4	1.16 E3	2.33 E3	2.82 E3
		II	1.58 E3	1.58 E3	1.58 E3	1.58 E3	1.58 E3
		III	---	---	---	1.58 E2	1.58 E2

*   A: Central region.
     B: Northern region related to Oyashio.

**   I: Fundamental distribution type based on a simplified diffusion formula.
     II: Uniform distribution.
     III: Retention in bottom layer.
     Concentrations under the condition, that total nuclide released was retained in the layer of 4.5-5.0 Km, were applied not only to 4.5-5.0 Km but also 3.0-5.0 Km.

*** Concentrations of nuclides were derived from the values of equilibrium level obtained by use of diffusion formula, assuming that 1 Curie of a single nuclide was continuously released annually.
Ex indicates E x 10^{-x}.

consumption/standing crop, instead of ration U in the equation (2).

Summarizing these, the following equations give the estimates of concentration of nuclide in biobox j.

$$R_j^{Total} = [CF]_j \cdot w \cdot R^W + \frac{\varphi \cdot \varepsilon}{\lambda_b} (\sum_k \alpha_k \cdot R_{ik}^{Total}) \qquad (10)$$

or

$$R_j^{Total} = [CF]_j \cdot w \cdot R^W + \frac{[CF]_j \cdot (1-w_j)}{\sum_k \alpha_k \cdot [CF]_{ik}} (\sum_k \alpha_k \cdot R_{ik}^{Total}) \qquad (11).$$

We utilized a computor in the calculation. First the concentration of nuclide in detritus, biobox D, at the bottom of trophic level was calculated and then the calculation was proceeded up and up the trophic level. For each biobox the concentration of nuclide was calculated using the equation (10) or (11).

Examples of computation

The above mentioned concept, model and equations were used to estimate the concentration of nuclides in commercial species in northwest Pacific where deep sea disposal of radioactive wastes are now planned. The results are shown below.

Since informations necessary to make this estimation, for example deep water current, abundance and distribution of organisms especially in deeper water, are very scarce, values of parameter and assumptions on model were so set as to give conservative results.

1. Target area

The northwest Pacific including supposed disposal area was simplified into a model ocean of 12,000km (E-W) × 6,000km (N-S) × 5km (deep) which has land boundaries, one extending north and south and the other extending east and west, 1,000km west and 2,000km north of the supposed dumping site, respectively. In this vast space, two representative areas, namely the central region and the northern region related to Oyashio, were taken as the object of study considering the present situations of main fisheries and fishing grounds in north Pacific. The central region is a water column of 100km square × 5km deep just above the dumping site and the northern region is a water column of the same volume situated 500km north of the dumping site.

2. A set of conditions

In this computation the conditions were given as follows.

i) Radioactive nuclides; Considering the nuclide composition of radioactive wastes which is now planned to be dumped at deep sea bottom, computation was made on three main nuclides, Cs 137, Sr 90 and Co 60.

ii) Nuclide concentration in sea water of each depth; Each nuclide was assumed to be released 1 Ci/year continuously and the concentration in water was calculated by the following formula.

$$C(x,y,z,t) = \int_0^t (q \cdot e^{-\lambda t} \cdot dt / 8(\pi t)^{3/2} \cdot (D_x \cdot D_y \cdot D_z)^{1/2} \cdot$$
$$\times \exp[(x^2/4D_x \cdot t + y^2/4D_y \cdot t + z^2/4D \cdot t)],$$

where $C(x,y,z,t)$ is nuclide concentration in water at the point of x,y,z at time t, q release rate of nuclide (1 Ci/year), D_x, D_y and D_z diffusion coefficient in the x-, y-, and z-direction, and λ physical decay constant 1/year. Here each nuclide is released in still-standing and homogeneous media at the flat bottom of 5 km deep. Though in the formula advection term is not included, the effect of advection were considered on selecting the values of diffusion coefficient and applying the nuclide concentration on further computation. As diffusion coefficients, 10^7 (cm^2/sec) for D_x and D_y and 200 (cm^2/sec) for D_z were chosen. In the computation the concentration at equilibrium condition was used and the values for the central region were those of the water column just above the released site and for the northern region related to Oyashio the averages of those of central region and those of the water column 500 km north of the released site. Actually water column was divided into five layers of 0-0.2, 0.2-1, 1-3, 3-5, 4.5-5 km corresponding to the inhabiting layers of each biological group and the average concentration of each layer was used in the computation. Further to this type of distribution, two more extreme types of distribution were considered. One is the case where all the released nuclide is to be retained in the deepest layer of 4.5-5 km and the other uniform distribution from deep to the surface considering large vertical transport of nuclide in case of upwelling etc. The concentrations used in the computation are given in Table I.

Table II. Tentative classification of biological groups as the unit of biological transportation of radionuclides in the North-West Pacific.

Biological group	Bathymetrical range, km	Central region	Northern region related to Oyashio
Detritus (D)	5.0 - 0 (1-4)*	Mainly biological remains but including living phytoplankton and nannoplankton in this report	
	5.0 - 3.0 (5)	Copepods, Chaetognaths, Decapod crustaceans, Mysids	Copepods, Chaetognaths, Ostracods
	3.0 - 1.0 (6)	Copepods, Chaetognaths, Medusae, Decapod crustaceans	Copepods, Chaetognaths, Ostracods, Euphausiids
Macroplankton (P)	1.0 - 0.2 (7)	Copepods, Chaetognaths, Euphausiids, Ostracods	Copepods, Chaetognaths, Ostracods
	0.2 - 0 (8)	Copepods, Chaetognaths	Copepods, Chaetognaths, Euphausiids, Medusae
Macroplankton (MP)1)	1.0 - 0.2 (9)	Mysids, Decapod crustaceans, Amphipods	
	5.0 - 3.0 (10)	Sternoptyx, Melamphaes, Anotopterus	
	3.0 - 1.0 (11)	Gonostoma, Cyclothone	Cyclothone
Micronekton (MN)	1.0 - 0.2 (12)	Myctophidae, Gonostoma, Vinciguerria	Lampanyetus
	0.2 - 0 (13)	Myctophidae, Chauliodus	Myctophidae, Tarletonbeania
Deepsea benthos (B)	5.0 - 4.5 2) (14)	Annelids, Amphipods, Bivalves, Ophiuroids, Asteroids, Holothuroids, Actinaria	Decapod crustaceans, Tunicates, Holothuroids, Echinoids, Asteroids, Mollusks
Shallow water benthos (SB)	1.0		
Deepsea bottom fish (BF)	5.0 - 4.5 (15)	Coryphaenoides	Coryphaenoides, Lionurus, Nematonurus
Shallow water bottom fish (DF)	0.2 - 0 (21)		Flatfish (Pleuronectiformes)
Squid (S)	1.0 - 0.2 (16)	Symplectoteuthis, Ommastrephes, Eucleoteuthis	Todarodes, Ommastrephes, Syplectoteuthis, Berryteuthis
Piscivorous fish (F)	0.2 - 0 (17)	Skipjack, Tunas, Marlins, Sharks	Yellow tail
Plankton feeding fish (PF)	0.2 - 0 (18)		Salmon, Cod, Saury, Skipjack, Herring, Sardine, Mackerel
Cetaceans (W)	0.2 - 0 (13,19)	Sperm whale and other small cetaceans	

1) Group of macroplankton which migrates vertically in the range of depth from 5 Km to surface.
2) In this model it is assumed that the bottom spreads within the depth from 5.0 to 4.5 Km.
*) Common numberings to Fig. 1 and Table 3, which are given to the both regional biological groups of every bathymetrical range.

iii) Model network of ecosystem; Organisms inhabiting in northwest Pacific were classified into 18 groups in the central region and 21 in the northern region as shown in Table II considering main inhabiting depth of each organism. These 18 and 21 groups in each region were considered to construct such a foodweb, a network of prey and predator as shown in Fig.1. Here predator was assumed not to feed on preys inhabiting in the upper layer so as to give conservative results.

iv) Biomass; Estimates of biomass of each group are given in Table III.

v) Feeding rate; Since informations are scarce on this parameter, three levels were assumed as high(H), medium(M) and low(L) on computation. The values of each level for each group are shown in Table IV.

vi) Parameters related to the metabolism of nuclide; In computation, concentration factor (CF), portion of uptake from water (w), absorption rate (ε) and turnover rate or biological decay constant (λ_b) are needed. These factors should be those for whole body since nuclide transport through ecosystem network is to be estimated. For the concentration factor Polikarpov (1966)[18], NAS (1971)[19] Ichikawa (1973)[20] and Shimizu (1973)[21] were referred. The CF value, however, is generally different from tissue to tissue and that for whole body is hardly found except for plankton. Therefore, when necessary, the values for each organ and tissue was were combined to get a whole body value according to body composition. As for fish the concentration factor for whole body was estimated from that for edible part which was used in the previous report of Nuclear Energy Safety Bureau (1976) [2] in order to be able to make a comparison of the results.

The parameters other than concentration factor have been usually obtained from the experiment. Rearing experiment, however, are quite few on organisms other than fish and shellfish, especially on plankton, and therefore, in the case of macroplankton the data obtained on euphausiids were quoted (Kuenzler 1969 [22], and Fowler et al. 1971[23]). The data on fish were sometimes applied to other organisms. Furthermore, necessary data on whales could not be found and parameters of metabolism in man (ICRP 1959)[24] were utilized.

The same values were used both in the central region and in the northern region, which are shown in Table V. As for the constant on the contribution of water (w), data could be hardly found and so in addition to the most probable value (L for Co 60, H for Sr 90 and M for Cs 137, c.f. Table 5) some different values were also applied to know the influence of changes in this parameter.

3. Results and discussions

The results of computation are given in Table VI. Here nuclide concentrations in commercial species are shown because of their importance in human consumption.

Two ways of calculation namely by the equation (10) and by the equation (11) were used in the present assessment. The former is based on dynamic processes of prey-predator relationship (hereafter referred as predation method) and the latter is, on the contrary, derived from static relationship expressed as CF ratio (CF ratio method). From the viewpoint of macroscopic and rather static analysis of the present paper, namely estimation of nuclide concentration in each biobox at steady state condition, CF ratio method is considered better than predation method either in principle or in reliability of parameters.

In the case of nuclide distribution in water of type I, nuclide concentration in piscivorous fish (F) estimated by CF ratio method using most probable values of w was compared with that which was obtained using concentration factors and concentration of nuclides in ambient water. This comparison, as is seen in Table VII, revealed that CF ratio method gave the almost same results for Sr 90, a little higher for Cs 137 and 3-5 times higher for Co 60. This means that the smaller the value of w, namely the larger the contribution of food in nuclide accumulation, the larger the influence of predation on prey inhabiting in lower layer where the nuclide concentration is higher.

When CF ratio method (I) is compared with predation method (II), the results are as follows;

Cs 137 (I) $\ll$ (II), Sr 90 (I) $\lesssim$ (II), Co 60 (I) $\gtrsim$ (II)..

Thus, predation method gives a quite higher for Cs 137 with high absorption rate.

When we examine the influence of changes in w and φ in CF ratio method and predation method, respectively, w has not so much influence. As for φ, in such a case as Sr 90 with low absorption rate and large w, change in φ gives no significant difference. However, in the case of Cs 137 with high absorption rate and Co 60 with small w influence of φ is quite large.

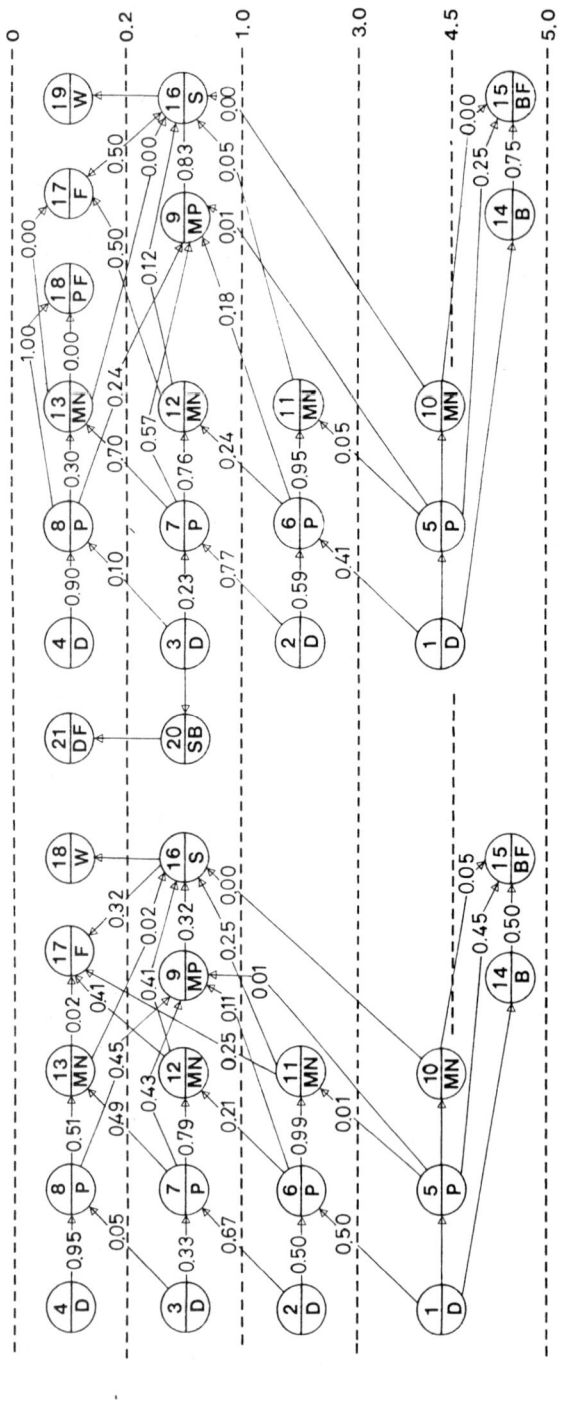

Figure 1. Simplified model of biological transportation of radionuclides in the North-West Pacific.

D: Detritus
P: Macroplankton
MP: Macroplankton
MN: Micronekton

B: Deepsea benthos
SB: Shallow water benthos
BF: Deepsea bottom fish
DF: Shallow water bottom fish

S: Squid
F: Piscivorous fish
PF: Plankton feeding fish
W: Cetacean

Representative or observed biological species and biomasses of these stratified biological groups can be found in Table 2 and 3, by the common numbers, which are shown in the above figure, to these tables. Mark of → indicates the relation from prey to predator, and the numerals written along this mark show the feeding ratio by which the predator depends on the individual biological groups. In this biological transportation model, it is assumed, in the case of predator feeds on plural number of preys, predator takes up the biological groups of prey with the ratio in proportion to their biomasses.

- 102 -

Table III. Estimates of biomass tentatively classified in consideration feeding habit and principally living zone.

Biological group	Bathymetrical range, km	Biomass, ton/(100km)^2			
		Central region		Northern region	
			Ref.		Ref.
Detritus (D)	5.0 - 3.0	(1) 3.2 E6	[1],[6]	(1) 3.3 E7	[6],[11]
"	3.0 - 1.0	(2) 3.2 E6	[1]	(2) 4.6 E7	"
"	1.0 - 0.2	(3) 1.6 E6	"	(3) 1.4 E7	"
"	0.2 - 0	(4) 3.2 E7	"	(4) 5.8 E7	"
Macroplankton (P)	5.0 - 3.0	(5) 1.0 E3	"	(5) 1.8 E4	[12]
"	3.0 - 1.0	(6) 5.8 E4	"	(6) 3.4 E5	"
"	1.0 - 0.2	(7) 2.2 E5	"	(7) 1.1 E5	"
"	0.2 - 0	(8) 2.3 E5	"	(8) 4.7 E5	"
" (MP)	5.0 - 0	(9) 1.4 E4	"	(9) 1.0 E5	"
Micronecton (MN)	5.0 - 3.0	(10) 1.1 E2	"	(10) 1 E1	"
"	3.0 - 1.0	(11) 1.1 E4	"	(11) 6.2 E3	"
"	1.0 - 0.2	(12) 1.8 E4	"	(12) 1.4 E4	"
"	0.2 - 0	(13) 1.0 E3	"	(13) 1 E1	"
Deepsea benthos (B)	5.0 - 4.5	(14) 1.1 E3	[7]	(14) 5.5 E4	[13],[14]
Shallow water bonthos (SB)	1.0 - 0.2	---		(20) 5.5 E5	
Deepsea bottom fish (BF)	5.0 - 4.5	(15) 1.1 E3	[1]	(15) 1.1 E4	**
Shallow water bottom fish (DF)	0.2 - 0	---		(21) 1.0 E4	[8],[15]
Squid (S)	1.0 - 0.2	(16) 1.4 E4	*	(16) 1.4 E4	*
Piscivorous fish (F)	0.2 - 0	(17) 4.8 E2	[8]	(17) 4.7 E2	[8],[16]
Plankton feeding fish (PF)	0.2 - 0	---		(18) 2.4 E4	[8],[17]
Cetaceans (W)	0.2 - 0	(18) 1.4 E3	[9],[10]	(19) 1.4 E3	***

Ex indicates 10^x

By use of the common numberings to Fig. 1 and Table II. presented as (x), representative or observed species of evry biological groups are shown in Table II, and position on the pathway of biological transportation of nuclides can be found in Fig. 1.

* estimated indirectly from whale stock
** no data, the value 10 times higher that of the central region used
*** the stock density was assumed the same as the central region

Table IV. Feeding rates of biological groups.

Biological group		Feeding rate, l/year		
		H	M	L
Macroplankton	5.0 - 3.0 km	109.50	36.50	18.25
	3.0 - 1.0	109.50	36.50	18.25
	1.0 - 0.2	365.00	109.50	36.50
	0.2 - 0	365.00	109.5	36.50
	5.0 - 0	109.50	36.50	18.25
Micronekton		36.50	18.25	3.65
Benthos		10.00	10.00	10.00
Deepsea benthos		10.00	10.00	10.00
Squid		73.00	36.50	18.25
Piscivorous fish		36.50	18.25	3.65
Cetaceans		14.60	14.60	14.60
Northern region related to Oyashio				
Macroplankton	1.0 - 0.2 km	109.50	36.50	18.25
	0.2 - 0	109.50	36.50	18.25
Plankton feeding fish		109.50	36.50	18.25
Shallow bottom fish		36.50	18.25	3.65

Table VII. Comparison of nuclide concentration in fish (F) of different calculation method

Locality	Nuclide	CF Whole body	Concentration in fish (F) Ci/ton		Ratio (1)/(2)
			(1)	(2)	
Central region	Cs 137	30	3.0 E14	2.5 E14	1.2
	Sr 90	20	1.7 E14	1.7 E14	1.0
	Co 60	1000	5.9 E13	1.2 E13	4.9
Northern region	Cs 137	30	2.5 E14	2.3 E14	1.1
	Sr 90	20	1.5 E14	1.5 E14	1.0
	Co 60	1000	3.0 E13	9.9 E14	3.3

Note: (1) CF ratio method, (2) obtained using CF and concentration of ambient water
Ex means 10^{-x}.

Table V. Values of parameters used for accumulation of radionuclides by the present biological transportation model.

Biological group	Absorption rate	Biological decay constant	Concentration factor	w H	w M	w L
^{60}Co						
Detritus	---	---	15×10^2	1.00	1.00	1.00
Macroplankton	0.3	109.5	7×10^2	0.10	0.10	0.10
Micronekton	0.1	1.46	10×10^2	0.50	0.10	0.01
Deepsea benthos	0.3	109.5	7×10^2	0.10	0.10	0.10
Shallow benthos	0.1	109.5	10×10^2	0.50	0.10	0.01
Bottom fishes	0.1	0.73	10×10^2	0.50	0.10	0.01
Squid	0.1	109.5	10×10^2	0.50	0.10	0.01
Fishes	0.1	0.73	10×10^2	0.50	0.10	0.01
Cetaceans	0.3	14.6	---	---	---	---
^{90}Sr						
Detritus	---	---	5×10^2	1.00	1.00	1.00
Macroplankton	0.6	109.5	5×10^2	0.90	0.90	0.90
Micronekton	0.05	7.2	5×10^2	0.99	0.90	0.50
Deepsea benthos	0.6	109.5	5×10^2	0.90	0.90	0.90
Shallow benthos	0.05	109.5	1	0.99	0.90	0.50
Bottom fishes	0.05	3.7	2×10^2	0.99	0.90	0.50
Squid	0.05	3.7	1	0.99	0.90	0.50
Fishes	0.05	3.7	2×10^2	0.99	0.90	0.50
Cetaceans	0.3	0.01	---	---	---	---
^{137}Cs						
Detrius	---	---	5×10^2	1.00	1.00	1.00
Macroplankton	0.3	36.5	2×10^2	0.80	0.50	0.20
Micronekton	0.9	3.65	3×10^2	0.80	0.50	0.20
Deepsea benthos	0.3	36.5	2×10^2	0.80	0.50	0.20
Shallow benthos	0.9	3.65	3×10^2	0.80	0.50	0.20
Deepsea bottom fish	0.9	3.65	3×10^2	0.80	0.50	0.20
Shallow bottom fish	0.9	1.83	3×10^2	0.80	0.50	0.20
Squid	0.9	3.65	3×10^2	0.80	0.50	0.20
Fishes	0.9	1.83	3×10^2	0.80	0.50	0.20
Cetaceans	1.0	1.83	---	---	---	---

Table VI. Radionuclides concentrations of biological groups obtained by applying varied values of parameters to biological transportation model, nCi/ton on wet basis.

Type of nuclides distribution	Biological group		FR-L w-L	FR-L w-M	FR-L w-S	FR-M w-L	FR-M w-M	FR-M w-S	FR-S w-L	FR-S w-M	FR-S w-S	w-L	w-M	w-S
	Central region in the North-West Pacific													
^{60}Co I	Squid	(16)	6.0 E4	5.1 E4	4.9 E4	1.3 E4	5.9 E5	4.4 E5	7.4 E5	1.7 E5	3.9 E6	2.5 E4	4.7 E4	5.4 E4
	Piscivorous fish	(17)	3.9 E3	3.5 E3	3.4 E3	5.2 E4	3.2 E4	2.7 E4	1.0 E4	2.5 E5	6.9 E6	2.4 E4	5.1 E4	5.9 E4
	Cetaceans	(18)	1.8 E4	1.5 E4	1.4 E4	3.7 E5	1.7 E5	1.3 E5	2.2 E5	5.0 E6	1.2 E6	7.5 E5	1.4 E4	1.6 E4
II		(16)	2.9 E3	2.0 E3	1.8 E3	1.1 E3	3.5 E4	1.8 E4	8.9 E4	1.9 E4	3.1 E5	1.6 E3	1.6 E3	1.6 E3
		(17)	1.0 E2	1.3 E2	1.2 E2	3.4 E3	1.4 E3	1.0 E3	1.1 E3	2.4 E4	4.1 E5	1.6 E3	1.6 E3	1.6 E3
		(18)	8.6 E4	5.9 E4	5.3 E4	3.3 E4	1.0 E4	5.3 E5	2.6 E4	5.6 E5	9.2 E6	4.7 E4	4.7 E4	4.7 E4
III		(16)	1.7 E3	1.7 E3	1.7 E3	1.5 E4	1.5 E4	1.5 E4	1.0 E5	7.9 E6	1.8 E5	7.7 E4	2.3 E3	2.8 E3
		(17)	1.2 E2	1.2 E2	1.2 E2	9.4 E4	9.4 E4	9.4 E4	1.8 E5	1.8 E5	1.8 E5	7.2 E4	2.6 E3	3.2 E3
		(18)	5.2 E4	5.1 E4	5.1 E4	4.4 E5	4.3 E5	4.3 E5	3.0 E6	2.4 E6	2.2 E6	2.3 E4	6.9 E4	8.3 E4
^{90}Sr I		(16)	3.4 E6	3.2 E6	2.3 E6	1.4 E6	1.3 E6	7.8 E7	1.1 E6	9.8 E7	5.4 E7	8.3 E7	8.4 E7	9.5 E7
		(17)	3.6 E5	3.4 E5	2.2 E5	2.2 E5	2.0 E5	1.1 E5	1.7 E5	1.6 E5	8.7 E6	1.7 E5	1.7 E5	2.0 E5
		(18)	2.9 E4	2.8 E4	2.0 E4	1.2 E4	1.2 E4	6.7 E5	9.2 E5	8.5 E5	4.7 E5	7.2 E5	7.3 E5	8.2 E5
II		(16)	6.8 E6	6.6 E6	4.5 E6	3.9 E6	3.0 E6	1.7 E6	2.5 E6	2.3 E6	1.3 E6	2.0 E6	2.0 E6	2.0 E6
		(17)	7.3 E5	6.8 E5	3.8 E5	4.9 E5	4.5 E5	2.6 E5	4.1 E5	3.7 E5	2.1 E5	4.0 E5	4.0 E5	4.0 E5
		(18)	5.9 E4	5.7 E4	3.9 E4	2.8 E4	2.6 E4	1.5 E4	2.2 E4	2.0 E4	1.1 E4	1.7 E4	1.7 E4	1.7 E4
III		(16)	1.1 E6	1.1 E6	1.1 E6	1.1 E7	1.0 E7	9.5 E8	1.3 E8	1.2 E8	9.1 E9	3.8 E10	6.0 E9	3.8 E7
		(17)	1.3 E5	1.3 E5	1.3 E5	1.1 E6	1.1 E6	1.1 E6	2.3 E8	2.3 E8	2.2 E8	1.1 E9	1.1 E7	1.2 E5
		(18)	9.9 E5	9.9 E5	9.6 E5	9.1 E6	9.0 E6	8.2 E6	1.1 E6	1.1 E6	7.9 E7	3.3 E8	5.2 E7	3.3 E5
^{137}Cs I		(16)	1.8 E2	1.7 E2	1.6 E2	1.9 E3	1.6 E3	1.4 E3	2.1 E4	1.5 E4	9.0 E5	2.6 E5	3.0 E5	3.8 E5
		(17)	1.2 E1	1.7 E1	1.1 E1	7.2 E3	6.2 E3	5.2 E3	1.7 E3	1.5 E4	8.7 E5	2.6 E5	3.0 E5	3.9 E5
		(18)	1.4 E1	1.4 E1	1.3 E1	1.5 E2	1.2 E3	1.1 E2	1.7 E3	1.2 E3	7.2 E4	2.1 E4	2.3 E4	3.0 E5
II		(16)	2.7 E2	2.5 E2	2.4 E2	3.0 E3	2.5 E3	2.0 E3	4.0 E4	2.7 E4	1.5 E4	6.0 E5	6.0 E5	6.0 E5
		(17)	1.8 E1	1.7 E1	1.6 E1	1.1 E2	9.6 E3	7.7 E3	3.9 E4	2.7 E4	1.4 E4	6.0 E5	6.0 E5	6.0 E5
		(18)	2.1 E1	2.0 E1	1.9 E1	2.4 E2	2.0 E2	1.6 E2	3.2 E3	2.2 E3	1.2 E3	4.8 E4	4.8 E4	4.8 E4
III		(16)	2.5 E2	2.5 E2	2.4 E2	2.1 E3	2.1 E3	2.1 E3	1.1 E4	1.1 E4	1.1 E4	1.3 E6	1.6 E5	6.3 E5
		(17)	1.6 E1	1.6 E1	1.6 E1	8.0 E3	7.9 E3	7.8 E3	1.0 E4	1.0 E4	1.0 E4	9.3 E7	1.5 E5	6.8 E5
		(18)	2.0 E1	2.0 E1	1.9 E1	1.7 E2	1.6 E2	1.6 E2	8.9 E4	8.6 E4	8.3 E4	1.0 E5	1.3 E4	5.0 E4

Northern region related to Oyashio in the North-West Pacific

^{60}Co			(16) Squid	1.4 E4	8.9 E5	7.9 E5	6.2 E5	1.9 E5	5.3 E5	1.2 E5	2.5 E6	1.8 E4	2.7 E4	2.9 E4
	I		(17) Piscivorous fish	1.1 E3	9.1 E4	8.6 E4	2.2 E4	1.0 E4	7.1 E5	1.6 E5	3.2 E6	1.5 E4	2.9 E4	3.0 E4
			(18) Plankton feeding fish	6.5 E3	6.1 E4	6.0 E4	1.3 E4	9.5 E5	7.7 E5	3.8 E5	2.9 E6	9.9 E5	9.9 E5	9.9 E5
			(19) Cetaceans	4.0 E5	2.6 E5	2.3 E5	1.8 E5	5.6 E6	1.6 E5	3.5 E6	7.3 E7	5.4 E5	7.9 E5	8.6 E5
			(20) Shallow water benthos	6.3 E5	2.3 E5	1.4 E5	6.3 E5	2.3 E5	6.3 E5	2.3 E5	1.4 E5	9.9 E5	9.9 E5	9.9 E5
			(21) Shallow water bottom fish	3.0 E4	1.0 E4	5.7 E4	1.7 E4	5.5 E5	7.4 E5	1.9 E5	6.6 E6	9.9 E5	9.9 E5	9.9 E5
	II		(16)	4.8 E4	2.2 E4	1.6 E4	3.4 E4	8.6 E5	3.1 E4	6.8 E5	1.3 E5	6.0 E4	6.0 E4	6.0 E4
			(17)	3.2 E3	1.9 E3	1.7 E3	8.9 E4	1.8 E4	4.2 E4	1.6 E5	3.2 E6	6.0 E4	6.0 E4	6.0 E4
			(18)	4.0 E3	3.7 E3	3.7 E3	8.1 E4	5.8 E4	4.7 E4	2.3 E4	1.8 E4	6.0 E4	6.0 E4	6.0 E4
			(19)	1.4 E4	6.4 E5	4.6 E5	1.0 E5	2.5 E5	9.3 E5	2.0 E5	3.8 E6	1.8 E4	1.8 E4	1.8 E4
			(20)	3.8 E4	1.4 E4	8.6 E5	3.8 E4	1.4 E4	3.8 E5	1.4 E4	8.6 E5	6.0 E4	6.0 E4	6.0 E4
			(21)	1.8 E3	6.1 E4	3.4 E4	1.0 E3	3.3 E4	4.5 E4	1.1 E4	4.0 E5	6.0 E4	6.0 E4	6.0 E4
	III		(16)	1.7 E4	1.7 E4	1.7 E4	1.4 E4	1.3 E5	9.1 E7	8.7 E7	8.6 E7	7.7 E7	7.8 E7	7.9 E7
			(17)	1.5 E3	1.5 E3	1.5 E3	1.1 E4	1.1 E4	2.1 E6	2.1 E6	2.1 E6	1.5 E5	1.6 E5	1.6 E5
			(19)	5.0 E5	5.0 E5	5.0 E5	4.0 E6	4.0 E6	2.7 E7	2.6 E7	2.6 E7	1.5 E5	1.5 E5	1.5 E5
^{90}Sr	I		(16)	3.1 E6	3.1 E6	2.7 E6	1.5 E6	1.4 E6	1.1 E6	9.9 E6	6.7 E7	6.7 E5	6.8 E5	6.8 E5
			(17)	2.5 E4	2.3 E4	1.5 E5	1.8 E5	1.7 E5	9.6 E6	1.4 E5	8.0 E6	6.7 E5	6.7 E5	6.8 E5
			(18)	1.0 E4	1.0 E4	9.4 E4	3.7 E5	3.5 E5	2.5 E5	2.4 E5	1.7 E6	6.7 E5	6.7 E5	6.8 E5
			(19)	1.7 E3	2.7 E2	2.4 E4	1.3 E4	1.2 E7	9.2 E4	8.6 E5	5.8 E5	6.7 E5	6.7 E5	7.7 E5
			(20)	9.4 E3	8.7 E7	5.6 E4	9.4 E7	8.7 E7	9.4 E7	8.7 E7	5.6 E5	7.7 E5	7.7 E5	7.7 E5
			(21)	1.6 E5	1.4 E5	8.0 E6	1.6 E5	1.4 E5	1.5 E5	1.4 E5	7.8 E6	1.5 E5	1.5 E5	1.5 E5
	II		(16)	5.8 E6	5.7 E6	5.0 E6	2.9 E6	2.7 E6	2.2 E6	2.0 E6	1.4 E6	1.6 E6	1.6 E6	1.6 E6
			(17)	4.8 E5	4.4 E5	2.8 E5	3.4 E5	3.4 E5	3.2 E5	2.9 E5	3.5 E5	3.1 E5	3.1 E5	3.1 E5
			(18)	2.1 E4	2.0 E4	1.9 E4	7.4 E5	7.1 E5	5.1 E5	4.8 E5	3.5 E5	3.2 E5	3.2 E5	3.2 E5
			(19)	5.0 E4	4.9 E4	4.3 E4	2.5 E4	2.4 E4	1.9 E6	1.7 E4	1.2 E4	1.4 E4	1.4 E4	1.4 E4
			(20)	1.9 E6	1.8 E6	1.1 E6	1.9 E6	1.8 E6	1.9 E6	1.8 E6	1.1 E6	1.6 E6	1.6 E6	1.6 E6
			(21)	3.2 E5	2.9 E5	1.6 E5	3.2 E5	2.9 E5	3.1 E5	2.8 E5	1.6 E5	3.2 E5	3.2 E5	3.2 E5
	III		(16)	9.1 E7	9.1 E7	9.0 E7	7.0 E8	2.7 E6	1.0 E8	1.0 E8	1.0 E8	2.5 E10	2.5 E9	2.2 E8
			(17)	3.1 E6	3.1 E6	2.8 E5	3.4 E5	5.0 E7	5.0 E9	5.0 E9	3.0 E7	2.5 E10	3.0 E8	7.4 E7
			(19)	7.8 E5	7.8 E5	7.8 E5	6.0 E6	6.0 E6	8.9 E7	8.9 E7	8.8 E7	2.2 E8	2.5 E7	1.9 E6
^{137}Cs	I		(16)	3.2 E3	2.9 E3	2.5 E3	4.5 E3	3.5 E4	1.1 E4	7.3 E5	3.6 E5	2.4 E5	2.5 E5	3.0 E5
			(17)	3.4 E2	3.1 E2	2.7 E2	2.8 E3	2.5 E3	1.5 E4	1.3 E4	5.1 E5	4.7 E5	2.3 E5	2.3 E5
			(18)	2.5 E2	2.3 E2	2.0 E2	4.5 E3	3.6 E3	1.8 E4	1.3 E4	8.4 E5	2.3 E5	2.3 E5	2.9 E5
			(19)	2.5 E2	2.3 E2	2.0 E2	3.6 E3	2.8 E3	8.8 E4	5.8 E4	2.9 E4	1.9 E4	2.0 E4	2.4 E4
			(20)	1.1 E3	1.1 E3	1.0 E3	1.1 E4	1.0 E4	1.1 E4	1.1 E4	1.0 E4	2.3 E4	2.3 E4	2.3 E4
			(21)	2.0 E3	1.9 E3	1.8 E3	1.0 E3	9.6 E4	2.2 E4	2.0 E4	1.8 E4	2.3 E4	2.3 E4	2.3 E4
	II		(16)	4.4 E3	3.8 E3	3.3 E3	7.2 E3	5.3 E4	3.4 E4	1.3 E4	6.2 E5	4.7 E5	4.7 E5	4.7 E5
			(17)	4.7 E2	4.1 E3	3.5 E3	2.2 E3	3.3 E3	2.8 E4	1.9 E4	8.8 E5	4.7 E5	4.7 E5	4.7 E5
			(18)	5.2 E3	4.7 E3	4.1 E3	9.1 E3	7.2 E3	5.4 E4	2.7 E4	1.7 E4	3.7 E5	3.7 E5	3.7 E5
			(19)	3.5 E2	3.0 E2	2.6 E3	5.7 E3	4.2 E3	2.7 E3	1.1 E3	4.9 E4	1.1 E3	1.1 E3	1.1 E3
			(20)	2.3 E3	2.2 E4	2.0 E3	2.2 E3	2.0 E4	2.3 E4	2.2 E4	3.7 E4	4.7 E5	4.7 E5	4.7 E5
			(21)	4.1 E3	3.9 E3	3.6 E3	2.1 E3	2.0 E3	4.5 E4	4.1 E4	3.7 E4	4.7 E5	4.7 E5	4.7 E5
	III		(16)	4.8 E3	4.7 E3	4.7 E3	4.0 E4	3.8 E4	2.6 E5	2.4 E5	2.3 E5	5.9 E7	7.1 E6	2.7 E5
			(17)	4.8 E3	4.8 E3	4.7 E3	2.2 E3	2.1 E3	3.2 E5	3.1 E5	2.9 E5	2.4 E7	4.7 E6	2.3 E5
			(19)	3.8 E2	3.7 E2	3.7 E2	3.1 E3	3.0 E3	2.0 E4	1.9 E4	1.8 E4	4.7 E6	5.7 E5	2.2 E4

- 107 -

Comparison of the central region with the northern region showed generally higher concentration in organisms in the former which is considered to reflect higher concentration of nuclides in water.

Difference in the type of nuclide distribution in water brought about different results. In the case of CF ratio method with higher w (large than 0.5), type II gave the highest result and I and III followed in this order. When w is small (less than 0.2), results changes into the following order III > II > I. In the case of predation method type II gave the highest result regardless of the value of φ. And I $\gg$ III was observed for Sr 90 with low absorption rate but for Cs 137 and Co 60 the difference between III and I is not remarkable.

Brief consideration was given above on the result of the computation but the model and parameters used in the present assessment were based on insufficient informations and therefore it is necessary to repeat the computation in accordance with the advance in our knowledge of ocean structure, ecology of deep sea and metabolism of nuclide in each species. Further in the present report nuclide distribution was assumed to be determined only by physical diffusion and not to be modified by the result of biological transport. On future assessment, together with this point, recycling of nuclide from organism to water by excretes and death should be considered.

Reference

[1] Fishery Agency (1975) Report of researches on deep sea disposal of solid radioactive wastes; survey on marine organisms 1972-1974 (in Japanese)
[2] Science and Technology Agency, Nuclear Energy Safety Bureau (1976) Report on the evaluation of environmental safety of test deep sea disposal (in Japanese)
[3] Ketchum, B.H. and B.T. Bowen (1958) Biological factors determining the distribution of radioisotopes in the sea Proc. 2nd U.N. Conf. Peaceful Uses Atom. Ener. Vol.18
[4] Feldt, W. (1967) The transport, dilution and reconcentration of radioactivity by marine biological mechanisms OECD/ENEA Seminar Lisbon
[5] Lowman, F.G., Rice, T.R. and F.A. Richards (1971) Accumulation and redistribution of radionuclides by marine organisms in "Radioactivity in the Marine Environment" NAS
[6] Aruga, S. (1970) Primary production in the ocean *Kaiyo Kagaku* (Marine Science) 2(6) (in Japanese)
[7] Filatova, Z. (1969) The quantitative distribution of the deep sea bottom fauna Biology of the Pacific Ocean II
[8] FAO (1975) Yearbook of Fishery Statistics Vol.40
[9] Doi, T. (1971) Stock assessment of sperm whale Bull. Tokai Reg. Fish. Res. Lab. No.66 (in Japanese)
[10] Ohsumi, S. (1974) Stock of whales in "Important Stock of Distant Waters Fishery" Fishery Agency (in Japanese)
[11] Handa, N. and K. Matsunaga (1977) Distribution of particulate organic matter and its decomposition processes in the marine environment, Preliminary report of the Hakuho maru crise KH-75-4
[12] Vinogradov, M.E. (1970) Vertical Distribution of the Oceanic Zooplankton (English Version)
[13] Okutani, T. (1972) Quantity of megarobenthos in shallow waters mainly in Sagami Bay and Tokyo Bay - its evaluation and problem Benthos Res. Ass. J. No.5/6 (in Japanese)
[14] Horikoshi, M., Imajima, H. and S. Gamo (1977) Biomasses of benthos IBP Synthesis Vol.14
[15] Takahashi, Z. and Y. Kitano (1974) Demersal fish stock in north Pacific in "Important Stock of Distant Waters Fishery" Fishery Agency (in Japanese)
[16] Kurogane, K. (1972) A trial of stock assessment of yellowtail in Pacific *Teichi* (Set Net) No.43 (in Japanese)
[17] Nakai, Z. (1962) Studies relevant to mechanisms underlying the fluctuation in the catch of the Japanese Sardine *Sardinops melanosticta* (T & S) Jap. J. Ichthyol. XI
[18] Polikarpov, G.G. (1966) Radioecology of Aquatic Organisms North Holland Pub.
[19] National Academy of Sciences (1971) Radioactivity in the Marine Environment
[20] Ichikawa, R. (1973) Accumulation of radioactive nuclides in fish in "Radioactivity and Fish" Koseisha-Koseikaku Pub. (in Japanese)

[21] Shimizu, M. (1973) Bioconcentration of radioactive materials in the environment Radioisotopes 22(11) (in Japanese)
[22] Kuenzler, E.J. (1969) Elimination of iodine, cobalt, iron and zinc by marine zooplankton in "Symposium on Radioecology" USAEC CONF-670503
[23] Fowler, S.W., Small, L.F. and J.M. Dean (1971) Experimental studies on elimination of zinc-65, cesium-137 and cerium-144 by euphausiids Mar. Biol. 8(3)
[24] ICRP (1959) Recommendations of the International Commission on Radiological Protection: Report of Committee II on Permissible Dose for Internal Radiation, ICRP Publication 2

This report is a summary of a part of the work on the evaluation of environmental safety of test deep sea disposal performed by the experts group on control technology of radioactive wastes of Nuclear Energy Safety Commission.

Discussion

A.W. VAN WEERS, Netherlands

Is there indeed a considerable lack of parameter values to be put into the biological transport model - for radioactive material on the bottom of the North Pacific ?

M. SHIMIZU, Japan

Our knowledge of oceanography and ecology of the deep sea environment is quite limited. In addition, data on the metabolism of radionuclides and/or stable element in deep sea organisms are also scarce. Nutritional physiology should differ in deep sea organisms from that of surface dwellers. But we do not have any knowledge of it. Therefore I should say that there is indeed a considerable lack of parameters to be put into the biological transport model.

OUTLINE OF BIOLOGICAL SURVEYS AROUND PLANNED WASTE DUMPING SITE OFF JAPAN
CONDUCTED BY TOKAI REGIONAL FISHERIES RESEARCH LABORATORY

T. KIDACHI*, K. HONJO**, T. OKUTANI*** and T. WATANABE*

* Tokai Regional Fisheries Research Laboratory, Tokyo, Japan
** Seikai Regional Fisheries Research Laboratory, Nagasaki, Japan
*** National Science Museum (Natural History Institute), Tokyo, Japan

ABSTRACT

The biological surveys were conducted by the R/V *Kaiyo-Maru* during 1972-1974 in and around the proposed dumping site in the Northwest Pacific Basin. The sampling was made by diverse kinds of gears to collect various size spectra of various ecological group of marine fauna. Some of biological result obtained by this survey is reviewed.

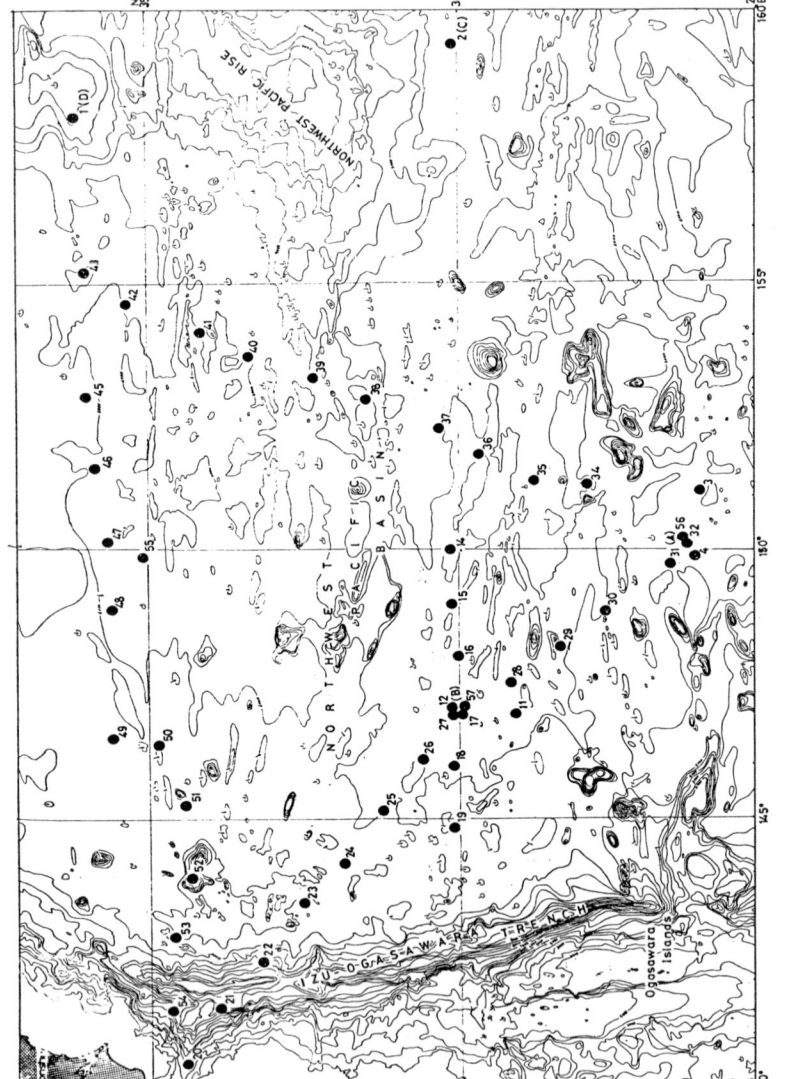

Figure 1. Biological stations occupied by the R/V Kaiyo-Maru during 1972 – 1974.

In response to a plan that solid waste of low activity level will be dumped in the Northwest Pacific Basin off Japan, Tokai Regional Fisheries Research laboratory, Fisheries Agency, carried out marine biological surveys on board the Research Vessels *Kaiyo-Maru* and *Soyo-Maru* around the proposed dumping site. Four stations, namely, Sts. A (26°N, 150°E), B (30°N, 147°E), C (30°N, 160°E) and D (36°N, 158°E), have been assigned to dumping sites. The purpose of the surveys of these two vessels has been in three folds: surveys on (1) fauna, (2) oceanography and (3) radioactivity background. The cruises conducted by the R/V *Kaiyo-Maru* during 1972-1974 covered all three branches of surveys. But, the R/V *Soyo-Maru* has been engaged in faunal survey, while the R/V *Kaiyo-Maru* has been devoted to oceanographic and radioactivity background surveys since 1977. The present paper will review the sampling plans and results of the surveys only from biological point of view.

The Coverage of Surveys

The surveys by the R/V *Kaiyo-Maru* during 1972-1974 were more concentrated to Sts. A and B. The routine macroplankton net hauls at regular interval and brief visits to Sts. C and D were also made. As the cruises of this ship were planned to be of over-all purposes, various gears that collect all size spectra of plankton (nanno-, micro-, macro- and megaloplankton), micronekton and epibenthos were employed.

In contrast to this, the new series of survey by the R/V *Soyo-Maru* is focused to faunal survey in which investigations of stratified distribution of macroplankton and species composition of megalobenthic fauna have been more emphasized. The geographical coverage by the R/V *Soyo-Maru* since 1977 has been not only the neighborhood of dumping site (Sts. A and B), but also some northern and southern extent of the Northwest Pacific Basin as well as a part of the Shikoku Basin.

The results of the 1972-1974 surveys were already made public (Fisheries Agency et al., 1975), but the data obtained by the surveys since 1977 have been on the way of analysis except preliminary results for 1977 and 1978 cruises were mimeographed as interim reports.

The contents of sampling and station plans for 1972-1974 and 1977-1979 surveys are shown in Table 1 and Figs. 1-2.

Table I. Biological sampling by the R/V *Kaiyo-Maru* (1972-1974) and the R/V *Soyo-Maru* (1977-1979)

Year	Vertical stratified plankton-net haul		Midwater horizontal opening-closing plankton-net tow		IKMT		Beam Trawl	Micro- and nanno-plankton (incl. suspension particles)		Routine macro-plankton-net haul
	No. of station	(No. of stratum)	No. of station	(No. of stratum)	No. of station	(No. of stratum)	No. of station	(No. of station)	(No. of stratum)	No. of station
1972	3	(12)	1	(3)	4	(5)	1	-		10
1973	4	(26)	1	(4)	2	(7)	1	3	(30)	33
1974	2	(16)	-		2	(8)	3	2	(14)	-
1977	2	(16)	2	(4)	-		14	-		16
1978	2	(16)	2	(5)	-		9	-		13
1979	7	(36)	2	(6)	-		11	-		29

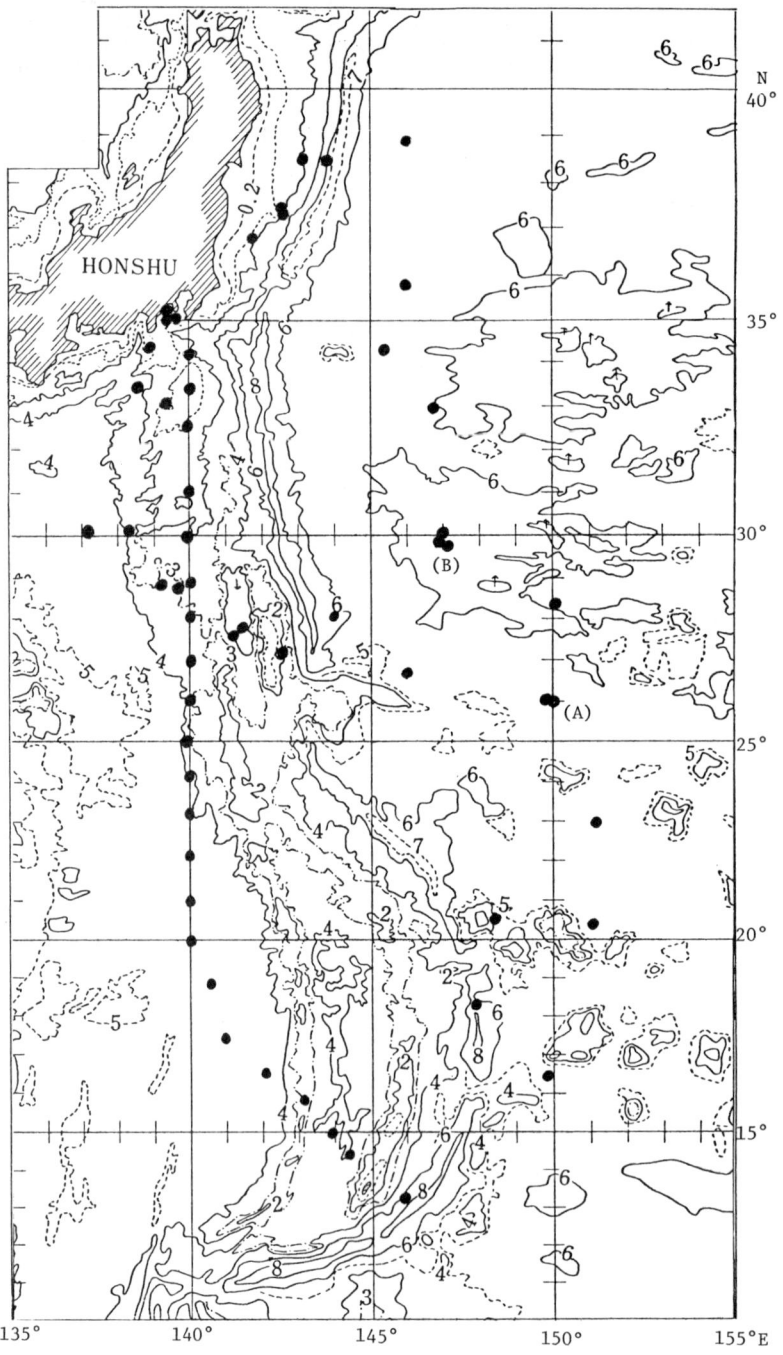

Figure 2. Biological stations occupied by the R/V *Soyo-Maru* during 1977 - 1979.

Summary of Biological Results for 1972-1974 Surveys

1. Macroplankton:

The standing crop of macroplankton in the surveyed area was unexpectedly large attaining some 53 g/1000 m^3 in the Kuroshio Extention and 98 g/1000 m^3 in the Countercurrent area. They are approximately equal to those in the neritic region of the Pacific coast of Japan. The vertical distribution of macroplankton biomass shows a tendency of gradual decrease down to 1000m-depth while abrupt decline deeper than 3000m-4000m. The biomass below 5000m attains only smaller than 1/1000 of the total biomass of the water column. Copepoda and Sagittoidea are dominant through all strata in wet weight. There is a certain discrepancy of dominant animal group in the present collection from the result obtained by Vinogradov (1968).

The numbers of copepod genera and species was minimum at 5000m-stratum and maximum between 500m and 2000m. The distribution of *Calanus sinicus*, which is the representative copepod in the neritic waters of Japan, ranged from 25°N in the south to 36°N in the north. The population of this copepod was found in the stratum between 150m and 3000m in the south of 30°N, while it shrunk in the south thereof where this species was found between 150m and 2000m. Subarctic copepods, *Calanus plumchrus* and *C. cristatus* appeared only in the north of 30°N at a depth of 500-3000m.

2. Microplankton:

Microplankton including nannoplankton and suspending organic particles were investigated during 1973 and 1974 by filtering 1000cm^3 of water taken with the Nansen bottle. The number of organisms (cells) was the most abundant in the surface layer and declined gradually with depth but again increased at depths of 3000m and 4000m. Such a tendency is quite different from that in netted plankton. Diatomacea found deeper than 1000m were mostly inactive cells. Dinoflagellata and Ciliata were more abundant shallower than 1000m than elsewhere. Some crustacean remains were particularly frequent in deeper layers.

3. Micronekton:

The micronektonic biomass taken by Isaacs-Kidd Midwater Trawl was the largest at 1000m stratum among 150m, 500m, 1000m and 3000m in any collected station. The micronektonic biomass of a station where some subarctic elements presented was twice larger than those from warm water area. The major components were fish, coelenterates, crustaceans and cephalopods. Fish is proportionally small in quantity at the 150m stratum.

Fish taken by IKMT comprised 112 species of 40 families and 11 orders, and more than 90% of population was occupied by two families Gonostomatidae and Myctophidae. Besides IKMT fish was taken by plankton net and beam trawl as well. Taking these catches into consideration, fish collected during 1972-1974 surveys was consisted of 131 species of 47 families and 15 orders, showing that this area yields quite rich piscifauna.

Some group of fish such as the Gonostomatidae shows ontogenic descend which was also demonstrated by the present study. According to the quantitative result of stratified collection of megaloplankton net, the fish population was the largest in the surface layer followed by upper bathypelagic zone (500-1000m) and minimized in abyssopelagic zone (below 3000m). The fish biomass expressed in wet weight was the largest in the upper bathypelagic zone followed by lower bathypelagic zone (1000-3000m) and very small in other strata. In qualitative point of view, the surface piscifauna contained some neritic-pelagic fish (mostly eggs and larvae) as well as intertidal species that are frequent in tropical coral reef. The occurrences of those fish will indicate that such an offshore region has a certain biological relation with insular neritic waters. Based on characteristics of bathymetrical distribution of species, fish under study were classified into seven categories. The analysis of the data made us concluded that mesopelagic fauna has a very close relation to the surface fauna, and upper bathypelagic fauna is consisted of not only endemic species of this zone but also visitors from both mesopelagic and lower bathypelagic zones functioning as a linkage between both zones. Some of near-bottom fish, such as Coryphaenoids, may play an important role in relaying lower bathyal and abyssopelagic zone through ontogenic and geographical migration. Such stepwise structure of dynamic distribution of micronekton will be the most important among biological transportation of substances and energy.

4. Benthos:

The small amount of megalobenthos was trawled from the basin with depths between 5740m and 6200m. (More intensive sampling in abyssal basin has been made by the R/V *Soyo-Maru* since 1977). Dominant taxa among trawl samples were Elasipodan holothurians and bivalve mollusks. Protobranchiate bivalves in the present collection were all widely distributed species. This suggests that the fauna around dumping site is involved in genetical dispersion for such an extensive geographical range.

Conclusive Remarks based on the 1972-1974 Survey

Through samplings by various kinds of gears, fauna of diverse size spectra and ecological units distributed in and around the proposed dumping site were clarified. The data came from pelagic sampling revealed that there is a quantitative discontinuity at around the depth of 2000m-3000m. Regardless the probable existence of stepwise linkage (biological transportation), nature and behavior of ecosystems above and below this boundary may be different from each other.
The origin and formation of organic particle from view point of its role in nutrition source of deepsea fauna and in concentrator of radionuclides in oceanic environment may be an important target for future study.

Zoogeography, quantitative distribution, compositional characteristics and other ecological aspects of deepsea fauna will become more evident after the material taken by the new series of faunistic surveys by the R/V *Soyo-Maru* will be worked out in the near future.

DISPERSION DE POLLUANTS D'UNE SOURCE AU FOND PROFOND
PAR DES TOURBILLONS A ECHELLE INTERMEDIAIRE

Kenzo Takano
Rikagaku Kenkyusho, Wako-shi
Saitama-ken, 351 Japon

et

M^{me}Sawa Matsuyama
Centre de Calcul, Université de Hosei
Koganei-shi, 184 Japon

RESUME

Quelques exemples numériques sont présentés pour la dispersion de particules rejetées au fond profond océanique et dérivant par l'advection due à des courants non permanents tourbillonnaires. Les courants sont régis par la force d'entraînement du vent et le flux de la chaleur de surface. Ils sont obtenus par un modèle mathématique de la circulation générale. On s'attache surtout à l'effet des tourbillons à échelle intermédiaire. Cet effet est remarquable. La dispersion se manifeste totalement différente si on les néglige.

ABSTRACT

Numerical examples are shown for the dispersion of particles discharged at the deep ocean bottom and drifting by the advection due to turbulent currents. The currents are governed by the wind stress as well as the surface heat flux. They are obtained by using a numerical model of the general circulation. Special attention is paid to the effect of the meso-scale eddies. This effect is so striking that the dispersion becomes quite different without taking it into account.

Position du problème

La présente note a pour objet de prendre un aperçu de l'-
effet des tourbillons à échelle intermédiaire sur la dispersion de
particules déchargées au fond océanique.
Rappelons que ces tourbillons dont l'existence a été bien
confirmée récemment [par exemple, 1] ont plusieurs dizaines de
kilomètres de diamètre et plusieurs dizaines de jours de période.
Alors que la dynamique des tourbillons n'est pas encore bien connue,
la plupart des énergies cinétiques de la mer s'y trouvent, ce qui
suggère leur importance non seulement dans la dispersion à échelle
intermédiaire mais encore dans la dispersion à grande échelle. Une
note précédente [2] a présenté un exemple de la dispersion de petites
masses d'eau dans un océan mondial au moyen d'un modèle mathématique
de la circulation générale. Celui-ci est plus rudimentaire que le
présent modèle à cause de la maille qui n'est pas suffisamment
affinée pour résoudre les tourbillons à échelle intermédiaire.
Les déchets radioactifs se dispersent par l'advection aussi
bien que par la diffusion turbulente. L'advection et la diffusion
turbulente ne sont bien définies que si l'échelle du phénomène dont
il s'agit est précisée. Dans le cas où s'impose le problème de la
dispersion de déchets radioactifs dans un océan, l'échelle du temps
est beaucoup plus longue que la période caractéristique des tourbillons, plusieurs dizaines de jours. A cet égard, les tourbillons y
interviennent en diffusion turbulente. C'est pouquoi nous nous
proposons, dans la dernière section, d'évaluer les coefficients de la
diffusion turbulente "virtuelle" due aux tourbillons à échelle intermédiaire.

Modèle mathématique

Pour commencer, on obtient les courants de base au moyen
d'un modèle mathématique de la circulation générale, en faisant les
hypothèses suivantes: 1° L'océan est borné par deux méridiens
espacés de 20,64° en longitude et par deux parallèles espacés de 18°
en latitude. La frontière sud est placée à 21°N. Il a 4000m de
profondeur partout. 2° Les courants s'établissent par la force
d'entraînement du vent et le flux de la chaleur de surface. La
première est donnée a priori. Quant au flux de la chaleur de surface,
il est proportionnel à la différence entre la température de l'eau de
surface à calculer et la température atmosphérique de "référence" qui
est, à son tour, donnée a priori. Ces deux paramètres externes ne
dépendent que de la latitude. 3° la densité de l'eau est une fonction
linéaire de la température.
L'hypothèse du flux de la chaleur est, bien que grossière,
justifiée en première approximation.
Les équations du mouvement s'écrivent alors:

$$\frac{\partial u}{\partial t} = -\frac{1}{\rho R \cos\varphi} \frac{\partial p}{\partial \lambda} + fv + A_M \nabla^2 u + k_M \frac{\partial^2 u}{\partial z^2}$$
$$- \frac{1}{R \cos\varphi} \left\{ \frac{\partial}{\partial \lambda}(uu) + \frac{\partial}{\partial \varphi}(uv\cos\varphi) \right\} - \frac{\partial}{\partial z}(uw) + \frac{uv}{R}\tan\varphi ,$$

$$\frac{\partial v}{\partial t} = -\frac{1}{\rho R} \frac{\partial p}{\partial \varphi} - fu + A_M \nabla^2 v + k_M \frac{\partial^2 v}{\partial z^2}$$
$$- \frac{1}{R \cos\varphi} \left\{ \frac{\partial}{\partial \lambda}(uv) + \frac{\partial}{\partial \varphi}(vv\cos\varphi) \right\} - \frac{\partial}{\partial z}(vw) - \frac{uu}{R}\tan\varphi ,$$

$$0 = -\frac{1}{\rho} \frac{\partial p}{\partial z} - g ,$$

$$V^2 = \frac{1}{R^2 \cos\varphi} \left\{ \frac{1}{\cos\varphi} \frac{\partial^2}{\partial\lambda^2} + \frac{\partial}{\partial\varphi} (\cos\varphi \frac{\partial}{\partial\varphi}) \right\},$$

où u, v et w sont les composantes de la vitesse vers l'est, le nord et le haut, respectivement, p la pression, ρ la densité, R le rayon de la terre, g l'accélération de la pesanteur, λ la longitude, φ la latitude, z positive vers le haut, t le temps, f le paramètre de Coriolis, A_M le coefficient de la diffusion horizontale et k_M le coefficient de la diffusion verticale.

L'équation de continuité est exprimée par:

$$\frac{1}{R\cos\varphi} \left\{ \frac{\partial u}{\partial \lambda} + \frac{\partial}{\partial\varphi} (v\cos\varphi) \right\} + \frac{\partial w}{\partial z} = 0.$$

L'équation de la chaleur est de la forme:

$$\frac{\partial T}{\partial t} = -\frac{1}{R\cos\varphi} \left\{ \frac{\partial}{\partial\lambda} (uT) + \frac{\partial}{\partial\varphi} (vT\cos\varphi) \right\} - \frac{\partial}{\partial z} (wT) + A_H \nabla^2 T + \frac{k_H}{\rho} \frac{\partial^2 T}{\partial z^2},$$

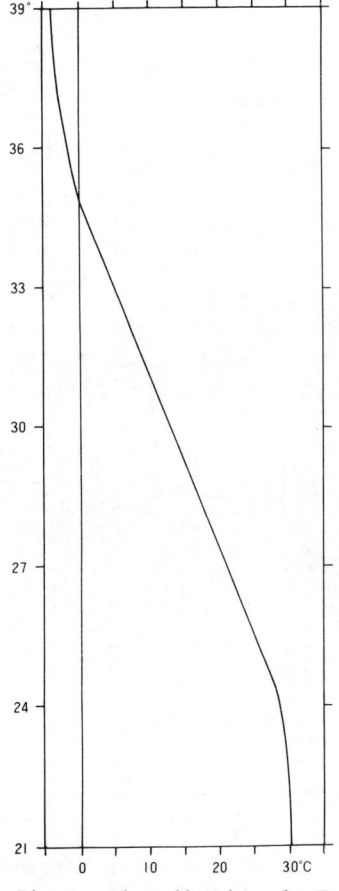

Fig.1. Distribution de T_A

où T est la température, A_H et k_H sont les coefficients de la diffusion et le paramètre δ est défini par:

$$\delta = \begin{cases} 1 \text{ dans le cas où } \frac{\partial \rho}{\partial z} \leq 0 \ , \\ 0 \text{ dans le cas où } \frac{\partial \rho}{\partial z} > 0 \ , \end{cases}$$

de sorte que, chaque fois que la stratification devient unstable, la diffusion verticale est faite indéfiniment intense pour restaurer la stratification neutre.

L'équation de la densité est approchée par:

$$\rho = \rho_0 (1 - \alpha T) \ ,$$

ρ_0 et α étant des constantes.
Cette relation à précision médiocre déforme à une certaine mesure la circulation générale [3]. On se permet tout de même de l'utiliser pour sa simplicité.

Les conditions à la surface sont exprimées par:

$$c \rho k_H \frac{\partial T}{\partial z} = \alpha (T_s - T_A) \ ,$$

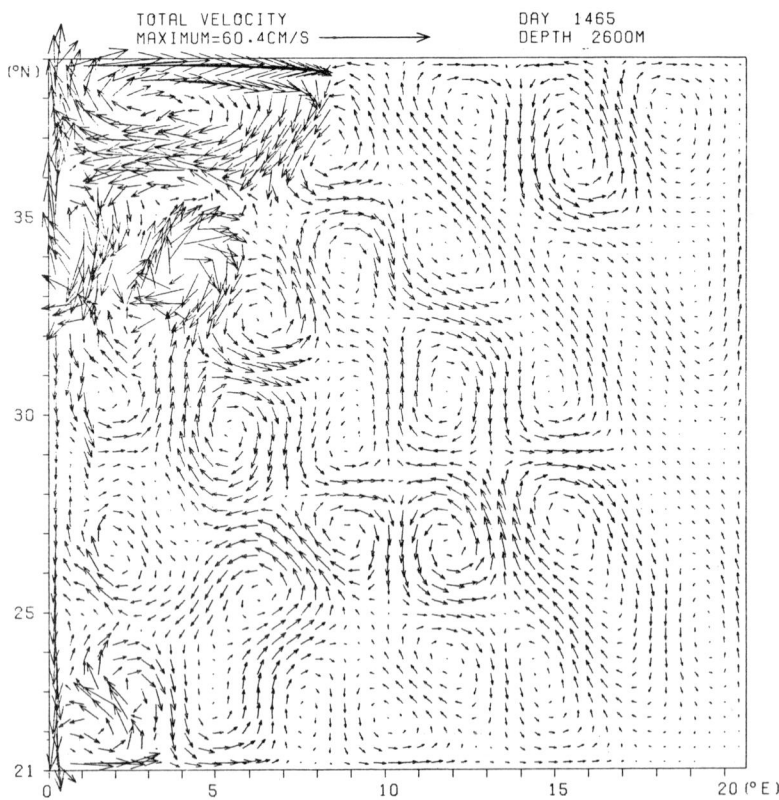

Fig.2. Distribution de la vitesse horizontale à 2600m de profondeur au jour 1485

$$\rho k_M \frac{\partial}{\partial z}(u,v) = (\tau_\lambda, 0),$$

$$w = 0,$$

où c est la chaleur spécifique de l'eau, T_S la température de l'eau de surface, T_A la température atmosphérique de référence et d une constante.

La composante vers l'est de la force d'entraînement du vent τ_λ est de la forme:

$$\tau_\lambda = -\tau_0 \cos\left\{\frac{\varphi - 21}{18}\pi\right\},$$

τ_0 étant supposée 2dyn/cm^2.
La composante vers le nord est supposée nulle. Le vent souffle donc vers l'ouest dans la moitié sud du bassin et vers l'est dans la moitié nord.

Il n'y a pas de frottement au fond et à la paroi verticale. Il n'y a ni de flux de la chaleur ni de flux de la masse d'eau à travers le fond et la paroi verticale.

Etant donné que le maillage et la méthode des différences finies sont déjà détaillés dans des notes précédentes [4][5], on se

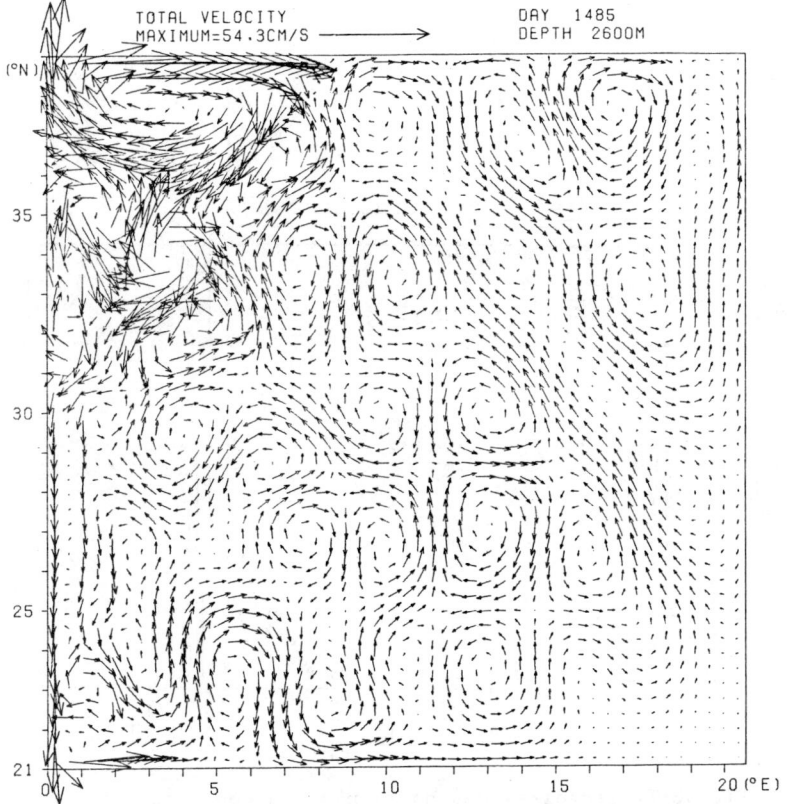

Fig.3. Distribution de la vitesse horizontale à 26oom de profondeur au jour 1485

borne à noter ici quelques points essentiels: Le pas d'espace est 0,43° en longitude et 3°/8 en latitude. Les composantes u, v et T sont calculées à 80, 400, 1600 et 2600m, tandis que la composante w est calculée à 240, 1000 et 2100m. Le pas de temps est une heure. Les valeurs numériques pour les paramètres principaux sont:

$$A_M = 10^6 cm^2/sec,$$
$$A_H = 2 \times 10^7 cm^2/sec,$$
$$k_M = k_H = 1 cm^2/sec,$$
$$d = 50 cal/cm^2/jour,$$
$$\alpha = 2,5 \times 10^{-4} /°C.$$

En partant de l'état initial où la température varie avec la latitude et la profondeur, l'intégration en temps des équations du mouvement et de l'équation de la chaleur s'avance pour 2315 jours. La vitesse et la température aux 986 dernier jours sont conservées à des intervalles de 2,5 jours pour les expériences ci-dessous. Bien que les paramètres externes moteurs soient permanents, la vitesse et la température produites par eux ne le sont pas. Les tourbillons se produisent spontanément et se mettent en mouvement. A titre d'indication, les figures 2 et 3 donnent la vitesse horizontale à 2600m à des intervalles de 20 jours. La longueur de la flèche est proportionnelle à la vitesse.

Ce sont les courants obtenus de cette manière qui mettent les particules polluantes en mouvement.

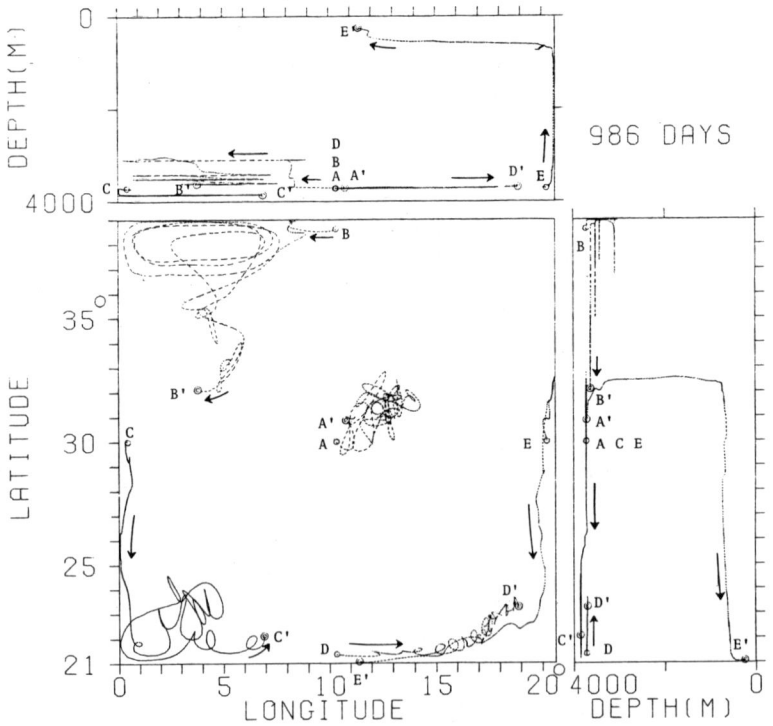

Fig.4. Trajectoires des particules au cours des 986 derniers jours. A, B, C, D et E sont les points de départ. A', B', C', D' et E' sont les points d'arrivée.

Expériences de la dispersion de particules

On suppose qu'une particule polluante quelconque ou une petite masse d'eau est déchargée de chacune de cinq sources A, B, C, D et E placées au-dessus du fond et qu'elle dérive par l'advection. Les positions géographiques de ces sources sont choisies plus ou moins arbitrairement. La diffusion et la désintégration sont négligées.

La figure 4 montre les cinq trajectoires au cours des 986 jours par la projection sur le plan horizontal en bas gauche, la projection sur le plan méridien en bas droit et la projection sur le plan perpendiculaire à la méridienne en haut gauche. Les points d'arrivée sont désignés par A', B', C', D' et E'. Les tourbillons sont si remarquables vis-à-vis des courants moyens en temps que les particules sont serpentées d'une manière frappante. La figure 5 donne, à titre indicatif, les trajectoires dues à l'advection par les courants moyens en temps.

Il est à noter que l'aire parcourue et polluée par les particules est beaucoup plus étendue dans la figure 4 que dans la figure 5. D'ailleurs, les points d'arrivée sont tout loin les uns des autres sauf D' et E', ce qui veut dire que le rôle joué par les tourbillons est tout différent de celui joué par les courants moyens. Le déplacement verticale n'est important que dans le voisinage de la paroi verticale, parce que la vitesse verticale est très petite en plein océan.

Un autre calcul porte sur des sources continues: trente et une particules sont déchargées une à une de la source A tous les dix jours au cours des 300 premiers jours. Les points d'arrivée au

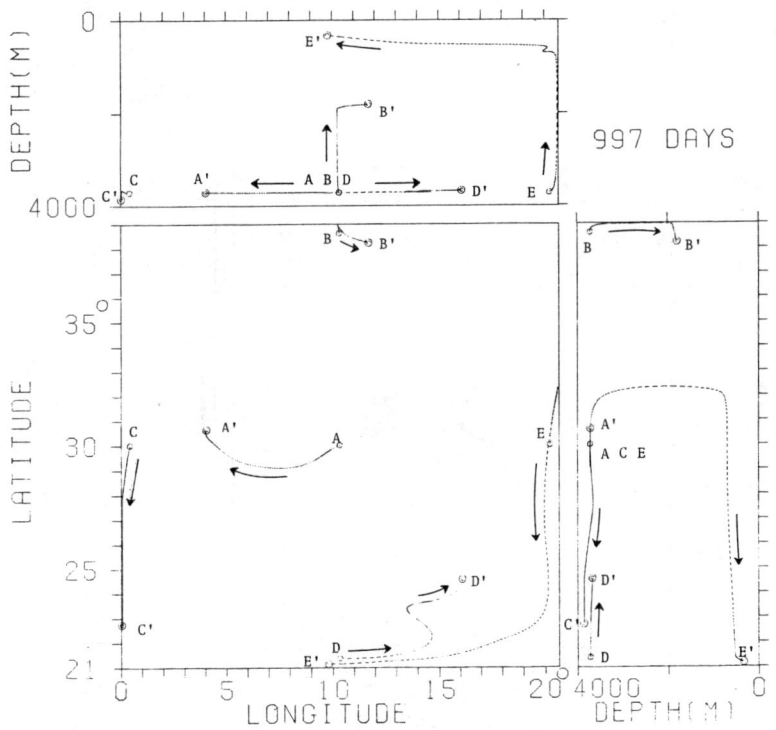

Fig.5. Trajectoires des particules par l'advection due aux courants moyens

dernier jour, 986e jour, sont illustrés dans la figure 6. Seuls les tourbillons transfèrent les particules. Les courants moyens sont omis. Les marques portées aux particules sont expliquées en haut droit. Par exemple, les cercles sont les six particules déchargées pendant 50 premiers jours. Le trait plein est la trajectoire par l'advection des courants moyens reproduite de la figure 5. Si c'étaient les courants moyens qui transfèrent les particules successivement déchargées, elles se trouveraient sur cette trajectoire, parce que les courants moyens sont indépendants du temps. On voit immédiatement que les particules soumises aux tourbillons se dispersent vers l'est sauf deux particules, tandis que les particules soumises aux courants moyens se dirigent vers l'ouest. Un autre point à signaler est que la distance entre la source et le point d'arrivée n'augmente pas toujours avec le temps.

Figures 7 et 8 montrent la dispersion des sources B et C. L'effet des tourbillons est aussi bien frappant que dans le cas de la source A.

Coefficients de la diffusion turbulente virtuelle

Les coefficients de la diffusion turbulente virtuelle résultante des tourbillons peuvent être définis par:

$$A_H^* \frac{1}{R^2 \cos^2\varphi} \frac{\partial^2 \overline{T}}{\partial \lambda^2} = - \frac{1}{R \cos\varphi} \frac{\partial}{\partial \lambda} \overline{(u'T')},$$

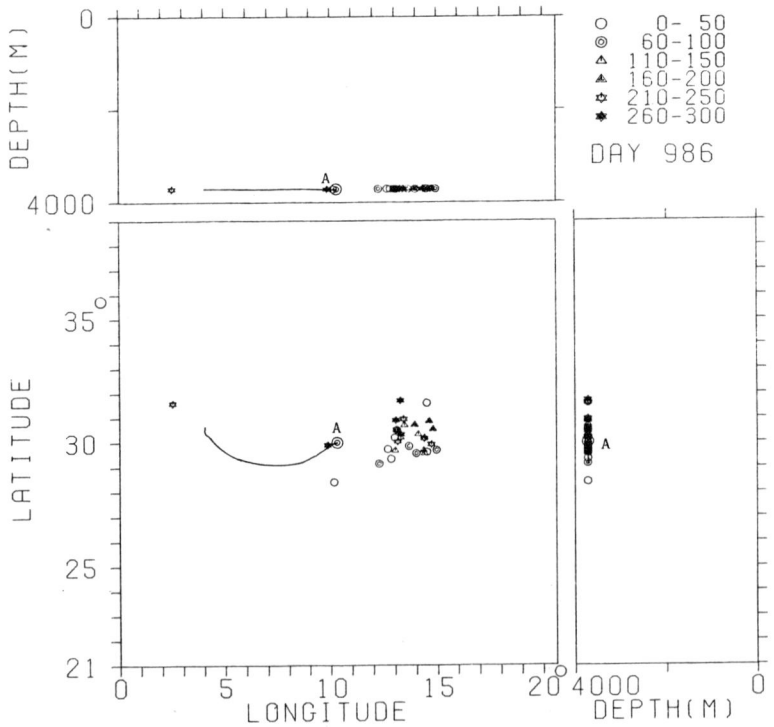

Fig.6. Particules dispersées de la source A par les tourbillons. Le trait plein désigne la trajectoire par les courants moyens reproduite de la figure 4.

$$A_H^{**} \frac{1}{R^2 \cos^2\varphi} \frac{\partial}{\partial\varphi}(\cos\varphi \frac{\partial}{\partial\varphi})\overline{T} = -\frac{1}{R\cos\varphi}\frac{\partial}{\partial\varphi}(\overline{T'v'\cos\varphi}) ,$$

$$k_H^* \frac{\partial^2 \overline{T}}{\partial z^2} = -\frac{\partial}{\partial z}(\overline{T'w'}) ,$$

où $\bar{u}$, $\bar{v}$, $\bar{w}$ et $\bar{T}$ sont les moyennes en temps et u', v', w' et T' sont les écarts des moyennes. Les données pour les 986 derniers jours permettent de calculer ces coefficients à tous les points de maille. La valeur de chaque coefficient est très variable d'un point à l'autre dans une large gamme. Le tableau I montre les maximums, les minimums et les moyennes globales.

Tableau I. Coefficients de la diffusion virtuelle (cm^2/sec)

	maximum	minimum	moyenne
A_H^*	$4,4 \times 10^{11}$	$-5,3 \times 10^{10}$	$1,8 \times 10^7$
A_H^{**}	$2,1 \times 10^{11}$	$-5,3 \times 10^{10}$	$7,3 \times 10^6$
k_H^*	$4,2 \times 10^2$	$-1,8 \times 10^3$	$0,4$

Les moyennes $1,8 \times 10^7$ et $7,3 \times 10^6$ sont près de 2×10^7 choisi pour A_H. La moyenne de k_H^*, 0,4, est aussi approché de k_H, 1,0. Ce

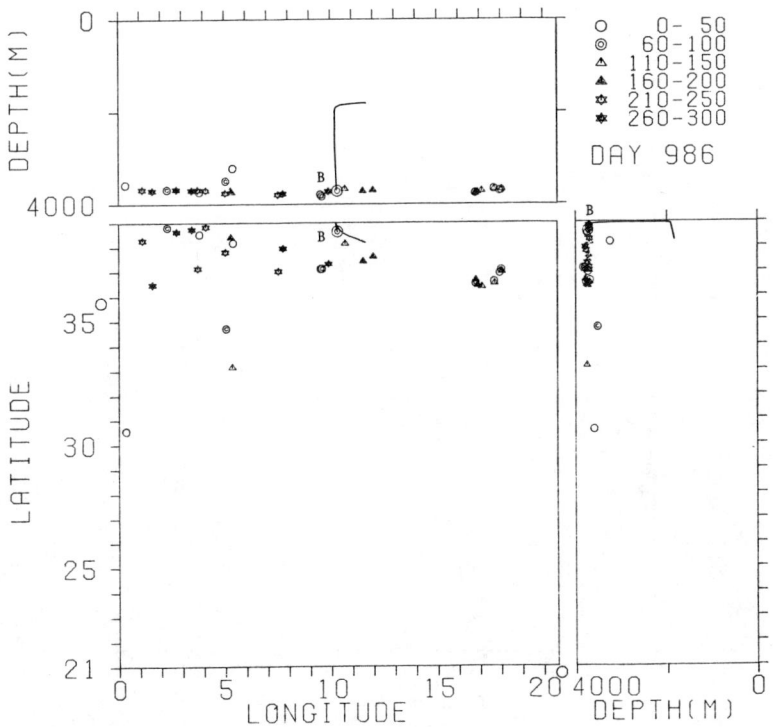

Fig.7. Particules dispersées de la source B par les tourbillons

n'est toutefois pas le cas du coefficient individuel. Le maximum et
la valeur absolue du minimum de chacun sont plus grands de trois ou
quatre ordres de grandeur que la moyenne. Le coefficient négatif
signifie que les tourbillons transfèrent la chaleur vers une région
plus chaude. La grandeur de sa valeur absolue souligne l'importance
de cette diffusion inverse.

Remarque

Les tourbillons dans la nature sont évidemment plus compli-
qués, probablement à cause de la complexité de la géometrie du bassin
océanique et de la force motrice. De plus, les tourbillons de modèle
altérés dans une certaine mesure par l'approximation médiocre de sont
l'équation de la densité. Tout cela n'empêchera pas de suggérer
l'importance de la recherche sur l'effet des tourbillons à échelle
intermédiaire. D'autres expériences numériques plus approfondies sont
en cours pour un océan plus large, à profondeur variable, régi par
la force motrice variable en temps.

Bibliographie

[1] The MODE Group:"The Mid-Ocean Dynamics Experiment", Deep-Sea
    Research, 25, 859-910 (1978).
[2] Takano,K.:"Exemples numériques de la dispersion de la pollution
    dans un océan mondial", La mer, 15, 196-204 (1977).

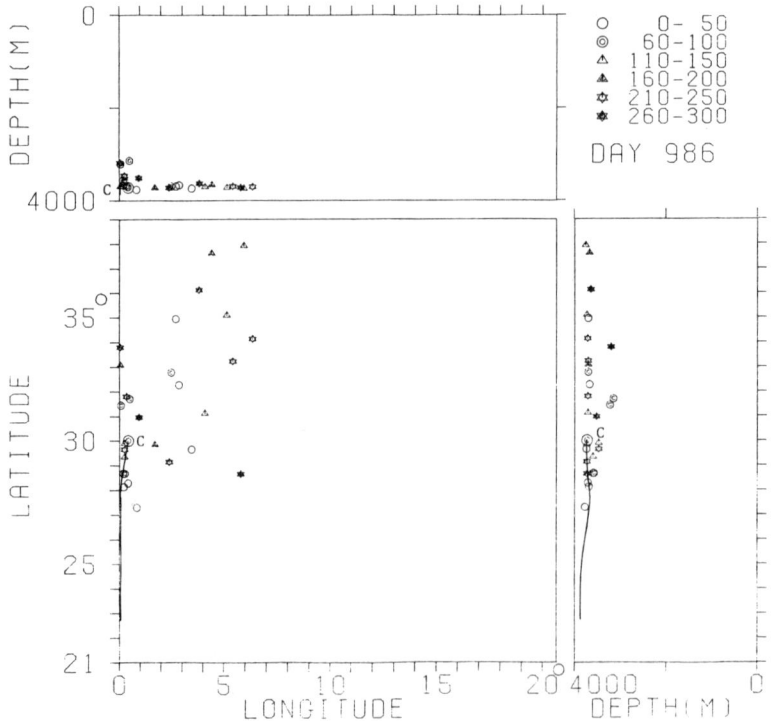

Fig.8. Particules dispersées de la source C
       par les tourbillons

[3] Takano,K.: "Effect of the approximation to the equation of state for sea water on the model general circulation", La mer, $\underline{16}$, 147-161 (1978).
[4] Takano,K.: "A general circulation model for the world ocean", Num. Simul. Weather and Climate, Tech. Rept. (8), Dept. Meteorol. Univ. California, Los Angeles, 1-46 (1974).
[5] Takano,K.: "Effect of a ridge on deep currents" (en japonais), Preserv. Mar. Environment, Rept. Spec. Project Res., Minist. Educ. Sci. Culture, 27-44 (1978).

Discussion

G. BRESSON, France

En conclusion, vous précisez que d'autres expériences numériques plus approfondies sont en cours pour un océan plus large. Est-ce que vous suggérez ces recherches ou bien sont-elles menées dans vos laboratoires ?

K. TAKANO, Japan

Nous avons presque terminé une série d'expériences numériques pour des océans plus larges à profondeur variable.

Session 3

Chairman - Président
Mr. A. AARKROG
(Denmark)

Séance 3

THE BEHAVIOR AND THE CHEMICAL FORMS OF METALLIC ELEMENTS DISSOLVED IN OCEAN WATERS

Y. Sugimura, Y. Suzuki and Y. Miyake*
Geochemical Laboratory, Meteorological Research Institute,
Koenji-kita, Suginami, Tokyo 166, Japan.
*Geochemistry Research Association, Koenji-kita, Suginami,
Tokyo 166, Japan.

ABSTRACT

It is important to study the chemical forms of metallic elements dissolved in sea water because the behaviors of these elements in the ocean are controlled by their chemical forms. According to our recent study on iron in sea water, it was found that a considerable part of iron is present in organic form. In order to confirm further this tendency, determination of metallic elements, such as Al, V, Mn, Fe, Co, Ni, Cu, Zn, Se, Mo, Ag, Cd and U dissolved in the western North Pacific surface and deep waters were carried out. In the separation of organic compounds of various metals, a new method using XAD-2 resin was employed. The results of determination are summarized as follows: (1) In case of Fe, Co, Cu and Cd more than 70 % is organic. (2) In case of Co and Cd, organic compounds prevail near the surface while they decrease in the deep. (3) In case of Al, V, Mn, Ni, Zn, Mo, Ag and U less than 40 % is organic. With respect to organic compound of metals, it is clarified that they are composed mainly of high polymer organic compounds, such as protein or protein-carbohydrate etc., with molecular weights ranging from 1×10^3 to 2×10^4. Therefore, we have to bear these facts in mind when behaviors and fates of radioactive elements dumped in the sea are considered or discussed in order to evaluate their effects on man and environment.

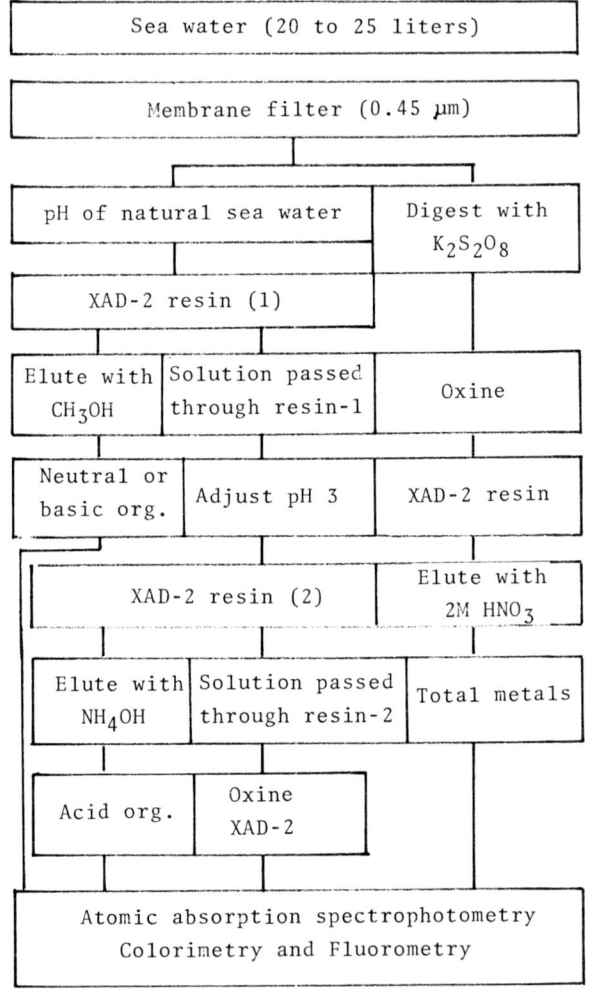

Fig. 1. A diagram of the analytical methods of organic and inorganic forms of metallic elements in sea water.

1. INTRODUCTION

It is important to study the chemical forms of metallic elements dissolved in sea water because the behaviors of these elements in the ocean are controlled by their chemical forms. Up to now there are plenty of studies on the effects of radioactive, metallic elements, both natural and artificial, in marine environment[1,2].

However, a most of them were carried out on the basis of simple assumption that metallic elements dissolved in sea water are present in inorganic forms either in ionic, molecular or complex compounds.

Recently the present authors reported the results of analytical study on iron and some other metallic elements dissolved in sea water [3,4]. As a result of investigation by using XAD-2 resin on which non-polar metal organic compounds can effectively be adsorbed, the present authors reached the conclusion that in sea water a most of metallic elements is present, more or less, in organic forms instead of inorganic forms as generally be regarded up to the present [5,6,7].

It is considered that metal organic compounds in sea water can be adsorbed on the XAD-2 resin at pH of the natural sea water, i.e., about 8 and also pH 3. The former is a neutral or basic metal organic compound and the latter is acid metal organic compound. In this connection, in our study on the adsorption capacity of organic compounds on XAD-2 resin, it was revealed that organic compounds with aromatic structure are quantitatively adsorbed on the resin, while those with aliphatic structure have lower affinity to the resin, and no inorganic compounds or ions are adsorbed [3].

In this paper, the present authors intend to present a report on the chemical forms and molecular weight of metal organic compounds in sea water collected in the western North Pacific Ocean.

2. METHODS OF ANALYSIS

The procedure of the analytical method of metal organic compounds dissolved in sea water was described in the previous report [4] as illustrated schematically in Fig. 1, on which some explanations are added as follows.

Immediately after sampling of sea water using a non-metallic sampler, a water sample is filtered through a membrane filter (0.45 μm of pore size, Nucleopore Filter) and metal organic compounds dissolved are separated by the successive adsorption through two sets of column containing XAD-2 resin at pH of about 8 and 3.

The metal organic compounds which are adsorbed on the resin are eluted by use of methyl alcohol, and a dilute solution of ammonium hydroxide of pH 10.

Determination of the amount of each metal in the effluent is carried out by the atomic absorption spectrophotometry, colorimetry and fluorometry. To determine a total amount of each metal dissolved in sea water, each metal ion is simultaneously adsorbed on a XAD-2 resin as in the form of metal oxinate, after oxidizing the organic matter in the solution by using potassium persulfate. Elution of metals which are adsorbed on the resin is done with 2 M HNO_3, and further analysis is carried out by using the same procedure as described above.

Table I. The average content of total metallic elements and the ratio of metal organic compounds to the total dissolved in surface waters in the western North Pacific

Element	Content of total (μg/l)	Ratio of organic form to total (%)		
		Neutral or basic (1)	Acid (2)	Sum of org. form (1)+(2)
Al	0.94 ± 0.33	14 ± 4	8 ± 2	21 ± 5
V	1.7 ± 0.3	46 ± 12	0	46 ± 12
Mn	0.25 ± 0.08	16 ± 9	0	16 ± 9
Fe	1.31 ± 0.37	30 ± 16	51 ± 21	82 ± 11
Co	0.05 ± 0.02	91 ± 13	0	91 ± 13
Ni	0.46 ± 0.24	22 ± 13	6 ± 2	28 ± 12
Cu	0.64 ± 0.21	66 ± 13	12 ± 5	78 ± 11
Zn	3.9 ± 1.5	27 ± 13	9 ± 4	37 ± 13
Se	0.08 ± 0.02	20 ± 6	0	20 ± 6
Mo	9.4 ± 1.3	10 ± 5	0	10 ± 5
Ag	0.09 ± 0.04	34 ± 17	0	34 ± 17
Cd	0.04 ± 0.02	64 ± 19	14 ± 10	78 ± 15
U	3.47 ± 0.13	0	8 ± 6	8 ± 6

Table II. Fractions (%) of metal organic compounds which belong to lipid, protein and carbohydrate

			Fe	Co	Cu	Zn	Se	Ag	Cd
Total dissolved (μg/l)			1.3	0.04	0.74	3.0	0.05	0.7	0.04
Ratio of organic form to total (%)		(1)*	38	100	72	23	45	15	100
		(2)**	52	0	11	0	0	0	0
Lipid	(1)		7	0	8	15	0	0	0
	(2)		1	0	5	6	0	0	0
Protein	(1)		24	75	27	46	80	40	0
	(2)		0	0	0	0	0	0	0
Carbohydrate	(1)		11	25	52	20	20	60	100
	(2)		57	0	8	13	0	0	0

(Sample: 35°03'N, 139°24'E, surface water).
* (1): neutral or basic metal organic compounds.
**(2): acid metal organic compounds

3. RESULTS AND DISCUSSION

The results of study on the average content of some metals dissolved in surface waters in the western North Pacific are shown in Table 1.

As seen in the Table, it is found that in surface layer, more than 70 % of the total content of Fe, Co, Cu and Cd are present in organic forms. It is to be noted that among these metals, about 60 % of Fe is associated with acid organic compounds, and the rest is involved in neutral or basic organic compounds. As to other metals such as Al, V, Mn, Ni, Zn, Mo, Ag and U, the surface metal contents of organic forms are less than 40 % of the total. In case of Se, up to 20 % of the total is in organic forms.

Fig. 2 shows examples of vertical distribution of Fe, Co, Cu and Cd in the western North Pacific with the observed data of water temperature, salinity and content of dissolved oxygen. In the vertical distributions, a most part of Fe and Cd is present in organic forms except at the depth of salinity minimum. Co is present mostly in organic forms in the surface, but in the deeper layers, a fraction of organic forms decreases to 30 to 40 %, though the total content of Co is nearly constant. As to Cu, more than 80 % is present in organic forms both in surface and deep layers, while in intermediate layer, the content of organic forms decreases to about 50 %.

It is important that of the metals such as Fe, Cu and Cd occurring in the ocean more than 50 % are present in organic forms regardless the depth except at salinity minimum. As to Al, V, Mn, Co, Ni, Zn, Se, Mo, Ag and U, about 10 to 30 % is present in organic forms anywhere in the marine environment.

To examine states of combination of metallic elements with three main groups of organic compounds, i.e., lipid, protein and carbohydrate, the content of metallic elements involved in each group was determined after separating organic matter into three categories by using liquid extraction and salting-out methods. The results of study are shown in Table 2.

As shown in the Table, among the metallic elements in organic forms, less than 20 % of Fe, Cu and Zn are associated with lipid fraction. More than 70 % of Co and Se in organic forms are involved in protein fraction, and 60 to 100 % of organic Fe, Cu, Ag and Cd are contained in the fraction of carbohydrate.

Since lipid, protein and carbohydrate are considered to be mainly composed of lipoprotein, protein and glycoprotein, it may be said that main parts of metal organic compounds dissolved in sea water are such as summarized as follows:

```
Fe    glycoprotein > protein > lipoprotein
Co    protein      > glycoprotein
Cu    glycoprotein > protein > lipoprotein
Zn    protein      > glycoprotein > lipoprotein
Se    protein      > glycoprotein
Ag    glycoprotein > protein
Cd    glycoprotein
```

In the next place, the molecular weights of these organic compounds were determined. Metal organic complexes which were obtained from 200 liters of sea water (collected at 0 m, 29°59'N, 137°07'E) by means of adsorption on XAD-2 resin, were separated from each other by using gel-filtration chromatographic method (Sephadex G-50, 1 cm dia. and 30 cm long) at a flow rate of 0.5 ml/min. Each 2 ml of effluent was collected successively in a test tube and it was subjected to the analysis of metallic elements, total amino acids

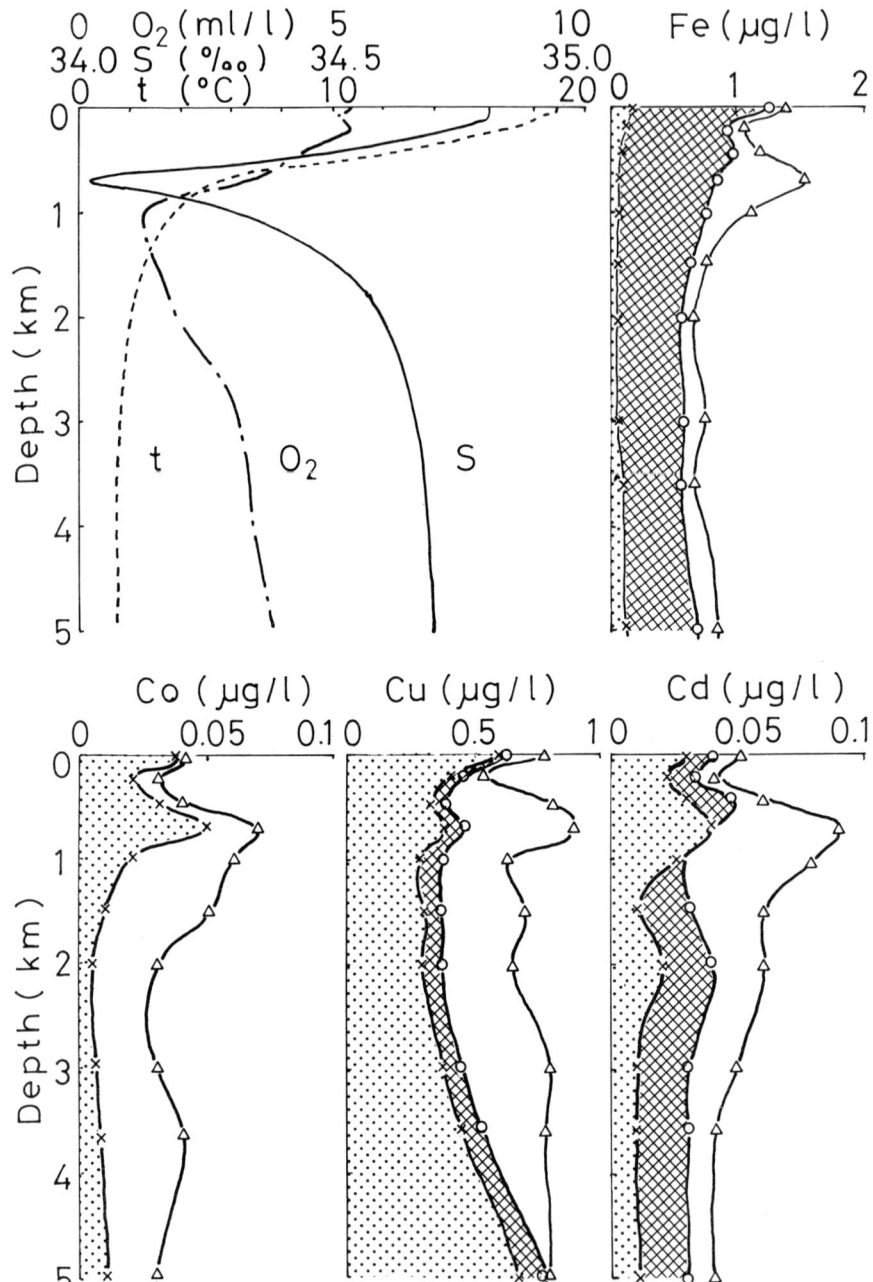

Fig. 2. Vertical distribution of Fe, Co, Cu and Cd, salinity, water temperature and the content of dissolved oxygen in the western North Pacific water (29°55'N, 146°57'E, 6,160 m).
—△— Total dissolved. ▨ Neutral or basic metal organic compounds, ▩ Acid metal organic compounds.

to represent proteinous matters and carbohydrate.

The relation between molecular weight and elution volume through the Sephadex column was determined by using several organic compounds such as myoglobin, cytochrome-C, insulin, cyanocobalamin etc., with known molecular weights ranging from 1.3×10^3 to 1.7×10^4.

For the assignment of organic compounds in each fraction of effluent the absorption of light at a wave length of 280 nm was measured. The results of UV absorption measurement are fairly in good agreement with the analytical results of metallic elements in respective fractions.

The results of the above determinations are shown in Fig. 3a and 3b, and Table 3. It is found that in the elution volume between 76 ml and 170 ml, there are six fractions in the neutral or basic metal organic compounds with molecular weight ranging from 1×10^3 to 2×10^4, and each fraction corresponds to a compound containing a certain metallic element.

Among iron organic complexes in the neutral or basic forms, a most part of iron is associated with a compound with molecular weight of 5×10^3 and the rest is contained in those with molecular weights of 1.7×10^4 and 1×10^4. As to organic complexes of Cu, Zn, Se and Cd, they are found mainly in the fractions with molecular weights ranging from 8×10^3 to 2×10^3. In case of Co, about a half is found in a fraction with a molecular weight of 1.3×10^3, and another half is present in the fractions with molecular weights ranging from 1 to 2×10^4.

With respect to the acid organic metal compounds, there are two fractions and Fe, Cu and Zn are involved in the fraction with a molecular weight of about 2×10^4.

In the proteinous part of the neutral or basic metal organic compounds, there are three fractions with molecular weights ranging from 1.7×10^4 to 8×10^3. All of Se and Cd are associated with these three fractions, and a most of Cu and a part of Fe, Co and Zn is involved in one or two of these fractions.

With respect to carbohydrate, there are two fractions with molecular weights of 8×10^3 and 5×10^3. A part of Cu, Zn, Se and Cd seems to associate with the former fraction, and a most of Fe and about a half of Zn to the latter.

As to the proteinous part of acid metal organic compounds, there are two fractions with molecular weights of about 2×10^4 and 8×10^3, in which Fe, Cu and Zn are associated with a fraction with a higher molecular weight. There is also a fraction of carbohydrate with a molecular weight of 1.7×10^4 to which a part of Zn is associated.

As a result of the above studies, it is concluded that metallic elements which are dissolved in sea water as minor constituents are not existing simply in inorganic ions or compounds, but, they are combined, more or less, with high polymer organic compounds, such as protein, protein-carbohydrate etc., with molecular weights ranging from 1×10^3 to 2×10^4. Therefore, we have to bear these facts in mind when behaviors and fates of radioactive elements dumped in the sea are considered or discussed in order to evaluate their effects on man and environment.

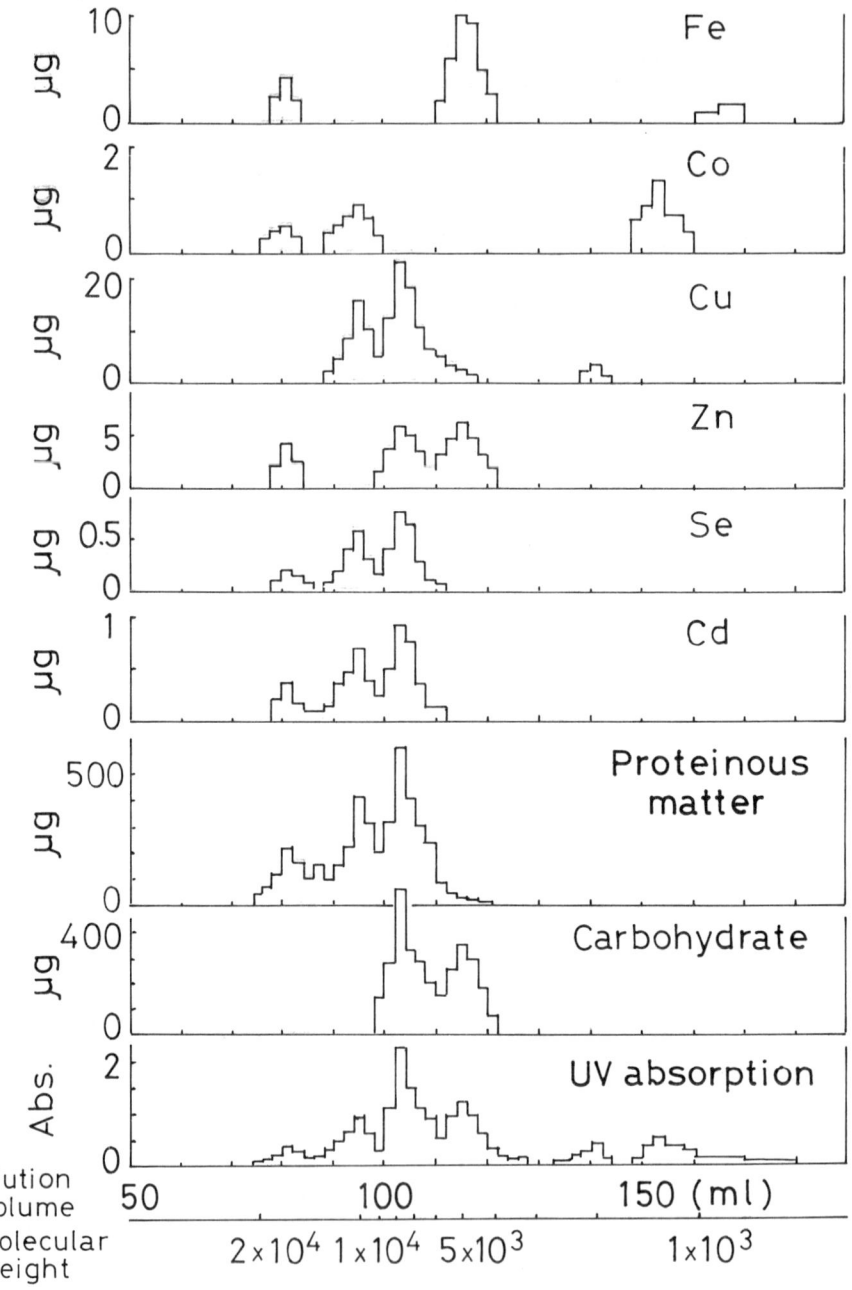

Fig. 3a. Separation of neutral or basic metal organic compounds through gel-filtration of Sephadex G-50

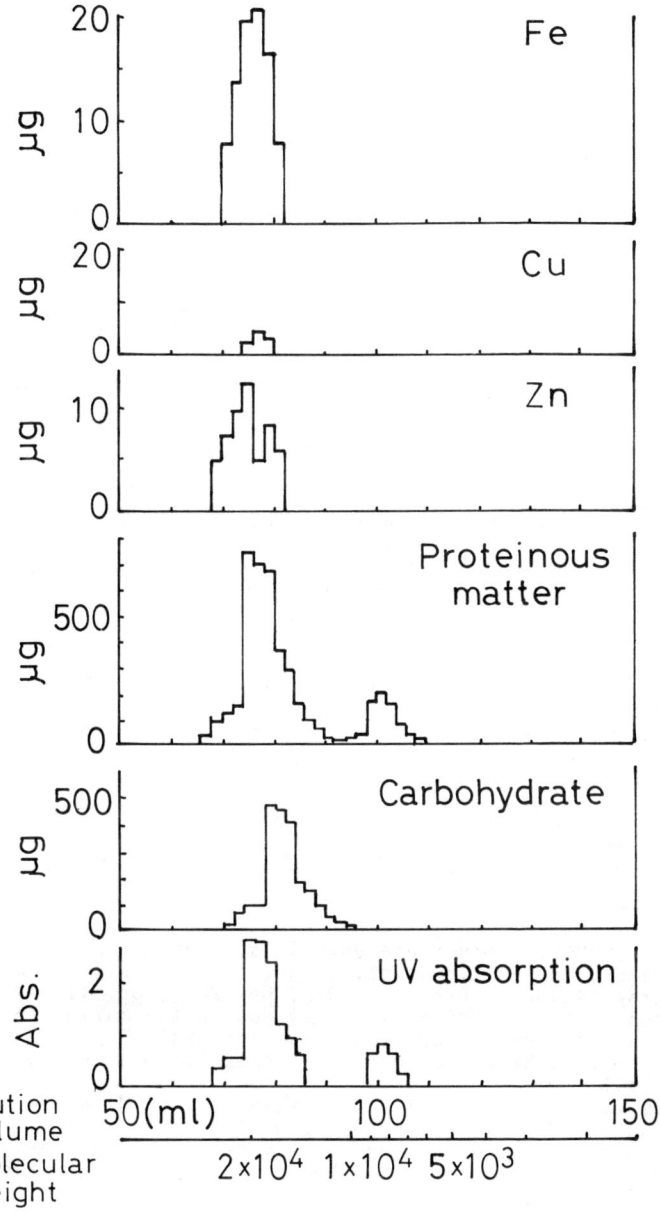

Fig. 3b. Separation of acid metal organic compounds through gel-filtration of Sephadex G-50

Table III. Distribution of molecular weights of metal organic compounds dissoved in surface water

	Fe	Co	Cu	Zn	Se	Cd
Total dissolved ($\mu g/l$)	1.24	0.03	0.57	4.8	0.10	0.03
Ratio of organic (1)*	29	100	80	17	24	67
form to total (%) (2)**	52	0	6	16	0	0

1: Molecular weight of neutral or basic metal organic compounds

	Fe	Co	Cu	Zn	Se	Cd
1.7×10^4	7 %	17 %	0 %	8 %	11 %	14 %
1×10^4	0	39	32	0	39	40
8×10^3	0	0	57	20	50	46
5×10^3	26	0	0	22	0	0
2×10^3	0	0	4	0	0	0
1×10^3	3	44	0	0	0	0

2: Molecular weight of acid metal organic compounds

	Fe	Co	Cu	Zn	Se	Cd
2×10^4	64 %	0 %	7 %	49 %	0 %	0 %

* (1): Neutral or basic metal organic compounds
**(2): Acid metal organic compounds
(Sample: 29°59'N, 137°00'E, surface water)

4. REFERENCES

[1]. Brewer, P. G.: Minor elements in sea water. In Chemical Oceanography, 2nd edition, Vol. 1. ed. by J. P. Riley and G. Skirrow, Academic Press, London, New York, pp. 415-496, 1975.
[2]. Burton, J. D.: Radioactive nuclides in the marine environment, In Chemical Oceanography, 2nd edition, Vol. 3. ed. by J. P. Riley and G. Skirrow, Academic Press, London, New York, pp.91-191, 1975.
[3]. Sugimura, Y., Y. Suzuki and Y. Miyake: The dissolved iron in sea water, Deep Sea Res., 25, 306-314, 1978.
[4]. Sugimura, Y., Y. Suzuki and Y. Miyake: Chemical forms of minor metallic elements in the ocean, J. Oceanogr. Soc. Japan, 34, 93-96, 1978.
[5]. Sillen, L. G.: The physical chemistry of sea water, In Oceanography, ed. by M. Sears, Amer. Assoc. Advance. Sci., Wash. D. C., pp. 549-581, 1961.
[6]. Stumm, W. and P. A. Brauner: Chemical speciation, In Chemical Oceanography, 2nd edition, Vol. 1. ed. by J. P. Riley and G. Skirrow, Academic Press, London, New York, pp.173-240, 1975.
[7]. Millero, F. J.: Thermodynamic models for the state of metal ion in sea water, In The Sea, Vol. 6, ed. by E. D. Goldberg et al., John Wiley, New York, pp. 653-693, 1977.

Discussion

<u>A.A. YAYANOS</u>, United States

There have been reports of bacteria that are small enough to pass through pores of 0.45 µm diameter. Have you attempted to analyze water filtered through smaller diameter pores, e.g. 0.2 µm ?

<u>Y. SUGIMURA</u>, Japan

We analyzed metal-organic compounds in filtrate through filters with different pore sizes, ranging from 0.02 µm to 0.45 µm, but there is no difference in the analytical results.

DISTRIBUTION OF Sr-90 AND Cs-137 IN DEEP WATERS AROUND JAPAN

Yutaka Nagaya and Kiyoshi Nakamura
Division of Marine Radioecology, National Institute
of Radiological Sciences, Nakaminato, Japan

ABSTRACT

Sr-90 and Cs-137 contents in deep waters around Japan were determined and the vertical distributions were examined.

The profiles of the radionuclides show a two layer distribution pattern and an exponential correlation is observed between the radionuclide contents and the depth, in both of the layers.

Generally speaking, the total amounts of Sr-90 and Cs-137 in the open waters of the northwestern Pacific seem to be well balanced with the cumulative fallout depositions, but in the regions where upwelling or evident currents are observed the total amounts of the radioisotopes in a water column are one half to one third of the fallout inputs on the surface.

Table I   Stations of Deep Water Collection.

Cruise	Stn.	Date	Latitude	Longitude	Depth (m)	Nos. of Samples
KH-76-4	L-1	10/09	28°57.2'N	135°26.3'E	5,030	1
(1976)	L-2	10/10	30°23.9'N	133°50.2'E	4,430	1
KH-77-3	LV-1	9/16	31°14.2'N	137°06.5'E	4,175	4
(1977)	LV-2	9/24	37°43.7'N	135°11.6'E	2,980	6
	LV-3	9/27	41°19.7'N	137°19.9'E	3,620	6
	LV-4	9/28	43°02.9'N	138°32.4'E	3,448	5
	LV-5	9/30	38°35.2'N	134°44.9'E	3,000	1
KH-78-1	LV-1	1/31	26°12.4'N	136°43.2'E	5,210	5
(1978)	LV-2	2/04	12°58.7'N	136°35.2'E	5,050	5
	LV-3	2/06	16°39.5'N	133°07.6'E	5,760	5
	LV-4	2/25	31°28.1'N	137°03.4'E	4,060	4
	LV-5	3/03	34°18.5'N	141°58.6'E	9,000	5
KH-79-3	LV-1	6/29	44°11.1'N	138°57.4'E	3,500	3
(1979)	LV-2	7/02	37°45.1'N	135°16.0'E	2,960	5
	LV-3	7/04	38°23.7'N	132°49.9'E	2,800	3
	LV-5	7/19	32°17.7'N	137°32.3'E	4,080	6
	LV-6	7/25	30°00.2'N	145°44.3'E	5,840	13

INTRODUCTION

Among the artificial radionuclides entered the seas and oceans, the soluble and long-lived fission products Sr-90 (half life 29.0 y) and Cs-137 (half life 30.2 y) are regarded as not only an important indicator of radioactive pollution of the marine environment, but are good water tracer. Their behaviour has been investigated since 1954.

For the northwestern Pacific including the seas around Japan, considerable observations on the horizontal and vertical distributions of the radioisotopes were reported [1,2,3,4,5,6,7,8,9,10,11,12]. The results indicate that recent Sr-90 and Cs-137 contents in the North Pacific surface water were approximately uniform along the latitudal zones, with some regional exceptions, and their horizontal distributions corresponded nearly to those of the radioactive fall-out. On the other hand, their vertical distributions showed remarkable decrease of the concentrations with increasing depth, and the contents in deep water were one or two orders of magnitude less than those in surface water.

Recently, in due course of the deep sea dumping of radioactive waste, the exact knowledge on Sr-90 and Cs-137 in deep waters are being required in respect to the radioactive background and the indices of radioanuclide behaviour in deep ocean. Unfortunately many of the results obtained in earlier observations were of surface or rather shallow subsurface waters. In addition, because of the technical difficulties of collecting a large volume sea water from a definite depth, to assure a high counting precision for extremely low radioactivity in deep layers, and also because of the uncertainty in results due to unsatisfactory sample collections, the reliable data of the radioisotope contents in deep waters were rather few.

A 250 liter - double barrel sampler was developed at the Ocean Research Institute, University of Tokyo[13] for the effective collection of reliable deep water samples. The authors have been collecting the deep water samples from the adjacent seas of Japan and the north-western Pacific since 1976, and have been determining their Sr-90 and Cs-137 contents. In this report, the results of our study will be discussed.

MATERIALS AND METHODS

As shown in Table I, 78 samples were collected from 17 stations on the cruises of the R/V Hakuho-maru, Ocean Research Institute, University of Tokyo in the seas around Japan. The vertical profiles of Sr-90 and Cs-137 contents were determined. The volume of sea water collected was 100 liters for surface above 100 m deep, and about 200 liters from the subsurface layer. The surface water up to 100 m deep was collected by use of a pumping system, and the subsurface water was obtained by a large volume sampler.

The principal structure of the sampler is two polyvinylchloride barrels (about 250 liters each in capacity) supported by a stainless steel frame. The water can be trapped in these barrels separately by use of two messengers. The upper and lower lids are closed by strong rubber bands and the lower lid is bolted tightly with three pins. Reversing thermometers and a pinger are also attached. Detail of the device is described elsewhere [13].

In order to obtain more data from each deep water sample, C-14, transuranic elements, Ra-226, Cs-137 and Sr-90 were extracted sequentially from the samples and in some cases the organic-chloride compounds were extracted also. Details of Sr-90 and Cs-137 analyses in laboratory were reported elsewhere [12]. Cross-contamination of the radionuclides during the extraction process on board and in laboratory analyses were examined and found to be less than 0.1 pCi per sample, i.e. nearly equivalent to or less than the counting errors of the deepest water samples. The beta ray radioactivity of Y-90, the daughter nuclide of Sr-90, and Cs-137 were measured by use of a low-background gas-flow counter. Chemical recoveries of the

Table II Comparisons of Integrated Fallout Depositions and Total Amounts of the Radionuclides in Water Column.

Region		Philippine Sea	Southeast of Boso Peninsula	Southeast of Shikoku	Sea of Japan
Latitude		13°-26°N	34°N	32°N	38°-43°N
Nos. of Station		3	1	4	4
Mean Depth of Observed Station		5,340 m	4,000 m*	4,117 m	3,349 m
Estimated Sr-90 Deposition in 1945-1977 (mCi/km^2)		55	91	91	108
Sr-90 in Water**	Upper	38	76	35	69
	Deep	14	12	8	5
	Total	52	88	43	74
Estimated Cs-137 Deposition in 1945-1977 (mCi/km^2)		88	146	146	173
Cs-137 in Water**	Upper	50	113	57	128
	Deep	11	14	12	2
	Total	61	127	69	130

* maximum depth of sample collection.

** expressed in mCi/km^2·column.

radionuclides were determined by atomic absorption spectrometry (Sr) and by gravimetric method (Y and Cs) respectively, and found to range 40 to 60 % for Sr-90 and Cs-137 approximately.

RESULTS AND DISCUSSION

All stations observed from 1976 to 1979 are shown in Figure 1, and classified for convenience into 6 regions -- the Philippine Sea (3 stations), the region east of the Izu-Ogasawara Ridge (1 station), the area southeast of Boso Peninsula (1 station), the seas south and southeast of Shikoku (2 and 3 stations) and the Sea of Japan (7 stations). Unfortunately, the results of the 1979 samples are not available yet. The vertical distributions of Sr-90 and Cs-137 contents in each region were obtained as shown in Figure 2 to Figure 5.

Generally speaking, the profiles of radioisotope concentrations could be separated vertically into two layers, i.e. upper and deep layers. In the upper layer the radionuclide concentrations decreased remarkably with increasing depth. On the other hand, the rates of variation in the deep layer were not so substantial in comparison to the upper layer and the concentrations were one to two orders of magnitude less than those in surface water. In the profiles, for example as shown in Figure 2, an exponential correlation is found between the radionuclide contents and the depth, in the upper and the deep layers respectively. Solid lines in the figures indicate the concentration-depth correlations of Sr-90 in both layers, and the dotted lines show those of Cs-137.

The depth of the boundary between the upper and deep layers tends to show regional variation. In the Philippine Sea, the boundaries lie around 1,000 m deep, but off the Boso Peninsula the depths of the boundaries are rather deeper (about 1,500 m), and the radionuclide concentration levels are higher than those in the Philippine Sea.

The stations in the area southeast of Shikoku clearly have different profiles of radioisotopes from the other regions, showing a rather higher concentration of radionuclides with increasing depth in the deep layer, although the boundary depth is similar to those off Boso Peninsula. This region is known as the Cold Water Mass area, and a temporary upwelling of cold water has been observed. The characteristic pattern of the radionuclide content profiles might be caused by the transport and upwelling of deep water having lower radionuclide concentrations, into the 1,000 m to 2,000 m layer from another region.

As shown in Figure 5, the boundary depths in the Sea of Japan are significantly deeper than those in the Pacific, especially in the case of Cs-137 the boundary practically reaches the sea bottom, reflecting a higher vertical mixing rate of water in the Sea of Japan in comparison with the Pacific.

In order to evaluate the transport and accumulation of the radionuclides in each region, the total amounts of Sr-90 and Cs-137 contents in a water column were calculated and compared with the estimated radioactive fallout depositions, as shown in Table II. Average integrated Sr-90 deposition observed in six cities, shown as little dots in Figure 1, in Japan [14] was used as the cumulative deposition in 30° to 40°N latitudal zone in the seas around Japan and the deposition of Sr-90 in each region was estimated from this value, applying the global distribution of cumulative Sr-90 fallout deposition between latitudal zones [15]. Cs-137 depositions were also estimated from the Sr-90 deposition and the Cs-137/Sr-90 ratio of 1.6 in the fallout [15].

Generally speaking, in the Philippine Sea and the area southeast of Boso Peninsula, the total amount of radionuclides is nearly equal to the fallout deposition, with a slight difference in Cs-137 contents (10 to 30 %). On the other hand, in the region southeast of Shikoku where temporal upwelling is observed, the total amount of the radioisotopes is less than half of the fallout input, suggesting the dilution of the radionuclide contents with low radioactivity

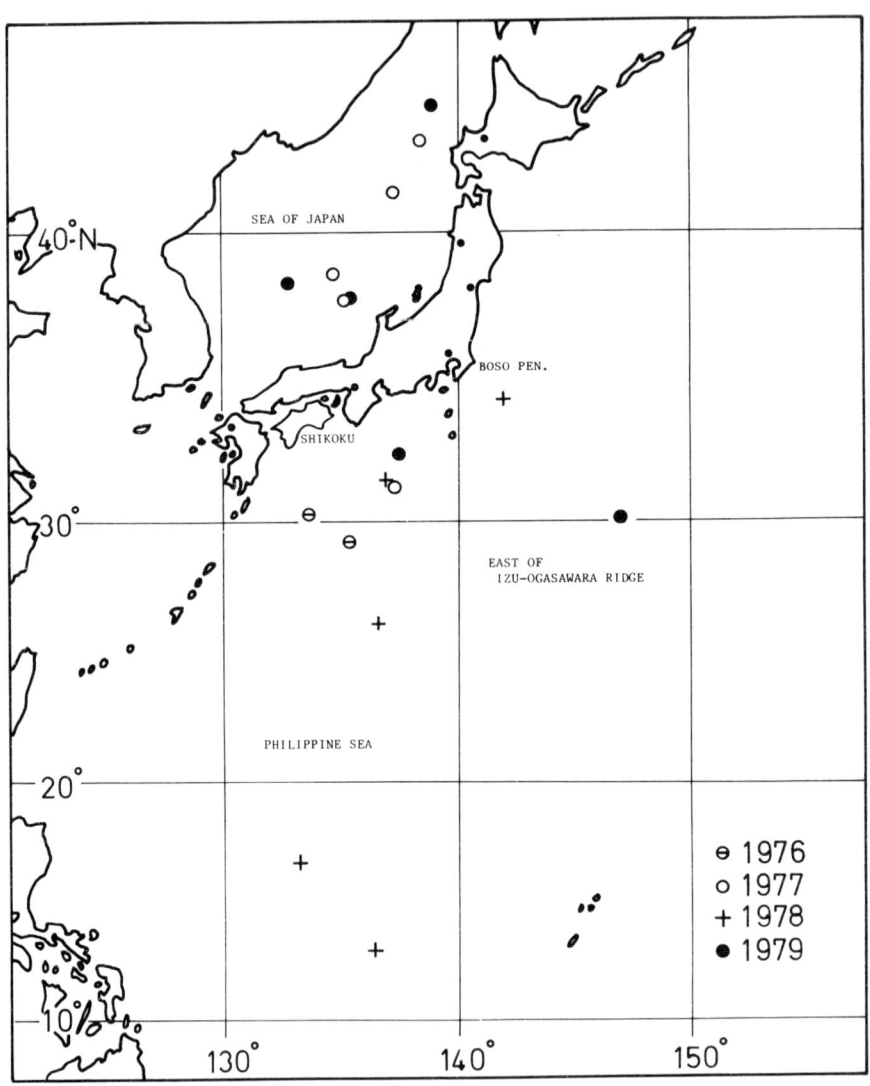

Figure 1 Sampling Stations.

deep water from another region. The radionuclide contents in the Sea of Japan are consistently less -- about 30 % -- than the fallout depositions. This tendency might be attributed to dilution or flow out of the radioisotope in the surface layer by the Tsushima Current and the others.

CONCLUSION

Sr-90 and Cs-137, originating from the radioactive fallout in the northwestern Pacific waters, show the vertical distributions consisting of two layers, upper and deep. Approximately 70 to 90 % of the radionuclides are retained in the upper layer. The depth of boundary of the layers varies regionally reflecting the differences in the progress of vertical mixing, and in the case of the Sea of Japan, where the vertical movement of water is reported to be intensive, the boundaries practically reach the sea bottom.

An exponential correlation is observed between the radioisotope concentrations and the depth, in both the upper and the deep layers.

Generally speaking, the total amount of Sr-90 and Cs-137 in the open seas of the northwestern Pacific seem to be well balanced with the cumulative fallout depositions, similar as reported in the North Atlantic[16], with a slight deviation in Cs-137. On the other hand, in the regions where upwelling or evident currents are observed the total amount of radionuclides in a water column are one half to one third of the cumulative inputs, suggesting the effects of dilution or flow out by the lower radioactivity waters from another region.

ACKNOWLEDGEMENT

The authors wish to express their hearty thanks to Dr. Y. Horibe, the other staff of the Ocean Research Institute, University of Tokyo, and the crew of the R/V Hakuho-maru of the Institute for their support in collecting samples. We are also grateful to Dr. H. Tsubota of Hiroshima University and other participating scientists in the cruises for their cooperation in sample collection.

REFERENCES

[1] Miyake,Y. et al.: "Vertical and Horizontal Mixing Rates of Radioactive Material in the Ocean", Report of the Scientific Conference on the Disposal of Radioactive Wastes, 167-173, IAEA-UNESCO, Monaco (1959).
[2] Miyake,Y. et al.: "Penetration of Sr-90 and Cs-137 in Deep Layers of the Pacific and Vertical Diffusion Rate of Deep Water", J. Radiat. Res., 3, 141-147 (1962).
[3] Popov,N.I. et al.: "Sr-90 in the Waters of the Pacific Ocean. I. Western Part and Adjacent Seas", Okeanologiya, 3, 666-668 (1963).
[4] Baranov,V.I. et al.: "Contamination of Oceans by Long-lived Radionuclides According to the Results of USSR Investigations", Proc. 3rd Intern. Conf. Peace. Use Atom. Energ., 14, 72-82 (1965).
[5] Nagaya,Y. et al.: "Some Fallout Radionuclides in Deep Waters around Japan", J. Radiat. Res., 6, 23-31 (1965).
[6] Folsom,T.R. et al.: "Distribution of Cs-137 in the Pacific", HASL-197, Part 1, 95-203, USAEC, New York (1968).
[7] Shirasawa,T.H. et al.: "Fallout Radioactivity in the North Pacific Ocean; Data Complication of Sr-90 and Cs-137 Concentrations in Sea Water", HASL-197, Part 1, 67-92, USAEC, New York (1968)
[8] Nagaya,Y. et al.: "A Study on the Vertical Transport of Sr-90 and Cs-137 in the Surface Waters of the Seas around Japan", J. Radiat. Res., 11, 32-43 (1970).

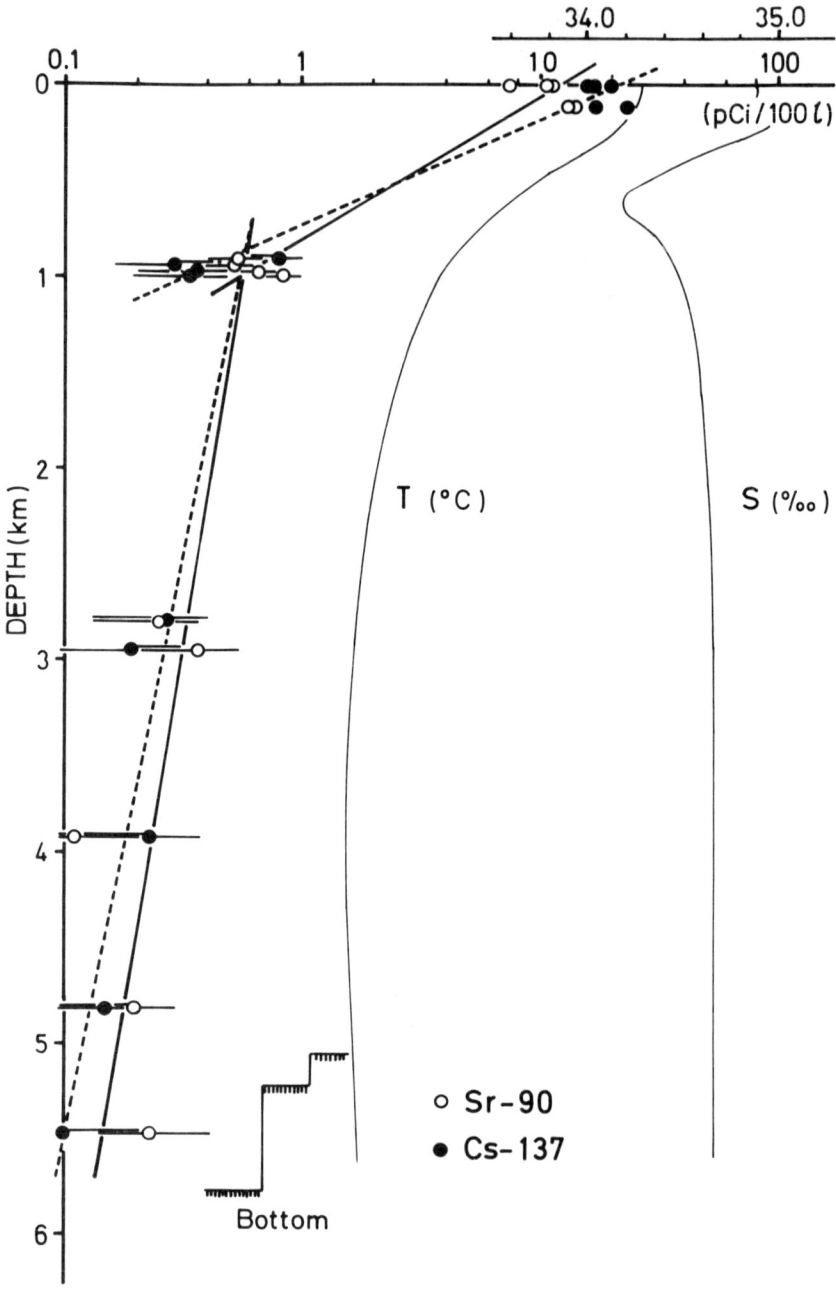

Figure 2 Vertical Distributions of Sr-90 and Cs-137 in the Philippine Sea (profiles of water temperature and salinity are of Stn.LV-3 in 1978).

[9] Volchok,H.L. et al.: "Oceanic Distribution of Radionuclides from Nuclear Explosions", Radioactivity in the Marine Environment, ed. by Seymour,A.H., 42-89, NAS-NS, Washington (1971).
[10] Shiozaki,M. et al.: "The Artificial Radioactivity in Sea Water", Researches in Hydrography and Oceanography, 203-249, Hydrographic Department of Japan, Tokyo (1972).
[11] Saruhashi,K. et al.: "Sr-90 and Cs-137 in the Pacific Waters", Rec. Oceanogr. Wks. Japan, 13, 1-15 (1975).
[12] Nagaya,Y. et al.: "Sr-90 and Cs-137 Contents in the Surface Waters of the Adjacent Seas of Japan and the North Pacific during 1969 to 1973", J. Oceanogr. Soc. Japan, 32, 228-234 (1976).
[13] Horibe,Y. et al.: "Development and Recovery of Moored Arrays of Instruments, Large-Volume Sampler, Auto-Analyzer", Environmental Marine Science, ed. by Horibe,Y., 184-208, Univ. Tokyo Press, Tokyo (1977). (in Japanese)
[14] Katsuragi,Y.: "Cs-137 and Sr-90 Deposition in Japan", Abstracts of Papers presented at the 20th Meeting on the Environmental Radioactivity Survey in Japan, 12-17, Nat'l. Inst. Radiol. Sci., Chiba (1978). (in Japanese)
[15] United Nations Scientific Comittee on the Effects of Atomic Radiation: "Radioactive Contamination due to Nuclear Explosions", Sources and Effects of Ionizing Radiation, 1977 Report, 165-222, United Nations, New York (1977).
[16] Kupferman,S.L. et al.: "A Mass Balance for Cs-137 and Sr-90 in the North Atlantic Ocean", J. Mar. Res., 37, 157-199 (1979).

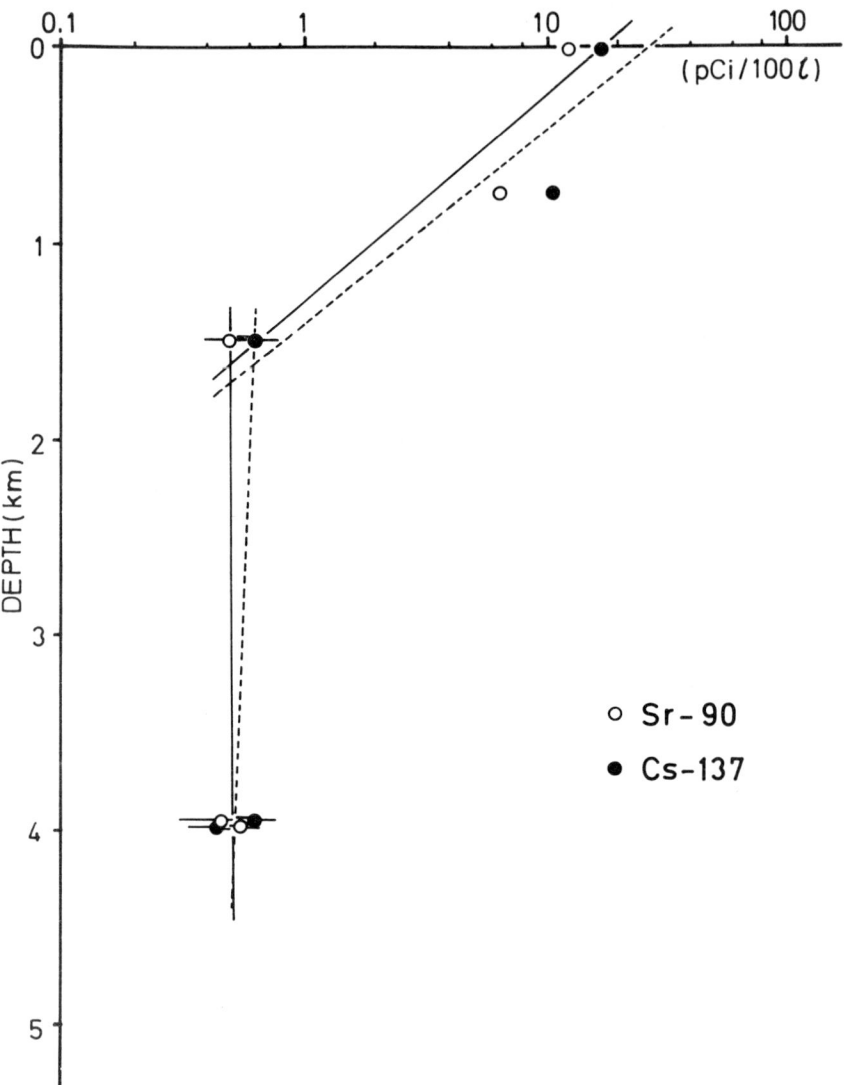

Figure 3　Vertical Distributions of Sr-90 and Cs-137 in the Area Southeast of Boso Peninsula.

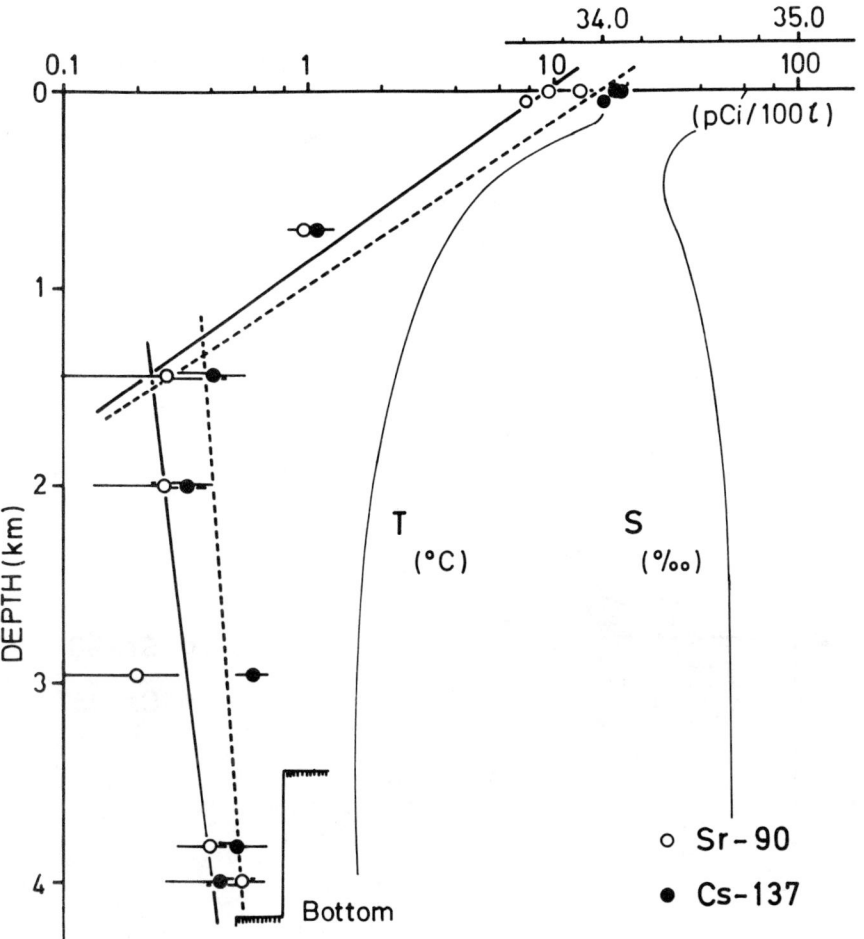

Figure 4 Vertical Distributions of Sr-90 and Cs-137 in the Sea Southeast of Shikoku (profiles of water temperature and salinity are of Stn.LV-4 in 1978).

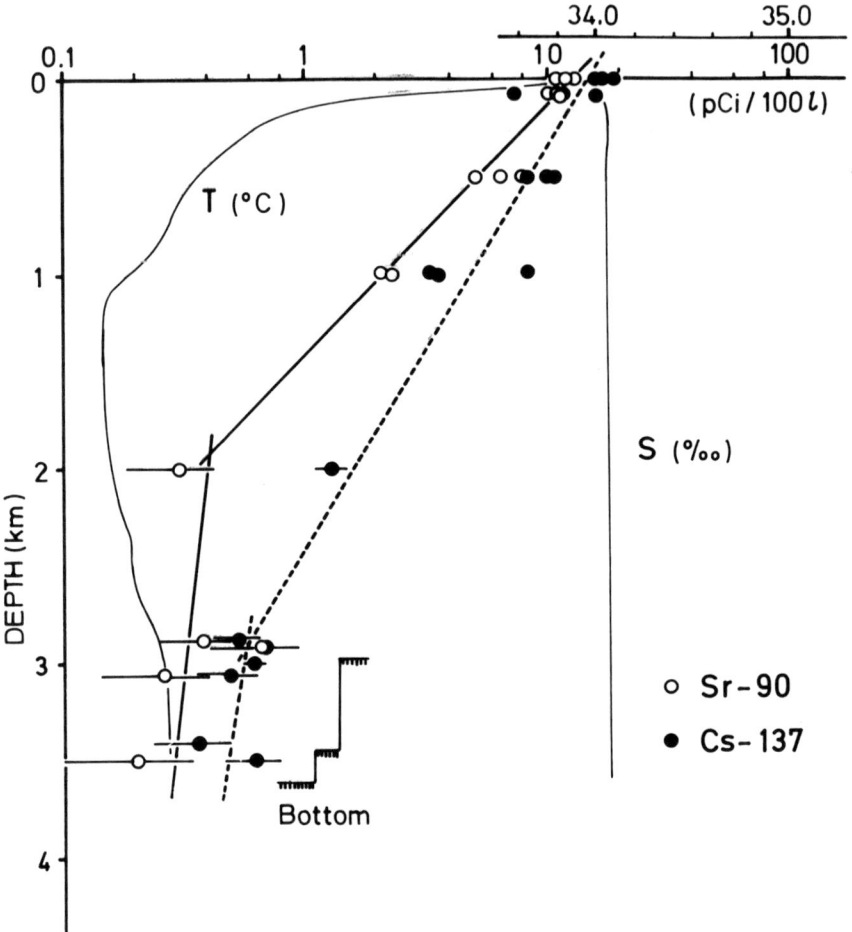

Figure 5 Vertical Distributions of Sr-90 and Cs-137 in the Sea of Japan (profiles of water temperature and salinity are of Stn.LV-3 in 1977).

Discussion

W.E. NOSHKIN Jr., United States

Table 2 shows a Cs-137/Sr-90 ratio in the deep water much different from 1.6, and very different from the value of the ratio in the upper water mass. In the sea of Japan the bottom water ratio is 0.4 compared to a surface ratio of approximately 1.6. Do you have an explanation for this difference ?

Y. NAGAYA, Japan

Sr-90 and Cs-137 concentrations in the deep water are extremely low, so the counting errors of radioactivity are relatively high. Therefore I could not have detected the statistically significant differences in Cs-137/Sr-90 ratio between the upper and deep water samples.

For the Sea of Japan, I have no ideas about the ratio of 0.4, but in Figure 5 the apparent boundary depth of the Cs-137 profile is much deeper than that of the Sr-90 profile ; in other words there are much smaller amounts of Cs-137 in the deep layer in comparison with those of Sr-90. I feel that a closer investigation is desirable on the difference in boundary depth between the radionuclide profiles. An additional sample collection was made in July 1979 from the Sea of Japan, and the results will be reported soon.

A. AARKROG, Denmark

I am not quite convinced that the Sr-90 and Cs-137 levels in the North Atlantic are well balanced with the cumulative fallout depositions. Generally speaking I think that the water collumn shows a surplus of radioactivity in the North Atlantic.

Y. NAGAYA, Japan

According to Kupferman et al. (16), Sr-90 and Cs-137 contents in the North Atlantic are well balanced with the total inputs of the radionuclides, which consisted of radioactive fallout and discharge from nuclear facilities. I think, in the open oceans, where the influence of the discharge is not significant, the radionuclides originate mainly from radioactive fallout and therefore their contents should be well balanced with the cumulative fallout depositions.

DISTRIBUTION ANOMALIES OF RADIO ISOTOPES IN DEEP SEA REGIONS OF THE NORTH ATLANTIC

H. Kautsky
Deutsches Hydrographisches Institut
Hamburg (Federal Republic of Germany)

Abstract

Exept in 1972, during the years 1966 to 1974 at all stations a decrease of the activity concentration values of the radioisotopes Cs 137 and Sr 90 from surface down to deeper layers could be observed.

However, at several stations taken along a line between the Biscay and the Azores, as well as in the northern North Atlantic at 60°, 70° and 72° N during the year 1972 we could observe in a depth region of about 750 to 1250 m an increase of the activity concentration values of Cs 137 up to two to eight times of the corresponding values in the surface water. The vertical distribution of Sr 90 does not indicate the same anomaly. Therefore it seems to us, that this concentration effect may be explained by biological processes.

During the years 1966 to 1974, repeated investigations concerning the vertical distribution of artificial radio isotopes in the Atlantic were carried out by the radiological working group of the Deutsches Hydrographisches Institut (DHI). Thereby, especially the Iberian Deep Sea Basin /¯1_/ - in which the actual European dumping region for packed, low-active wastes lies - as well as regions of the northerly North Atlantic /¯2, 3_/, and the western Mediterranean Sea /¯4_/, were investigated.

The activity concentrations, in the water, of the isotopes Cs 137 and Sr 90 which originate from the atomic bomb experiments and enter the sea with the global fallout, were measured. In no case, thus far, could we find activity values in the deep-sea water indicative of a liberation of measurable amounts of the aforesaid isotopes from the radioactive wastes stored on the sea bottom. The amounts of waste dumped in the Iberian Basin to date compared with the surrounding watermass are also still so negligible that one could reckon with a possibility of detection of artificial radio isotopes possibly set free from those wastes - at the most - in the immediate vicinity of stored containers.

The limits of detection of our analytical methods for Cs 137 lie at about 0.005 pCi/l (5 fCi/l) with about ± 5 % 1σ propagated error, and about 0.01 pCi/l with less than ± 10 % 1σ propagated error for Sr 90. Of course, in order to attain that, we must work off 400 l seawater per analysis.

The measurement values' deviation range between individual stations, lying comparatively close to one another, and also in the repetition of a station within the period of a few days - in individual cases - can be surprisingly large.

However, in general, one can state from the measurements already carried out, that certain deviations of the Cs 137 and Sr 90 activity concentration values in the upper water layer down to about 200 m depth can occur on both sides, which give an indication of the layering of the water in this region. Below these depths - from 1972 onwards, only from about 500 m depth - a stronger decrease in concentration occurs down to about 1,000 m depth. In the depth below 1,000 m to 1,500 m, the values then stay fairly constant. The advance of the fallout products, transported from the surface, into greater depths takes place only very slowly in the sea areas investigated.

Altogether, the Cs 137 values in the surface water of the eastern North Atlantic have decreased from somewhat above 0.3 pCi/l in 1966 to about 0.10 pCi/l (range from 0.08 to 0.12 pCi/l) in 1978. The corresponding values for Sr 90 amounted to somewhat above 0.2 pCi/l in 1966, for 1972 0.08 to 0.1 pCi/l.

Apart from one exception in 1970, at a station in the western Mediterranean (37°02' N; 00°01'30" W), only in 1972 at several stations in the Iberian Basin (Fig. I) and in the northern North Atlantic between 50° N and 72° N (Fig. II) were we able to measure extraordinarily high Cs 137 activity concentration values in a depth range between 500 m and 1,500 m. In individual cases, these were several times higher than those surface concentrations which were available at the same time. In fact, in one case in the Iberian Basin, the value was about twice as high as the highest 1966 surface water measurement in the same area (0.62 to 0.34 pCi/l). The corresponding surface value at that station lay at 0.14 pCi/l at a quarter of the depth value. At 70° N, with 0.14 pCi Cs 137/l in the surface water, we could in fact ascertain at 1,000 m depth 1.1 pCi Cs 137/l; that is, a concentration that was about eight times higher than that of the surface.

We can provide no explanation for this phenomenon. Insofar as measurement values for Sr 90 are available from the same stations, these show the normal decrease of the activity concentration values between surface and depth; that is, no enrichment tendency can be recognized (Fig. II). This observation speaks for the possibility that biological processes are responsible for these enrichment phenomena of Cs 137. Sr 90 practically will not - on the other hand Cs 137 can - accumulate in plankton or other marine organisms to about a factor of 100. It is astounding that, in the 1974 measurement values, absolutely no indication on this distribution anomaly which occured in 1972 was to be recognized.

The investigations specified here refer to the distribution of fallout products only; that is, to transport mechanisms from the sea surface down to the depths. Nevertheless, they illustrate a clear warning of how different radio isotopes with relatively negligibly different biological and chemical behaviour can react under comparatively simple and foreseeable conditions. One could presumably assume a similar, probably essentially larger complexity also for transport processes from the bottom regions upwards to the surface regions of the sea.

For that reason, in all considerations concerning the dumping of radioactive wastes in the sea, whether in packed form or by means of an open discharge of waste solutions in the near-coastal sea, one must pay careful attention to the peculiarities of the isotopes of each individual element with reference to its chemical resp. biological behaviour in the marine environment. I am convinced of the fact that, even with clearly obvious circumstances, in many cases one will arrive at surprising and unexpected results.

References

1. Kautsky, H., Koltermann, K.P., Prahm, G.: "Iberische Tiefsee, hydrographische und radiologische Untersuchungen", Deutsches Hydrographisches Institut, Meereskundliche Beobachtungen und Ergebnisse 45, Hamburg (1977).

2. Kautsky, H.: "Distribution of Radioactive Fallout Products in Atlantic Water between 10° S and 81° N during the Years 1969 and 1972", Deutsche Hydrographische Zeitschrift 30, 216-227 (1977).

3. Kautsky, H.: "Distribution of Radioactive Fallout Products in the Water of the North Atlantic and the Barents Sea during the Year 1972", Isotope Marine Chemistry, in press.

4. Kautsky, H.: "Die Vertikalverteilung radioaktiver Falloutprodukte im westlichen Mittelmeer in den Jahren 1970 und 1974, Deutsche Hydrographische Zeitschrift 30, 175-184 (1977).

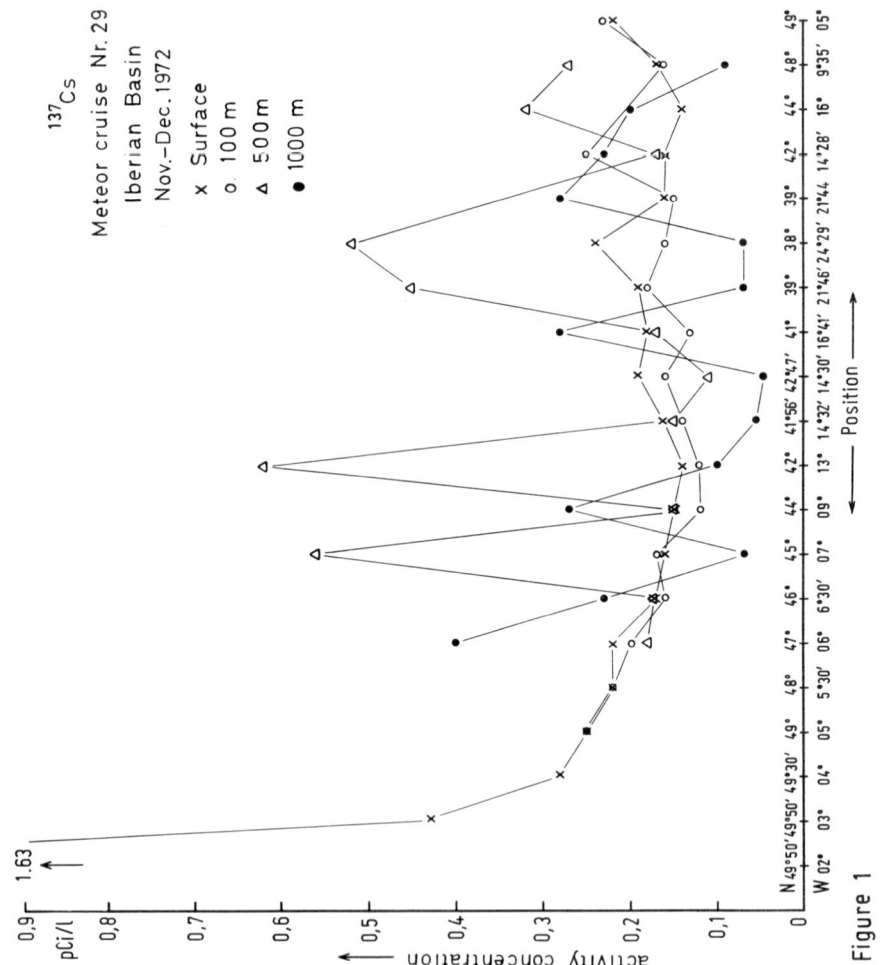

Figure 1

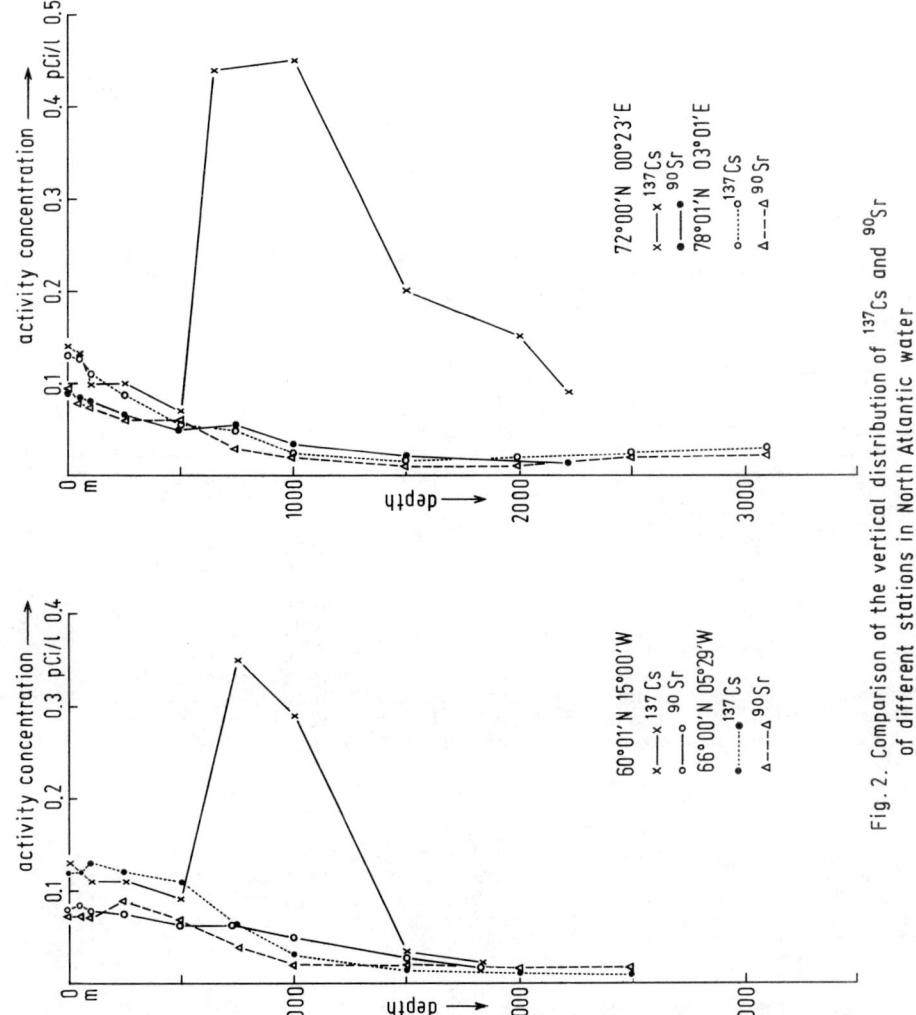

Fig. 2. Comparison of the vertical distribution of ^{137}Cs and ^{90}Sr of different stations in North Atlantic water

Discussion

W.L. TEMPLETON, United States

Between 1969 and 1971 the discharges of Cs-137 from the French reprocessing plant at Cap de La Hague increased by about an order of magnitude (Paper n° 17). Is there any evidence to suggest that the increase at depth you measured in 1972 could be a result of this "spike" release ?

A.M. ORTINS DE BETTENCOURT, Portugal

Do you think that the abnormal caesium values could be related to previous dumping operations in the ocean ?

Y. NAGAYA, Japan

I have never observed such high intermediate maxima, as you have shown, in Cs-137 profiles in the North Western Pacific. Have you any idea of the source of these high values ?

H. KAUTSKY, Federal Republic of Germany

As I have already stated in my presentation we do not know the reason for this accumulation. In my opinion, this is a temporal accumulation of fallout products caused by biological processes. But this cannot be proved. This observation was only made once in 1970 in the Mediterranean, and on a second occasion in 1972 at some stations (about 15 to 18) in the North Atlantic (Iberian Basin and along the Greenwich Meridian between 50°N and 80°N). I personnaly do not believe the possibility that these enhanced activities originate from land based sources (for example in the Mediterranean). At one station, sampled twice within one week in the Iberian Basin, on the first occasion a normal decrease of Cs-137 values from surface down to depth was observed, but on the second occasion a heavy enrichment at a depth of around 500 to 750 m was observed. This phenomenon is therefore not a permanent one. It can change within a short time and appears at different locations. If you take a line of three stations you may observe an enrichment at the first and the third one, whereas at the second one, lying in between, a normal decrease of activity between surface and deep water regions may be observed.

I feel the question is not the source of the origin of the nuclides, the question is : what mechanisms cause these local enrichment effects ? In my opinion the source is the widespread artificial radioactivity originating from bomb tests, or even from other releases of radioactive wastes into the sea at any given point in the world.

I does not seem possible to decide the exact origin of the very low activity concentrations in different widespread sea areas (with the exception of some defined small regions like the Irish Sea where the source of the main bulk or radioisotopes found therein is well known).

We have found this enrichment phenomenon in 1970 in the Western Mediterranean as well as in 1972 in the North Atlantic in the region of the Iberian Basin and up to 80°N. Therefore, in any case, it seems impossible to me that "clouds" of higher activity concentration values of Cs-137 only (not of Sr-90 !) should be transported from the few reprocessing plants situated on European coasts to very different sea regions far away from the source.

There must exist a defined, locally arising mechanism causing this enrichment phenomenon in depths of about 500 to 1500 m at very different places. To date we have proven explanation for this mechanism.

Y. MIYAKE, Japan

You showed us the vertical distribution of radioactive materials in which the gradient of the content of radioactive material with depth is much smaller in deeper layers. This means that the rate of penetration of these nuclides is much greater in deeper layers and the eddy diffusion coefficient must be much larger. We estimated 200 cm^2/sec as the eddy diffusion coefficient by means of the vertical profile of Sr-90 in the Western North Pacific in 1960, which is much larger than that considered hitherto (0.1-1 cm^2/sec).

I would like to ask your opinion of this problem.

H. KAUTSKY, Federal Republic of Germany

This question cannot be clearly answered. I have never made any calculation of diffusion coefficients. We have not enough values to come to a clear answer.

The existing values are widespread at different stations. My statement with regard to fairly constant values of Cs-137 in depths below 1500 m is just a simplification of an average measurement compilation. In reality you can observe a wide variation of the gradients at different stations. I would not dare to make any calculations from these values. We cannot determine from the very few stations available the range of possible variations.

For example (1972)

0-250 m	0.12-0.12	0.12-0.12	0.13-0.088
250-500 m	0.12-0.10	0.12-0.11	0.088-0.052
500-1000 m	0.10-0.049	0.11-0.030	0.052-0.025
1500-depth (~3000 m)	0.039-0.0054	0.013-0.010	0.016-0.030

Ref. H. Kautsky. Distribution of Radioactive Fallout Products in Atlantic Water between 10°S and 57°N during the years 1969 and 1972. Deutschen Hydrographischen Zeitschrift Band 30, Heft 6 (1972) 216-227.

and Ref. : 1 to 4 of my paper.

PLUTONIUM MOBILIZATION FROM SEDIMENTARY SOURCES TO SOLUTION IN THE MARINE ENVIRONMENT

V. E. Noshkin and K. M. Wong
Lawrence Livermore Laboratory
Livermore, California (United States)

ABSTRACT

Inventories of plutonium radionuclides greatly in excess of global fallout levels persists in the benthic environments of Bikini and Enewetak Atolls. It now appears that the atolls have reached a chemical steadystate condition with respect to the partitioning of $^{239+240}$Pu between solution and solid phases of the environment.

The mobilized $^{239+240}$Pu has solute-like characteristics, passes rapidly and readily through dialysis membranes, has adsorption characteristics similar to those of fallout plutonium in the open ocean, and exists in solution primarily as some oxidized +5 or +6 chemical species. Water-column profiles of $^{239+240}$Pu taken outside the atolls show a plutonium excess in the deep water mass. This remobilized $^{239+240}$Pu possibly originates from the contaminated sediments previously deposited on the outer slopes of the atolls and surrounding basins.

"Work performed under the auspices of the U.S. Department of Energy by the Lawrence Livermore Laboratory under contract number W-7405-ENG-48."

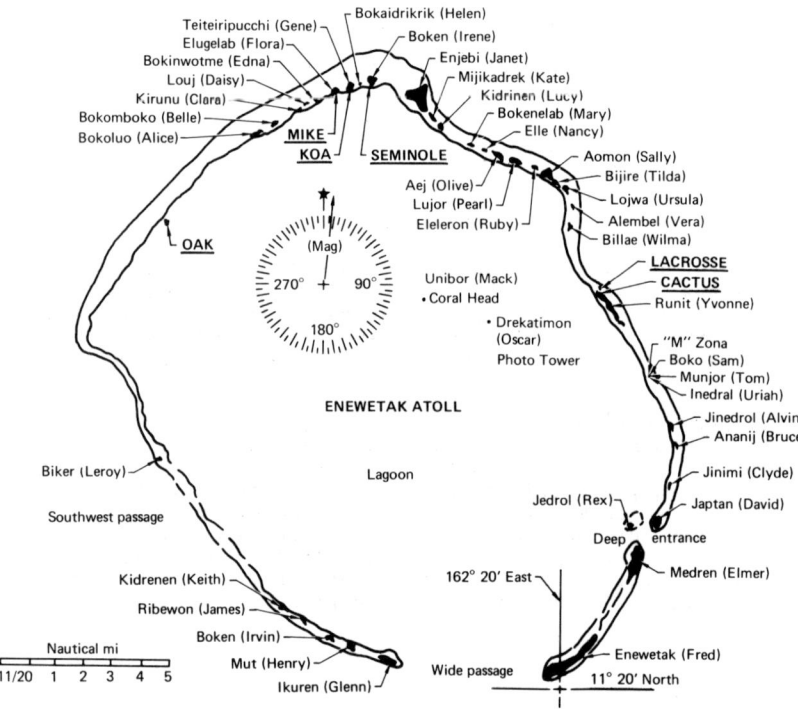

Fig. 1. Enewetak Atoll with names and locations of islands and the six nuclear craters.

INTRODUCTION

Planning for sea-bed disposal or deposition of radiological wastes (either high or low level) must necessarily account for the various processes, reactions, and rates that influence the fate of specific long-lived radionuclides in the event of their release to the sedimentary environment. Many of the processes leading to release of radionuclides and the migration pathways for these activities from buried and/or disposed containment systems leading back to man have been defined [1, 2, 3] but are still imperfectly understood.

Significant quantities of plutonium isotopes and other transuranics will be among the longer-lived, toxic radionuclides associated with radiological waste materials. It is therefore essential to understand by what processes, if any, and at what rates plutonium, subjected to various chemical, physical, and biological disturbances on the sea floor, migrates back to the oceanic water column.

During the past few years we have been conducting studies at Enewetak and Bikini Atolls to better define the environmental physical, chemical, and biological transport mechanisms and fate of the transuranics and other long-lived radionuclides in the aquatic environments. The radionuclides were introduced to the environments during testing of nuclear devices by the United States at these Pacific Atolls between between 1946 and 1958 and were subsequently deposited to the lagoon sediments in association with settling particulate material. These radionuclides are studied mainly to evaluate their impact on critical processes essential for the establishment and continuity of life at the atolls and partly because we recognize these studies can provide data of some significance related to understanding mobilization and migration of plutonium and other radionuclides from oceanic sedimentary deposits to the water column and back to man. The radiological studies at the atolls are therefore germane to problems related to the disposal of transuranic and other radioactive wastes in the ocean. In this paper we discuss in-situ results related to the partitioning of plutonium between solid sources and sollution in the atoll environment. A great deal of similarity has been found in the aquatic characteristics and behavior of plutonium at Enewetak and Bikini. For example, the rates of plutonium mobilization and atoll residence time are very similar. The results from one atoll have great value in predicting transuranic behavior at other Pacific Atolls and, as we will show, in other contrasted marine environments. Results from Enewetak Atoll will be emphasized and supplemented by data from Bikini when it is necessary to clarify the interpretation of data.

Enewetak Atoll and Plutonium Inventory

Enewetak Atoll consists of 39 islands on an elliptical coral reef encompassing a lagoon with an area of 931 km^2. The islands, which were given alphabetic code names during the U.S. occupancy, and several land marks including the locations of craters formed by nuclear tests are identified and shown in Figure 1. The islands, which make a total land area of approximately 6.9 km^2 are situated on a reef 84 km^2, in area. The average depth of the lagoon is 47.4 m; the maximum depth is 60 m.

The U.S. moritorium on testing began on 31 October 1958 and marked the end of all nuclear testing at Enewetak and Bikini Atolls. The fallout history plus other activities during and after the testing period produced a very heterogenous distribution of radionuclides in the lagoon sediments. Today

TABLE I

Summary of Mean and Range of $^{239+240}$Pu Concentrations in Seawater
From Locations in the North Equatorial Pacific Ocean

Location & Region Sampled	Month/Year Sampled	No. of Locations Sampled	Mean $^{239+240}$Pu (pCi/m^3)^a Soluble	Particulate	Total
Enewetak Atoll					
Lagoon surface samples	11/72	29	--	--	29.0 (0.4-96)
Lagoon bottom samples	11/72	6	--	--	44.0 (10-75)
Lagoon surface samples	8/74	48	23.6 (1.4-65)	18.6 (0.8-125)	42.2
Lagoon bottom samples	8/74	23	28.5 (3-69)	18.9 (0.9-36)	47.4
Ocean reef samples	10/75	9	114 (3-644)	N.A.^{b}	
Lagoon reef samples off N. Yvonne Island	10/75	9	94.0 (72-120)	N.A.	
Lagoon surface samples 2 km offshore the inner Atoll perimeter	5/76	19	17.0 (2.1-31)	12.6 (1.6-67)	29.6
Lagoon bottom samples 2 km offshore the inner Atoll perimeter	5/76	10	14.0 (0.7-26)	15.0 (2.6-43)	29.0
Ocean reef samples	5/76	6	26.0 (10-70)	58.0 (6-167)	
Ocean reef samples	10/76	3	55.0 (28-94)	220 (71-774)	
Lagoon reef samples	11/78	5	17.0 (3-46)	N.A.	
Bikini Atoll					
Lagoon surface samples	11/72	10	40.4 (3.9-79)	12 (0.1-42)	
Lagoon bottom samples	11/72	7	40.8 (8.6-64)	71 (5-460)	
Lagoon surface samples	2/77	18	52.0 (27-84)	N.A.	
Lagoon bottom samples	2/77	8	44.0 (13-104)	N.A.	
Lagoon reef samples	11/78	8	29.0 (7-50)	N.A.	
Kwajalein Atoll					
Lagoon surface samples	5/75-10/76	10			0.46 ± 0.15
Lagoon bottom samples	5/75-10/76	3			0.59 ± 0.27
Wotho Atoll					
Lagoon reef samples	11/78	3			0.35 ± 0.16
Rongerik Atoll					
Lagoon reef samples	11/78	4			0.32 ± 0.20

TABLE I (Continued)

Location & Region Sampled	Month/Year Sampled	No. of Locations Sampled	Mean $^{239+240}$Pu (pCi/m^3)^a		
			Soluble	Particulate	Total
Bikar Atoll					
Lagoon reef samples	11/78	2			0.32 ± 0.04
Equatorial Pacific					
1-5 miles W & S of Bikini	10/72	4			15 ± 6
26-90 miles W of Bikini	7/78	3			3.0 ± 0.8
1 mile S of wide pass Enewetak	11/72	5			4.8 ± 3.0
1 mile S of wide pass Enewetak	4/76	3			5.3 ± 3.0
2 miles W of Enewetak	10/76	9			1.7 ± 0.4
North equatorial Pacific	10/72-7/78	14			0.38 ± 0.12

a—Values in parenthesis represent the range in concentrations encountered at locations sampled.
b—Not analyzed.

quantities of long-lived fission products such as ^{137}Cs, ^{90}Sr, ^{155}Eu, and others; activation products such as ^{55}Fe, ^{60}Co, and ^{207}Bi; and transuranics such as ^{238}, ^{239}, ^{240}, ^{241}Pu and ^{241}Am persist in the atoll's environment. The largest inventory of plutonium at Enewetak and Bikini is found associated with the lagoon sedimentary components. Analyses of over 100 grab and core samples collected from Enewetak lagoon in 1972 defined the areal distribution of plutonium in the sediment and showed that approximately 250 curies are associated with the surface 2.5 cm layer and the inventory to a depth of 16 cm in the sediment column is estimated to be 1200 curies.

Highest concentrations are associated with the sediments from the northwest quadrant of the lagoon roughly 2-3 km east of the islands of Alice and Belle (see Figure 1) and several km southwest of Mike and Koa craters. A second region of relatively high concentrations is in the sediments off the shore of Yvonne Island. Most of the plutonium inventory in the surface sediments can be separated roughly from the lesser contaminated deposits by a line extending from the southwest passage to the island of Tom (Monjor), which is south of Yvonne on the eastern reef. The surface $^{239+240}$Pu concentrations north of this line range from 2 to 170 pCi/g dry weight; those south of this line are less than 2 pCi/g. The average concentration in the lagoon sediment determined from the samples collected during 1972 is 5.2 pCi/g dry weight. Bikini sediment inventories were estimated from substantially fewer data than were available for Enewetak, and future results from Bikini might alter the following estimates of the plutonium inventory: 309 curies in the surface 2.5-cm-thick layer and 1470 curies to a depth of 16 cm in the sediment column; the mean activity in the first 16 cm of sediment is 9.5 pCi/g dry weight of sediment.

Following the last nuclear test at Enewetak in 1958, the residual radionuclides deposited to the lagoon water either settled to the bottom or remained as dissolved or particulate species in the water and were eventually discharged to the North Equatorial Pacific by the prevailing exchange of water between the ocean and the lagoon. It has been generally stated that, following introduction to the surface layer of the ocean from any source term, plutonium radionuclides become associated with particles and settle rapidly to the sea floor where they remain immobile after deposition. Accepting this argument, the concentrations of $^{239+240}$Pu and other long-lived radionuclides in the lagoon water column during any year subsequent to 1958 should then only equal the temporal fallout levels in the north equatorial Pacific surface water.

Concentrations in the Water

A considerable number of lagoon water samples have been collected and analyzed for $^{239+240}$Pu by this laboratory since 1972. Several studies are in progress at the atolls that require data on concentrations in lagoon water so the number of samples and location sampled during any year were predicated by the requirements of the program. Table I summarizes arithmetic mean $^{239+240}$Pu concentrations and the range in concentrations detected in filtered water samples collected during different periods from the regions of Enewetak and Bikini indicated. Also shown for comparison are concentrations in surface water collected 1-2 miles outside (west and south) Bikini and Enewetak Atolls, surface concentrations in the Equatorial Pacific 25-90 miles directly west of Bikini, concentrations in lagoon water at other Marshall Island Atolls, and levels in the surface North Equatorial Pacific water well away from the Atolls. The range in values in the lagoon shown in Table I indicates there are significant spatial and temporal differences in concentrations in the water. However, wherever and whenever water was sampled in the lagoon or on the reef, the $^{239+240}$Pu concentrations in solution and in association with particles greatly exceeded the 0.3-0.5 fCi/l fallout background levels in the Equatorial Pacific surface waters. These results are a direct indication that $^{239+240}$Pu is mobilized to solution from the solid phases of the environment. Concentrations in the perimeter surface samples and in the surface water at distances west of Bikini, show there is a flux of mobilized $^{239+240}$Pu continuously advecting from the lagoon to the Equatorial Pacific water mass. Small but measurable amounts of $^{239+240}$Pu are continuously mobilized to solution from sources within the atoll, resuspended to the water column for subsequent redistribution both within and outside the atoll, and are concentrated by all lagoon organisms. The term "soluble" plutonium herein refers to the quantity of plutonium in water samples that passes through filters of stated diameters. A considerable number of tests have been run these past years using filters of different pore sizes. We have found that a 1 micron filter, normally used to remove particulates, is as efficient as 0.45- and 0.2-micron filters. We have been unable to identify the species of plutonium in solution but we have identified many of its characteristics. We find, for example, it is present in the lagoon water in more than one valence state; it has solute like characteristics and passes readily through dialysis membranes; less than 6% of the plutonium in solution is found associated with organically bound material; the quantity mobilized to solution on the windward reef is relatively inert and highly complexed when compared to the highly exchangeable species in the lagoon; the species in the lagoon has exchange characteristics similar to fallout levels in the open ocean; dissolved plutonium released on the reef has been traced for considerable distances by a plutonium radionuclide balance that involves the change in the ^{238}Pu:$^{239+240}$Pu ratio in the water; the dissolved plutonium moves in solution apparently without interacting rapidly with sediment deposits during transport; ^{238}Pu behavior is similar to $^{239+240}$Pu, and although we find some similarities between the characteristics of ^{241}Am and Pu$^{239+240}$, there are sufficient differences in properties that set the behavior of the two transuranics apart.

Rates of Plutonium Loss from the Lagoon

A complete description of the biological, physical, and chemical interactions that are potentially capable of mobilizing $^{239+240}$Pu from the solid sedimentary sources at a coral atoll and of the processes moving $^{239+240}$Pu to and within the lagoon water mass are beyond the scope of this report. We take a less sophisticated approach in the form of radionuclide budgets to attempt an interpretation at a rather simple level of the rates and processes affecting $^{239+240}$Pu mobilization. However, by following this procedure we have been able to determine that some processes and mechanisms dominate the mobilization and redistribution of $^{239+240}$Pu at the atolls; some are insignificant and some appear important but are difficult as yet to test experimentally.

There is a significant quantity of environmental data that now shows that the amount of $^{239+240}$Pu mobilized to solution in the lagoon has been relatively constant for perhaps as long as the past 13 years. The lagoon water mass, with the dissolved inventory of $^{239+240}$Pu, is constantly replaced by ocean water containing only background levels of $^{239+240}$Pu. In order to maintain a constant, elevated $^{239+240}$Pu level in the lagoon water, the rate of plutonium removal from the atoll must be balanced by a continuous flux of dissolved plutonium from sources within the atoll. Horizontal and vertical concentration gradients prevail in the lagoon water mass during any period sampled. To attempt a material balance, it is necessary to assume that a reasonable, average concentration in the lagoon can be derived from the results during those periods, when sufficient samples were obtained from the lagoon for analysis. Results from Enewetak during November 1972, July-August 1974, and April-May 1976 and from Bikini during November 1972 and January-February 1977 show that this is a reasonable assumption. During November 1972 the average concentration of plutonium (soluble plus particulate) in the lagoon water at Enewetak was 32 fCi/l. In 1974 and 1976, the average concentrations of soluble $^{239+240}$Pu were 24 and 16 fCi/l, respectively, and the particulate concentrations represented 46 and 43%, respectively, of the average total concentrations of $^{239+240}$Pu. Assuming there is, at any time, 44% of the total plutonium in association with the particulate phase, we estimate that in November 1972 10 fCi/l of $^{239+240}$Pu was in association with suspended particulates and 22 fCi/l was the average soluble concentration. The differences in the soluble concentration determined during these 3 periods is not considered significant. It should be emphasized that the average concentration is based on results each year from a different number of samples. These samples were quite often taken from different lagoon locations. When concentrations were determined in water samples from lagoon stations previously sampled, more than 3 times out of 4 the agreement in measured values was excellent. Therefore, in spite of sampling different water masses during different seasons of different years, the average quantity of $^{239+240}$Pu in solution, based on results from different numbers of samples, has been reasonably constant, at least since 1972. During 1972 and 1977, the average concentrations of soluble $^{239+240}$Pu in Bikini lagoon water were 40 and 49 fCi/l, respectively. No significance is attached to these differences and, as at Enewetak, we assume the standing average amount of $^{239+240}$Pu in the lagoon water mass at anytime has been constant. At Bikini in 1972, the average concentration of $^{239+240}$Pu associated with lagoon particulates represented 48% of the total concentration in the water. This average is very similar to the mean percentage now measured twice at Enewetak.

With the appropriate dimensions for each lagoon, the average concentrations in solution at Enewetak and Bikini convert to a plutonium standing inventory of 0.9 and 1.3 curies, respectively. The average inventories associated with lagoon suspended particulates are 0.7 and 1.1 curies, respectively. At Enewetak the quantity in solution represents only 0.36% of the $^{239+240}$Pu inventory in the sediment measured to a depth of 2.5 cm in the sediment column and 0.075% of the total inventory to a 16 cm depth. At Bikini, the mean soluble inventory of $^{239+240}$Pu in the lagoon water is 0.40% of the 309 curies estimated in the surface 2.5 cm layer of sediment and 0.086% of the 1470 curies inventory to a 16 cm depth. Particulate inventories in the water column are proportionally smaller fractions of the respective sediment inventories. The amount of $^{239+240}$Pu mobilized and found at any time in solution at the atolls represents a very small fraction of the inventory in the major atoll reservoirs.

Radiological data from different biological indicators show that the lagoon- water residence times vary between 118-170 days with a reasonable average being 144 days. However, physical circulation data indicate that the lagoon water mass, on the average, is exchanged with the open ocean at a much more rapid rate [4,5]. A resolution of these differences is in progress, but for the present, the slower rate of 144 days will be taken to represent the rate at which the $^{239+240}$Pu in solution is exchanged between the lagoon and open ocean. A large fraction of the $^{239+240}$Pu associated with the lagoon particulates is associated with resuspended sediments. Very little sedimentary material escapes from the lagoon, and resuspended bottom material probably settles out again on the lagoon floor close to its point of origin. Therefore, no plutonium associated with particulate material is assumed lost from the lagoon.

Using the residence time of 144 days and the average soluble inventories of plutonium, 2.8 and 3.2 curies, respectively, are discharged annually to the open ocean from Enewetak and Bikini Atolls. If the inventory to 16 cm in the sediment column (1200 curies) is the reservoir for the mobilized $^{239+240}$Pu at Enewetak, then the mean life for $^{239+240}$Pu in this reservoir is 435 years. The mean life is computed from Equation (1), where dn/dt is the annual rate of plutonium depletion, N -- represents the total sedimentary inventory considered, and t is the mean life.

$$\bar{t} = \frac{N}{dn/dt}$$

(1)

At Bikini, the computed mean life for the $^{239+240}$Pu in the sedimentary reservoir, to a depth of 16 cm in the sediment column, is 460 years. Although the inventory of $^{239+240}$Pu at any time in these lagoon water masses represents a small fraction of the sedimentary inventory, if the mobilization processes continue at the same rate, $^{239+240}$Pu will be depleted from the sedimentary environments in a geological time span that is short compared to the radiological half-lives for $^{239+240}$Pu.

Exchange of Plutonium Between Sediment and Seawater

Since $^{239+240}$Pu has been found in solution at the atolls, there must be some release of $^{239+240}$Pu from the contaminated sediments to the water phase as a consequence of desorption or loss by some other means. A number of laboratory and in-situ experiments have been conducted with contaminated sediments from the lagoons to arrive at a value for the distribution coefficient (K_d) for plutonium. Laboratory studies involved measuring the fraction of $^{239+240}$Pu desorbed from the sediments while field studies involved the collection and analysis of interstitial water and sediment. Different size fractions from different geographical regions and from different depths within the sediment column were used in desorption experiments with uncontaminated seawater. The range in K_d values for the different lagoon sediments was between 0.5×10^5 and 3.6×10^5 with an average K_d value for $^{239+240}$Pu of 2.3×10^5.

Nelson and Lovett [6], recently studied plutonium distributions in the Irish Sea. They found that the average K_d for Pu (+3 or +4) in seawater was 24.9×10^5 while the average K_d for the oxidized forms of Pu (+5 or +6) in seawater was 0.15×10^5. The method outlined by Nelson and Lovett [6] was followed to separate the reduced from oxidized forms of dissolved $^{239+240}$Pu in the lagoon seawater at Enewetak and Bikini. Analysis of three lagoon water samples showed that on the average, 92% of the dissolved $^{239+240}$Pu was in the oxidized (+5 or +6) form while the remaining 8% represented a reduced (+3 or +4) state of dissolved $^{239+240}$Pu. These results demonstrate that different oxidation states of $^{239+240}$Pu are capable of coexisting in the lagoon water.

Also using the K_d values for the oxidized and reduced states of $^{239+240}Pu$ from Nelson and Lovett [6], and the fractions of oxidized and reduced $^{239+240}Pu$ in seawater at Enewetak, we have been able to correctly predict our average measured K_d value of 2.3×10^5. The value of the distribution coefficient for any element relates to the activity on solids in equilibrium with a quantity in water ($K_d = pCi\ g^{-1}/pCi\ ml^{-1}$). If two species coexist in a seawater/sediment system, then an apparent K_d would fall between the two respective values as shown for the different oxidation states of plutonium by Equations (2) and (3).

$$\frac{pCi\ g^{-1}(+3\ or\ +4) + pCi\ g^{-1}(+5\ or\ +6)}{pCi\ ml^{-1}(+3\ or\ +4) + pCi\ ml^{-1}(+5\ or\ +6)} = K_d\ apparent \qquad (2)$$

$$\frac{K_{d(+3\ or\ +4)} \times pCi\ ml^{-1}(+3\ or\ +4) + K_{d(+5\ or\ +6)} \times pCi\ ml^{-1}(+5\ or\ +6)}{pCi\ ml^{-1}(+3\ or\ +4) + pCi\ ml^{-1}(+5\ or\ +6)} = K_d\ apparent \qquad (3)$$

Substituting the respective K_d values and the fractions of the oxidized and reduced plutonium in seawater at Enewetak into Equation (3) results in an apparent K_d of 2.1×10^5, which is in excellent agreement with our average measured value. The range in K_d values encountered with sediments from the lagoon may merely reflect the quantity of oxidized and reduced plutonium species capable of disassociation from the particular lagoon sediments tested.

The average concentration of $^{239+240}Pu$ in the surface 16 cm of sediment at Enewetak and Bikini are, as previously given, 5.2 and 9.5 $pCi\ gm^{-1}$, respectively. If the surface sediment layer is forcibly mixed with overlying lagoon water, mixing of the interstitial water with dissolved $^{239+240}Pu$ would take place. A fraction of the exchangeable $^{239+240}Pu$ held on the suspended material will also be liberated during each mixing. There are several other identified processes that may cause mixing of the interstitial lagoon fluids and the dissolved $^{239+240}Pu$ with the lagoon water. Assume that a simple chemical balance is continuously maintained between the $^{239+240}Pu$ in the sedimentary sources and the interstitial and interface water. The rate at which the interstitial and interface water mix within the lagoon is balanced by a flux of undersaturated lagoon water which, then rapidly equilibrates with the exposed sediments. This process simulates a huge continuous batch extraction in which we allow the $^{239+240}Pu$ retained on the solid sediments to be continuously in equilibrium with the overlying water column, and the desorption mechanism follows the law of mass action applicable to an ion-exchange type equilibrium. Then average concentrations of $^{239+240}Pu$ in the lagoon water at any time should relate to the K_d value and the average sediment concentrations. Computed inventories and average concentrations in the water, using the basic equation relating K_d to concentrations in the water and sediment, are 0.99 curies and 23 fCi/l, respectively, at Enewetak, and 1.3 curies and 41 fCi/l at Bikini. There is general agreement between the average quantities of $^{239+240}Pu$ predicted and those measured in solution during each of the periods of 1972, 1974, 1976, and 1977 at Enewetak and Bikini. For many reasons, it may be argued that this agreement is fortuitous. Nevertheless, the general agreement found between computed and the average concentrations in both lagoons between 1972 and 1977 measured several times supports the contention of a steady state condition and demonstrates the general usefulness of this simple model in predicting long term average concentrations in the lagoon water.

In summary, small quantities of $^{239+240}$Pu are found to be continuously mobilized to solution from the sedimentary sources at Enewetak and Bikini Atolls. The plutonium is slowly being depleted from the atoll reservoir and is discharged to the surface waters of the North Equatorial Pacific. $^{239+240}$Pu mobilized to solution has solute-like characteristics, is available for uptake by organisms, and different valence states are capable of coexisting in solution. It will require more than 400 years to mobilize the entire inventory of $^{239+240}$Pu from these atoll sediments. Although this period of time is long compared to our lifespan, it is small compared to the radiological half-life of plutonium.

The ratio of the average concentration in lagoon water at Bikini and Enewetak is approximately 1.8. This value is nearly identical to the ratio of the mean sediment concentrations of 9.5 and 5.2 pCi gm^{-1}. The computed mean lives for $^{239+240}$Pu in the sediments of both lagoons are essentially the same.

The average quantity of $^{239+240}$Pu mobilized to the water column is proportional to the average concentration in the lagoon sediments. Therefore, geological mean lifetimes for plutonium in sediments from any atoll will be similar to the mean life as determined at Enewetak and Bikini, and the average concentration in the overlying water column will be proportional to the respective concentration in the sediment column considered. A simple mass action model can be used to estimate the quantity of $^{239+240}$Pu capable of dissociation from the sediments to solution.

Concentrations in Equatorial Pacific Waters

During June and July 1978, we participated in a joint oceanographic cruise to regions of the Equatorial Pacific with V.T. Bowen of Woods Hole Oceanographic Institution. Water, sediment, and plankton samples were collected, and several other ancillary experiments were conducted. The following represents a summary of plutonium water-column inventories in samples we collected for analysis. Some of these conclusions may be revised after the results from both laboratories are critically compared.

Inventories of the $^{239+240}$Pu to different depths for five stations, along with the calculated range, in the North Equatorial Pacific are given in Table II. Stations 2 and 4 are 270 to 300 miles to the east and northeast, respectively, of Bikini Atoll. Station 9 is approximately halfway between Bikini and Enewetak Atolls. Station 11 is 15 miles west of the northwest reef of Enewetak and Station 13 is located about 212 miles northwest of Enewetak. Between the surface and 1000 m depth, there is a slight east to west gradient in the $^{239+240}$Pu inventory, but the mean values from all stations fall within the average inventory of 1.28 $\pm$ 0.10 mCi/km^2. This quantity alone in the upper 1000 m exceeds the estimated global fallout levels delivered to these latitudes [7]. The inventory of plutonium in the 1000 m of water at Stations 9 and 11 are essentially identical to the quantities in the upper 1000 m of water at Stations 2 and 4. Plutonium concentrations from 3 to 10 times higher than fallout levels were previously detected in surface water samples collected west of Bikini and west of Enewetak. An estimated 3 curies of plutonium are remobilized annually from the lagoon sedimentary deposits and exchange with the north equatorieal surface waters. This annual input to the surface layers of the North Equatorial Pacific is so rapidly

TABLE II

Inventories of 239+240Pu (mCi/km^2) in filtrates from water columns at stations in the North Equatorial Pacific.[a]

Station Number	Location	239+240Pu (mCi/km^2) to various depths (m)					
		0-1000	0-2000	0-3000	0-4000	0-5000	Bottom
2	270 mi E of Bikini 11° 'N 170°00 'E	1.19 ± 0.07	2.06 ± 0.14	2.38 ± 0.16	2.57 ± 0.17		(0-4300) 2.62 ± 0.21
4	300 mi NE of Bikini 13°40 'N 170° 'E	1.20 ± 0.06	1.89 ± 0.11	2.18 ± 0.15	2.35 ± 0.23	2.43 ± 0.25	(0-5431) 2.48 ± 0.27
9	95 mi W of Bikini 11°40 'N 163°36 'E	1.23 ± 0.07	2.24 ± 0.15	2.67 ± 0.20	3.00 ± 0.25	3.28 ± 0.33	(0-5004) 3.28 ± 0.33
11	15 mi W of Enewetak 11°37 'N 161°48 'E	1.35 ± 0.09	2.31 ± 0.18	3.04 ± 0.27			(0-3850) 4.34 ± 0.37
13	212 mi NW of Enewetak 13°30 'N 154°00 'E	1.41 ± 0.15	2.37 ± 0.20	2.82 ± 0.24	3.20 ± 0.27	3.67 ± 0.30	(0-5757) 4.01 ± 0.34

[a]These inventories are calculated from the activity of the filtrate only. An additional 10 to 25% of the plutonium activity usually found in the particulate fraction (collected on 1 μm filter cartridges) was not included in these inventories.

diluted and horizontally transported westward that essentially no change in the plutonium inventory is detectable within the upper 1000 m of water west of the atolls when compared to inventories in this layer east of the atolls.

At Stations 2 and 4 the average inventory of $^{239+240}$Pu between 1000- and 4000-m depths is 1.27 mCi/km^2. At Stations 9, 11, and 13 the plutonium inventory within this depth interval is 1.77 $\pm$ 0.26, 3.00 $\pm$ 0.38, and 1.79 $\pm$ 0.31 mCi/km^2, respectively. These amounts are statistically different from those in the deep water east of the atolls. They show that plutonium may be remobilized to the deeper waters west of these Pacific test sites, possibly from the contaminated sediments previously deposited on the outer slopes of the atolls and surrounding basin. Concentrations in sediments from the region of Station 11 greatly exceed expected fallout concentrations. If the mobilization rate of plutonium from sedimentary deposits outside the atolls does not differ from that within the atolls, the bottom waters of the North Equatorial Pacific must move substantially slower than the surface layers to account for the increased inventories at depth from the atoll source term. Station 13 is 212 miles west of Enewetak, and excess plutonium is still evident in the deep water mass. The areal extent of contaminated bottom water and the fate of the plutonium introduced to the deep ocean are unknown. We can, however, safely assume that plutonium in the deep water, originating from the test series at Enewetak and Bikini, is present at least 212 miles from the atolls' source term and moved with the bottom water to this distance during the last 20-30 years.

References

1. I. Aoyana, M. Yamamoto, and Y. Inoue, "Evaluation of the Radioactive Waste Disposal into the Deep Sea", Health Phys., 33, 227 (1977)

2. C. Machida, A. Ito, A. Matsumoto, S. Sakata, T. R. Nagakura, and H. Abe, "Developments and Studies for Marine Disposal of Radioactive Wastes", Management of Radioactive Wastes from the Nuclear Fuel Cycle, Vol. II, IAEA, Vienna, 339 (1976).

3. D. R. Anderson, W. P. Bishop, V. T. Bowen, J. P. Brennan, W. N. Caudle, R. J. Detry, T. E. Ewart, D. E. Hayes, T. R. Heath, R. R. Hessler, C. D. Hollister, K. Keil, J. A. McGowen, R. W. Rohde, W. P. Schimmel, C. L. Schuster, A. J. Silva, W. H. Smyrl, B. A. Taft, and D. U. Talbert, "Release Pathways for Deep Sea bed Disposal of Radioactive Wastes", Impacts of Nuclear Releases into the Aquatic Environment, IAEA, Vienna, 483 (1975).

4. S. Smith, Hawaii Institute of Marine Biology (private communication), 1979.

5. W. S. VonArx, "Circulation Systems of Bikini and Rongelap Lagoons", U.S.G.S. Professional Paper No 260-B. Govt. Printing Office, Washington, D.C. (1954)

6. D. M. Nelson and M. B. Lovett, "Oxidation State of Plutonium in the Irish Sea", Nature, 276, 599 (1978).

7. C. P. Hardy, P. W. Krey, and H. L. Volchok, "Global Inventory and Distribution of Fallout Plutonium", Nature, 241, 444 (1973).

NOTICE

"This report was prepared as an account of work sponsored by the United States Government. Neither the United States nor the United States Department of Energy, nor any of their employees, nor any of their contractors, subcontractors, or their employees, makes any warranty, express or implied, or assumes any legal liability or responsibility for the accuracy, completeness or usefulness of any information, apparatus, product or process disclosed, or represents that its use would not infringe privately-owned rights."

Discussion

G.A.M. WEBB, United Kingdom

A caution on the extrapolation of these results for long term predictions in the widespread deep ocean - especially the 400 year clean-out time. There will be an effect from fresh sediment deposition and movement to deeper sediment layers. Also the K_d definition implies an equilibrium process ; with time the concentration in seawater will increase and the concentration in sediment will decrease - both of these will tend to lengthen the effective residence time.

W.E. NOSHKIN Jr., United States

A 400 year clean-up time only applies to Enewetok and Bikini Atolls. As long as fresh ocean water continues to exchange with the lagoon and its dissolved plutonium, the K_d concept remains valid. With time, the concentration in the lagoon seawater will change in proportion to the sediment levels. As the average sediment levels decrease, the average concentrations in the lagoon water will decrease. Barring catastrophic events, the present distributions and inventory of plutonium in the lagoon sediments will be only slightly altered during the next few decades. K_d values determined in a variety of sediments differ very little from our measured average. Considering the difference in the types of environmental samples tested in the literature it is striking that the K_d for plutonium differs so little. We feel our results can be used with a reasonable degree of accuracy to predict bottom or interstitial water concentrations at other contaminated sedimentary environments, knowing the existing sediment levels.

Y. SUGIMURA, Japan

According to our results from a study, already published, on the distribution of Pu in the Pacific Ocean, the content of Pu in subtropical surface water is in good agreement with your results in the N. Equatorial water.

Concerning the vertical distribution of Pu in the subtropical region in the Western North Pacific, a subsurface maximum was observed around 600 m depth, which is approximately the same depth of distribution as given in your report.

With respect to the soluble and particulate form of Pu in surface water, we found that 10 to 50 % of the Pu is associated with the particulate form.

W.E. NOSHKIN Jr., United States

Agreement between measured low level plutonium concentrations in sea water samples, obtained by different laboratories, using different techniques, is very satisfying since our efforts have provided each of us with a valuable, unanticipated, intercomparison of analytical results.

CONSIDÉRATIONS SUR LES CHAÎNES ALIMENTAIRES

M.C.Vaz Carreiro et A. Ortins de Bettencourt,
L.N.E.T.I. - Département de Protection et Sureté Radiologique
Sacavém, Portugal

Résumé

Des références sont faites au sujet des caractheristiques d'adaptation des communautés benthoniques à la vie des grands fonds, des migrations verticales de différents organismes et des chaînes alimentaires. On considère aussi les conséquences de l'immersion des fûts.

Il en ressort la possibilité que ces matériaux radioactifs ont d'atteindre la surface, soit par des courants d'upwelling, soit à travers les chaînes alimentaires habituelles, ou soit encore par une nouvelle exploitation des produits de pêche à de plus grandes profondeurs.

Abstract

Adaption characteristics of the benthic communities to life at the abyssal sea floor, vertical migrations of different organisms and food chains are referred. Consequences of drums dumping are also considered.

It comes out that radioactive materials may reach the surface by upwelling, throughout the usual food chains or through a new exploitation of the fishing resources at greater depths.

1. INTRODUCTION

Depuis plusieurs années on procède à l'immersion de déchets radioactifs dans l'Océan Atlantique nordeste et on peut observer que les activités immergées, bien que restant au dessous des taux maxima reccomandés par l'AIEA, viennent en s'accroîssant au long des dernières opérations.

En dépit des études déjà effectuées par différents pays et organisations internationales sur ce sujet, il faut remarquer que celles-ci sont, pour la plupart, théoriques et que dans certains domaines les informations sont encore insuffisan - tes et parfois même contradictoires.

Étant donné que les phénomènes en cause ont un temps d'inertie long, les conséquences pouvant atteindre des générations lointaines dans l'avenir, aucune pos sibilité de sous-estimer le risque ne doit être laissée au hasard.

Le but de cette courte contribution est donc d'attirer l'attention sur les chaînes alimentaires et sur autres possibilités de transfert des radioéléments immergés jusqu'à la surface et d'encourager la discussion sur ce sujet.

2. RÉFÉRENCES AUX COMMUNAUTÉS BENTHONIQUES ET PÉLAGIQUES DES GRANDS FONDS

Les régions abyssales qui correspondent à la plaine à faible inclinai - son, qui va de 3000 à 6000 - 7000 m, ont des conditions physiques particulières, dont sont connues:

- la décroissance de la température avec la profondeur, sans variation sai sonnière;

- l'homogénéité de la salinité, avec des valeurs plus faibles qu'en surfa - ce;

- la distribution irrégulière de l'oxygène dissous, avec diminution aux abords du fond;

- les courants faibles (1/3 à 1/2 de noeud), avec influence possible des courants de turbulence jusqu'à de grandes profondeurs;

- le substrat constitué par des sédiments mobiles, rarement dur;

- la pression élevée, environ 100 kg/cm^2 par kilomètre de profondeur.

D'autre part, sont mal connues des caractéristiques telles que les cou - rants verticaux, la vitesse et la direction des courants de fond et les vitesses de sédimentation.

En ce qui concerne le biota, les grands fonds sont caractérisés par une certaine stabilité et par l'adaptation de la faune à une faible activité métaboli - que, ayant aussi des adaptations spéciales.

Les animaux benthoniques des fonds abissaux ont des caractéristiques tel les que:

- réduction des formations calcaires, par exemple, des coquilles des mol - lusques; dans le cas des poissons, les déformations du squelette pour - raient être une conséquence de l'insuffisance en vitamine D (en relation avec l'obscurité);

- gigantisme par rapport à des formes proches non abyssales, atribué à l'action de la pression hydrostatique sur le métabolisme;

- corps plus aplati et appendices plus élargis que ceux des formes proches des régions moins profondes.

En ce qui concerne le problème de la nourriture, la base de la pyramide alimentaire dans les profondeurs abyssales ne peut être occupée que par des bactéries ou des détritus organiques. La tendance actuelle est d'admettre que le rôle alimentaire des bactéries peut être plus important que celui des détritus organiques, puisque les bactéries semblent être très nombreuses à la surface des grands fonds.

Outre les animaux qui se nourrissent de bactéries ou de détritus il y a les macrophages qui, étant donné la faible abondance des proies disponibles, ont des adaptations spéciales qui leurs permettent de les retenir juste en les appercevant (par exemple, une très grande bouche, une bouche extensible et l'esophage élargi).

On pense à présent que, plus que la retombée passive de matière organique, il y a un transport actif des organismes pélagiques dû aux migrations, qui sont la conséquence des variations d'ilumination, du cicle biologique ou des habitudes alimentaires. Il y a une vraie chaîne de migration qui transporte de la matière organique pas à pas, dès les riches couches de l'epiplancton jusqu'aux grands fonds océaniques, laquelle peut alimenter les organismes, soit benthoniques, soit pélagiques des grandes profondeurs. C'est le cas du Zooplancton dont les migrations peuvent atteindre les 5000 m et plus.

D'autres groupes d'organismes, comme les Bryosoa, Crinoidea, Echinoidea, Ophiuroidea e Holothuroidea, peuvent également vivre jusqu'à 5000 m de profondeur et plus; l'holothuria Epidia glacialis a déjà été rencontrée entre 50 et 9000 m.

Outre les formes benthoniques on trouve également dans les fonds abyssaux des formes pélagiques comme des poissons. Les Poissons Myctophides (poissons lanterne), peuvent vivre de 2 m environ jusqu'à 3000 m et les crabes rouges de 300 à 3000 m.

Plus importants, à notre avis, sont les céphalopodes, largement distribués dans les océans, des exemplaires de la même espèce ayant été pris à 15 m et aussi à 5000 m. On ignore s'il s'agit d'une migration, mais des migrations verticales de 2500 m ont déjà été enregistrées.

Les habitudes alimentaires des céphalopodes ne sont pas très bien connues, mais ils se nourissent de poissons, de copépodes, d'euphausides, etc.. Par ailleurs, beaucoup d'espèces différentes de céphalopodes ont été rencontrées dans les estomacs d'un grand nombre de prédateurs (oiseaux, phoques, poissons, comme le thon, et baleines), représentant donc un maillon important dans plusieurs chaînes alimentaires.

L'importance croissante des céphalopodes comme source de protéines et l'ampleur de ses migrations confère donc un intérêt particulier aux études visant une meilleure connaissance de ce groupe, nommément des facteurs de concentration.

3. RÉFÉRENCES AU TRANSPORT VERTICAL

On pense, à présent, que le mélange dans le fond de l'océan et le transport vertical vers la surface peuvent avoir lieu surtout aux abords des pentes abruptes, la thermocline ne constituant pas de barrière effective aux courants verticaux.

Il est ainsi important d'étudier le transport entre les différentes couches, non seulement dans la région d'immersion, mais également aux limites de la zone océanique.

L'étude de l'upwelling vertical de profondeur dans les limites océaniques est assez difficile mais d'une très grande importance, car il peut permettre un court-circuit dans le transfert au long des chaînes alimentaires biologiques et donc un apport de concentrations relativement élevées de matières radioactives jusqu'à des zones de pêche.

Shepherd dans son modèle de dispersion attire l'attention sur l'effet d'advection verticale qui fait accentuer le transport de l'eau de profondeur jusqu'à la surface.

En ce qui concerne le temps de permanence des eaux profondes, il n'y a pas d'accord; en effet on parle de 500 années pour l'Atlantique Nord, mais on admet aussi qu'il peut être beaucoup plus réduit.

Les doutes existants sur le transfert vertical réel de la radioactivité justifient:

- a) que dans les modèles les plus récents, l'on ait prit, pour la surface les concentrations prévisibles, plus élevées, pour les eaux profondes,
- b) que l'on considère la possibilité d'un transfert rapide vers la surface, en discutant la validité d'un programme de contrôle radiologique,
- c) que l'on fasse des recherches plus poussées dans ces domaines.

4. CONSÉQUENCES PRÉVISIBLES DE L'IMMERSION DES FÛTS DE DÉCHETS RADIOACTIFS

Par l'observation des photographies du fond des océans, on sait que presque tous les objets durs sont couverts d'organismes fixes filtrants et de prédateurs et commensaux de ces groupes fixes, dépendant de facteurs tels que la nature du substract, les courants, etc.. Les objets métaliques ont tendance à être couverts de crinoides pédonculés, éponges, actinies et corails. L'immersion d'un grand nombre de fûts, métaliques ou non, peut donc changer la biologie benthonique du site et même la dynamique des courants.

Quand les fûts sont en partie enfouis dans les sédiments, une attaque électrolytique est à craindre dans la zone de séparation, étant donné que la surface est soumise à des conditions soit d'oxydation, soit de réduction, selon qu'elle est en contact avec l'eau ou les sédiments. Ceci peut encore conduire à des taux de fixation variables pour les différents radioéléments par les sédiments.

Certains des radionucléides libérés par les fûts semblent, de toute façon, demeurer immobilisés par les sédiments, pouvant atteindre des concentrations parfois très élevées, ce qui rend critique la question de savoir comment et dans quelle mesure ils sont disponibles pour le biota benthonique. En effet, même sans oublier les differences de comportement dûes aux formes physico-chimiques des radioéléments, il y a certaines contradictions dans l'interprétation des phénomènes, comme quand c'est le cas, par exemple, de savoir si la principale voie d'absorption des actinides par les organismes benthoniques consiste en une contamination à travers des sédiments, des matières en suspension ou bien d'autres maillons de la chaîne alimentaire.

La perte du contenu par les fûts de déchets radioactifs, en plus de la contamination, donne lieu à une augmentation sur place de carbone organique (papiers, gants en caoutchouc, etc.) et de complexes organiques spécifiques, ce qui amène certainement une augmentation de la nourriture disponible pour les bactéries marines et pour les organismes qui habitent les sédiments, entrainant, comme conséquence, une augmentation des populations qui peuvent donc devenir fortement contaminées.

L'observation des populations benthoniques des lieux d'immersion déjà utilisés et de leur contamination actuelle serait, donc, du plus grand intérêt pour la compréhension de ces phénomènes et la confirmation des modèles théoriques.

5. CONSIDÉRATIONS FINALES

Étant donné que les communautés biologiques vivant aux fonds océaniques, dans la région d'immersion, peuvent être soumises à une radioactivité plus élevée que celle admise pour les organismes consommés par la population, il faut prendre en considération la possibilité d'un court-circuit qui les amèneraient plus di - rectement jusqu'à l'homme. Ainsi, par exemple, des chaînes alimentaires interliées d'organismes à migrations verticales, comme les céphalopodes et les euphausides, peuvent éventuellement conduire à une contamination rapide des organismes de surface qui font partie de l'alimentation humaine, bien que la probabilité en soit réduite, étant donné que le transport de la nourriture en mer se fait d'une façon préférentielle vers le fond.

La pêche des céphalopodes étant en dévelopement, ces Mollusques peuvent devenir une voie non négligeable de transfert des matières radioactives immergées jusqu'à l'homme. Selon les statistiques, des bateaux portugais en ont pêché en 1977, 6131 tonnes, dont la plus grande partie provient de l'Atlantique Nordeste et on en a importé 1874 tonnes de provenance inconnue.

D'autres espèces d'eaux profondes comme les poissons lanterne et les crabes rouges ont, par ailleurs, déjà été identifiées comme des organismes marins potentiellement importants pour une consommation future par l'homme.

Les poissons migrateurs méritent également une référence, comme c'est le cas du thon, dont les céphalopodes constituent un des aliments. Ces poissons migrent de l'Atlantique nord jusqu'à la Méditerranée et ils utilisent normalement des régions où il y a des vortex (qui peuvent correspondre à un transport vers la surface) pour se reposer et se nourrir, pendant leurs migrations, en particulier en période de frai. Ceci se passe, pour les thons, dans les régions de l'Atlantique nordeste proches du site d'immersion, ce qui justifierait leur contrôle.

On observe actuellement une tendance à élargir les zones de pêche, surtout de descendre jusqu'à 1500 - 2000 m de profondeur, une fois que les couches jusqu'à 500 m sont déjà surexploitées. À Madère on pêche depuis longtemps le poisson sabre noire, Aphanopus carbo, qui habite à environ 2000 m de profondeur et se nourrit surtout d'une espèce de crevette géante et de céphalopodes, pouvant ainsi constituer un bon indicateur biologique de pollution radioactive d'origine profonde, du moment que l'on en connaisse la radioactivité de référence.

Dans ces brèves considérations, il nous semble raisonable de conclure le besoin de bien identifier les communautés benthoniques, en n'oubliant pas la colonisation sur et autour des fûts immergés, de déterminer les voies alimentaires significatives pour le transfert depuis les grands fonds jusqu'à l'homme et encore de prendre en considération les nouvelles pratiques de pêche prévisibles à de plus grandes profondeurs, qui pourront donner lieu à de nouvelles voies de transfert.

Il faudrait, en outre, étudier les facteurs de concentration des organismes des grands fonds et en particulier des céphalopodes et les coefficients de transfert des proies aux prédateurs.

RÉFÉRENCES

"Criteria for Selection, Management and Surveillance of Ocean Dumping Sites".
Technical Document, IAEA, 1979.

"The Oceanographic Basis of the IAEA Revised Definition and Recomenda tions Concerning High-level Radioactive Weste Unsuitable for Dumping at Sea".
IAEA - 210, 1978.

Moore, H.B.,
Marine Ecology
John Wiley & Sons, Inc. London, 1966

"Radioactivity in the Marine Environment"
Prepared by the Panel on Radioactivity in the Marine Environment of the Committee on Oceanography. National Research Council. National Academy of Sciences. 1971

Pérès, J. - M.
"La Vida en el Océano"
Ediciones Martinez Roca, S.A. Barcelona 1976

Estatísticas de Pesca
Instituto Nacional de Estatística, Lisboa.1977

Session 4

Chairman - Président
Mr. A. AARKROG
(Denmark)

Séance 4

DONNEES RADIOECOLOGIQUES
CONCERNANT LE SITE MARIN DE LA HAGUE

J. Ancellin et P. Bovard
Commissariat à l'Energie Atomique
Institut de Protection et de Sûreté
Nucléaire, Département de Protection
Centre d'Etudes Nucléaires de
Fontenay-aux-Roses (France)

Résumé

De nouvelles observations radioécologiques faites dans l'environnement de l'usine de traitement de combustibles irradiés de La Hague (Nord-Ouest du Cotentin) ont permis d'apporter des compléments aux études déjà réalisées en ce qui concerne le transfert des radionucléides rejetés en mer et les niveaux de radioactivité relevés dans le secteur Nord-Ouest du Cotentin et à distance sur les côtes de la Manche.

Parmi les radionucléides rejetés par l'usine, le ruthénium-106 et le cérium-144 constituent la part la plus importante de la radioactivité artificielle γ observée dans les espèces et les sédiments. Leur dispersion, ainsi que celle du plutonium, a tendance à être relativement limitée par rapport à celle d'un radionucléide se présentant sous forme essentiellement soluble comme le césium-137. Certains apports côtiers seraient d'autre part plus accentués au Sud-Ouest qu'à l'Est-Sud-Est du point de rejet. Enfin les niveaux de radioactivité observés et leur incidence sur le plan sanitaire permettent de conclure à une bonne concordance avec les résultats de l'étude prévisionnelle effectuée avant la mise en fonctionnement des installations.

Le fonctionnement de l'usine de traitement de combustibles irradiés située à La Hague, dans le nord-Cotentin, a commencé en 1966 et a correspondu, après une étude prévisionnelle effectuée avant les premiers rejets en mer, à la mise en oeuvre d'un programme de mesures et d'observations régulièrement échelonnées dans le temps, tant au titre de la surveillance qu'au titre des recherches radioécologiques sur les transferts des radionucléides dans le milieu océanique.

Nous nous proposons de rendre compte ici - plus spécialement dans le domaine radioécologique - de diverses données qui viennent compléter les résultats déjà publiés de travaux consacrés au site de La Hague [1] et qui s'appliquent notamment aux espèces et sédiments de la zone soumise à l'influence des rejets en mer. Cet exposé sera précédé d'un rappel concernant les caractéristiques et conditions d'utilisation du site ainsi que les principaux résultats de l'étude prévisionnelle.

1. LE SITE

1.1 Caractéristiques naturelles

Au point de vue hydrodynamique, le site se caractérise par la grande amplitude des marées et par l'existence de courants importants - de l'ordre de 6 à 8 noeuds ou plus - dans le détroit du Raz Blanchard situé entre la pointe nord-ouest du Cotentin et l'Ile d'Aurigny.

En ce qui concerne la morphologie et la sédimentologie littorales, la côte est constituée de falaises rocheuses entrecoupées d'anses sableuses, telles que l'anse d'Ecalgrain et celle de Vauville, où se produisent des apports sédimentaires. Au large les fonds sont constitués de cailloutis, graviers ou sable graveleux et correspondant à des zones où les conditions hydrologiques empêchent la sédimentation.

Les peuplements biologiques sont surtout remarquables par l'abondance, sur les supports rocheux, des algues - notamment <u>Laminaria digitata</u> et <u>Fucus serratus</u>, dont la biomasse va de 2,5 à 10 kg/m^2 - et par l'abondance des balanes (crustacés fixés).

1.2 Utilisation du milieu par l'homme

La récolte des produits maritimes dans le nord-ouest du Cotentin comporte :

. <u>la pêche en mer</u> qui se pratique à bord de petits bateaux artisanaux, soit aux casiers (homards, crevettes, crabes et araignées de mer), soit aux lignes de fond (congres) ou aux lignes trainantes (maquereaux), soit, plus rarement, au chalut (poissons plats) ou à la drague (coquilles Saint Jacques). Les quantités capturées sont de l'ordre de 15 tonnes par an dans la partie ouest de la côte (de Vauville au Cap de La Hague) et de l'ordre de 180 à 200 tonnes par an dans la partie nord (du Cap de La Hague à Cherbourg); elles comprennent 75 à 80% de poissons, 15 à 20% de crustacés, 5 à 10% de mollusques. Le temps de présence en mer et de manipulation des engins de pêche par un pêcheur a été évalué à 2.000 heures par an en moyenne.

. <u>la pêche à pied</u> qui se pratique surtout en période d'été pour <u>la récolte de l'algue Chondrus crispus</u> (extraction d'alginates) dont les apports s'élèvent à environ 400 tonnes par an. Le temps moyen consacré par un pêcheur à cette récolte a été évalué à 300 heures par an. Il existe également une pêche de coquillages (pourpres, ormeaux) et de crustacés relativement peu importante.

En ce qui concerne la <u>distribution des produits maritimes</u>, le marché en gros de Cherbourg (7.000 à 8.000 tonnes/an expédiées en notable partie vers des centres de consommation situés en dehors de la région) est surtout alimenté par les apports de chalutiers et cordiers travaillant sur des lieux de pêche se trouvant à une assez grande distance du Cotentin (Mer celtique entre autres). Par contre il existe un marché de détail - pour les besoins des consommateurs locaux - qui est pour une part importante approvisionné par la pêche artisanale s'exerçant dans la zone côtière de Vauville à Cherbourg, englobant le secteur des rejets.

L'<u>alimentation</u> des populations locales a fait en 1965-1966 l'objet d'une enquête portant sur trois secteurs géographiques : l'agglomération cherbourgeoise (groupe de population n° 1), la zone située à l'intérieur des terres (groupe n° 2), la zone côtière où l'on distingue une population rurale (groupe n° 3) et une population constituée par les familles de pêcheurs pratiquant la pêche artisanale (groupe n° 4). Pour ce dernier groupe qui est le plus grand consommateur de produits marins, la consommation moyenne journalière de ces produits était la suivante :

. poissons   48 grammes
. crustacés 19   "
. mollusques 12   "

Enfin sur le plan des <u>loisirs</u>, le nord-ouest du Cotentin fait l'objet d'une certaine fréquentation des estivants, notamment les plages de l'anse de Vauville, de la baie d'Ecalgrain et d'Urville-Nacqueville près de Cherbourg. Le temps moyen de présence d'un estivant sur l'estran a été évalué à 100 heures par an.

1.3. Niveaux de radioactivité avant le fonctionnement des installations

Les mesures de radioactivité du milieu marin effectuées avant la mise en marche de l'usine en 1966 se rapportent pour la plupart à l'activité bêta totale.

Ces mesures ont notamment été effectuées sur les algues - qui, par suite de leur pouvoir d'accumulation des radionucléides, se révèlent être parmi les meilleurs indicateurs de la pollution radioactive. Pour les trois espèces suivantes : <u>Corallina officinalis</u>, <u>Fucus serratus</u> et <u>Chondrus crispus</u>, la radioactivité naturelle - due principalement au potassium 40 - était de l'ordre de 4.000 à 8.000 picocuries par kilo frais. A cette radioactivité naturelle se superpose celle, variable, provenant de la pollution par la radioactivité artificielle. En 1963-1964 la radioactivité bêta totale (y compris celle du potassium 40) était de l'ordre de 10.000 à 12.000 picocuries par kilo frais pour <u>Fucus serratus</u> et <u>Chondrus crispus</u>. Cette radioactivité, à laquelle contribuaient très largement les retombées atmosphériques, a marqué ensuite une tendance à la diminution en corrélation avec une diminution du nombre des explosions nucléaires. L'incidence des retombées était toutefois encore notable à une certaine profondeur dans les sédiments au cours des dernières années [5].

2. RESULTATS DE L'ETUDE PREVISIONNELLE : DOSES D'IRRADIATION CALCULEES

L'étude prévisionnelle a consisté, entre autres, à recueillir les informations voulues sur le trajet et l'importance des courants et à évaluer la capacité de dilution de l'effluent à

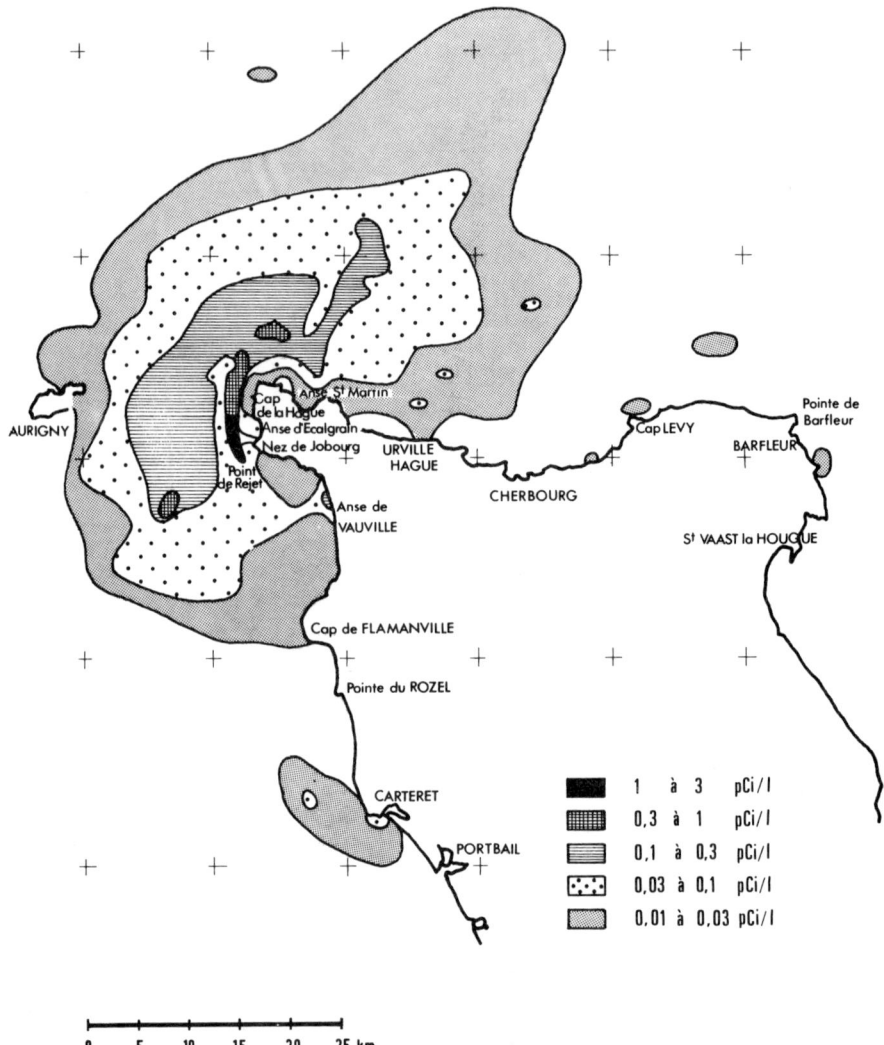

Figure 1. Carte prévisionnelle des concentrations moyennes pour un rejet de 1 curie/jour. (Extraite de /1/).

laquelle on pouvait s'attendre dans la zone des rejets, en recourant à des injections de colorant sur modèle réduit et in situ. Les données obtenues ont conduit à estimer qu'une dilution de 10^7 devait intervenir à 6-8 km du point de rejet et ont permis d'établir une carte des taux moyens prévisibles de radioactivité (isoconcentrations) dans le secteur marin des rejets, calculés pour un rejet de 1 curie/jour, (Fig. 1).

Les calculs prévisionnels de doses d'irradiation externe et interne appelées à être délivrées - pour un débit déterminé de rejets radioactifs - aux différents groupes de population ont été établis à partir de données concernant essentiellement : a) les conditions de dilution précédemment mentionnées (cf. carte d'isoconcentration), b) les facteurs de concentration des radionucléides chez les espèces marines consommées, c) les quantités consommées par jour et par habitant, d) les durées de séjour des pêcheurs et des estivants dans le milieu appelé à être soumis à l'influence des rejets [1].

Pour les pêcheurs (groupe le plus exposé) et pour un rejet prévisionnel de 45.000 curies par an, les valeurs d'irradiation exprimées en fractions de limite de dose (rapport entre la dose calculée résultant des rejets et la dose admissible pour les "personnes du public") ont été les suivantes :

. Irradiation interne par consommation alimentaire
  Tractus digestif : $1,52.10^{-2}$
  Organisme entier : $0,45.10^{-3}$

. Irradiation externe
  Tractus digestif : $1,25.10^{-3}$
  Organisme entier : $3,25.10^{-3}$

. Irradiation totale
  Tractus digestif : $1,64.10^{-2}$
  Organisme entier : $0,37.10^{-2}$

3. SITUATION EN COURS DE FONCTIONNEMENT DE L'USINE

Les principaux radionucléides rejetés en mer, outre le tritium, sont les suivants : ruthénium 103 et 106, césium 134 et 137, strontium 89 et 90, antimoine 125, cérium 144, zirconium-niobium 95.

Parmi les autres radionucléides rejetés figurent en quantité faible ou très faible : cobalt 60, argent 110m, zinc 65.

On trouvera dans le tableau I le relevé en curies et par radionucléide des quantités rejetées en mer entre 1966 et 1977.

La surveillance s'exerce d'une manière continue sur les différents constituants du milieu marin : eau, sédiments, espèces. Elle porte sur l'ensemble du Cotentin, les secteurs côtiers sélectionnés étant au nombre de 12; s'y ajoutent plusieurs secteurs situés au large.

Les résultats de la surveillance n'ont mis en évidence l'influence des rejets qu'à partir de 1968. L'évolution de l'activité des éléments surveillés a été en relation directe et en bon accord avec l'importance des rejets d'eaux résiduaires. L'examen des résultats montre également une nette influence de la proximité de l'usine, malgré une dilution immédiate des effluents au voisinage de l'émissaire atteignant 10^6, comparable à celle trouvée lors des expériences de rejets de colorants [2].

TABLEAU I

Rejets annuels en mer (en Ci) de l'usine de traitement de combustibles irradiés de La Hague

Année	^{3}H	^{103}Ru ^{103}Rh	^{106}Ru ^{106}Rh	^{134}Cs	^{137}Cs	^{89}Sr	^{90}Sr ^{90}Y	^{144}Ce ^{144}Pr	^{95}Zr	^{95}Nb
1966			389		197	28	28			
1967			1 326		443	17	11			
1968			1 627		768	21	37			
1969			1 434		545	7	19			
1970	1 657	92	5 409	374	2 409	10	107	25	74	
1971	2 113	187	7 712	1 298	6 556	21	447	355	72	
1972	2 280	450	7 571	166	890	68	868	145	322	
1973	2 967	116	7 104	227	1 872	33	510	177	23	
1974	7 598	241	14 518	244	1 513	233	2 817	1 123	411	39
1975	11 120	410	22 425	115	931	235	2 029	566	309	360
1976	7 132	60	15 004	177	939	21	1 078	157	92	270
1977	8 958	31	14 591	258	1 372	73	1 965	135	20	53
										35

Année	^{125}Sb	^{60}Co	^{110m}Ag	^{65}Zn	Autres radioéléments	^{239}Pu+^{238}Pu	Uranium (en kg)	Total en curies (uranium non compris)
1966						0,03	190	642,03
1967						0,32	280	1 797,32
1968						0,85	244	2 453,85
1969						0,36	126	2 005,36
1970	25					0,64	193	10 610,64
1971	68				428	3,91	205	20 038,91
1972	492				1 206	1,79	2 576	13 867,09
1973	1 789	11,3	0,3		613	2,19	1 346	16 158,79
1974	1 862	0,3	1,3	1,47	1 287	15,69	11,3 (*)	32 877,96
1975	1 943	5,9	0,5	0,04	1 940	7,07	6,2 (*)	43 050,51
1976	1 971	5,4	0,5	0,21	2 684	4,23	5,6 (*)	26 430,54
1977	1 481	6,0	0,7	0,60	736	6,46	10,4 (*)	28 932,66
			0,6					

(*) Uranium et autres émetteurs alpha (hormis le plutonium) en curies.

Au point de vue des conséquences sanitaires de ces rejets, selon les premières estimations établies sur la base des niveaux de radioactivité observés et en tenant compte des divers modes d'utilisation du milieu, l'exposition à laquelle est soumis le groupe critique (pêcheurs) représenterait moins de 1% de la dose admissible définie par la Commission Internationale de Protection Radiologique [2].

4. RESULTATS D'OBSERVATIONS RECENTES DANS LE DOMAINE RADIOECOLOGIQUE

Les données récentes qui viennent s'ajouter à celles déjà utilisées pour établir un bilan de la radioactivité du site [2] proviennent essentiellement des résultats de mesures effectuées par le Laboratoire de Radioécologie Marine de La Hague et qui ont fait l'objet de trois principales publications [3], [4], [5].

Ces données appellent à effectuer un certain nombre de constatations sur la dispersion des radionucléides dans le secteur des rejets et à distance, sur les niveaux de radioactivité observés à proximité de l'émissaire et sur les conséquences qui en découlent sur le plan sanitaire.

4.1 Dispersion des radionucléides

4.1.1 Diminution de la radioactivité en fonction de la distance.

D'une façon générale et au moins en ce qui concerne les espèces, on constate qu'une décroissance de radioactivité se produit assez régulièrement en fonction de la distance à laquelle on se trouve de l'émissaire. Cette décroissance est cependant sensiblement plus marquée pour les radionucléides ayant tendance à se présenter sous forme particulaire - tels que le ^{144}Ce, le ^{106}Ru, le ^{239}Pu - que pour des radionucléides solubles, ou considérés comme tels, comme le césium 137. Le tableau II indique le rapport entre la valeur A de radioactivité massique près de l'émissaire et la valeur B trouvée à une distance de 100 km (Baie du Mont Saint Michel) pour trois espèces marines : deux algues, Fucus serratus et Fucus vesiculosus et un mollusque, Gibbula sp. On voit qu'à cette distance la radioactivité des espèces étudiées diminue d'un facteur 5 à 10 pour les trois premiers radionucléides mentionnés, alors qu'elle ne diminue que d'un facteur 2 environ pour le césium 137 par rapport à la radioactivité observée non loin du point de rejet. Cette observation est évidemment à rapprocher de celles faites sur le site de Windscale où, en comparant les teneurs dans l'eau du ruthénium, du plutonium et du césium, il apparait que ce dernier radionucléide s'élimine à distance moins rapidement que les deux premiers de la masse d'eau [6], [7].

Il n'est pas sans intérêt de faire remarquer à ce sujet que les calculs prévisionnels de dilution, basés sur le comportement de colorants solubles, ne s'appliquent qu'imparfaitement au cas de radionucléides à formes particulaires.

4.1.2 Cas de l'antimoine 125

Les teneurs de l'antimoine 125 chez l'algue Pelvetia canaliculata sont au contraire plus élevées à une certaine distance qu'à proximité de l'émissaire, aussi bien à l'ouest qu'à l'est de celui-ci (tableau III). Ce phénomène s'expliquerait par les particularités de l'hydrolyse de l'antimoine 125 lequel évoluerait très lentement au cours de son transfert dans le milieu marin d'une forme non adsorbable vers une forme partiellement absorbable, plus aisément fixée [5].

TABLEAU II
(d'après [3] et [4])
Gradient de diminution de radioactivité
en fonction de la distance du point de rejet
(Rapport entre la radioactivité massique A d'une espèce déterminée près du point
de rejet et la radioactivité massique B de cette même espèce à 100 km de distance)

	^{238}Pu ^{239}Pu	^{144}Ce	^{106}Ru	^{137}Cs
Fucus vesiculosus			A/B = 5,2	A/B = 1,3
Fucus serratus	A/B = 6,6	A/B = 5,1	A/B = 9,2	A/B = 2,4
Gibbula sp.			A/B = 7,6	

TABLEAU III
(d'après [4])
Teneurs en antimoine 125 chez l'algue Pelvetia canaliculata,
en rapport avec la distance du point de rejet
(moyenne observée de Sept. 1976 à Fev. 1977)

Station	km	Teneur en ^{125}Sb de Pelvetia en pCi/kg frais
Roscoff	180	-
Cancale	110	169
Dielette	15	33
Herquemoulin	5	24
Point de rejet	0	-
Goury	5	76
Omonville	20	167
Fermanville	50	233
Saint Vaast	80	198

TABLEAU IV
(d'après [4])
Radioactivité de quatre espèces-témoins,
prélevées mensuellement à la station de Goury (secteur du point de rejet),
entre 1976 et 1978. (valeurs moyennes en pCi/kg frais)

	^{144}Ce	^{106}Ru	^{137}Cs	^{40}K
Fucus serratus	370	3 680	58	9 000
Fucus vesiculosus	340	2 600	65	8 250
Patella sp. (chair)	380	3 472	≠ 0	2 000
Gibbula sp. (chair)	1 160	19 350	≠ 0	1 755

4.1.3 Dissymétrie de répartition de la radioactivité par rapport au point de rejet.

Sur les figures 2 et 3 sont reportées les valeurs moyennes de radioactivité des sédiments fins littoraux pour le ruthénium 106 et le cérium 144. On voit d'après ces données qu'une limite moyenne de radioactivité de 5 pCi/g pour le ruthénium 106 passe à l'est de l'émissaire vers Barfleur (50 km du point de rejet), alors qu'au sud-ouest cette même limite se situe vers la Baie de Saint Brieuc (130-140 km du point de rejet). Une constatation analogue peut être faite pour le cérium 144 : la limite de radioactivité de 2 pCi/g pour ce radionucléide passe à l'est entre Fermanville et Barfleur (40 km du point de rejet), et au sud-ouest également au voisinage de la Baie de Saint Brieuc (130-140 km du point de rejet). D'après ces données les apports en région côtière des deux radionucléides en question seraient plus prononcés vers le Golfe normano-breton au sud-ouest qu'en Baie de Seine à l'est-sud-est. Des études complémentaires sont nécessaires pour déterminer dans quelle mesure cette situation (qui n'est pas observée pour l'antimoine 125, le zirconium 95 et le césium 137) est liée à des conditions hydrologiques particulières et compte tenu qu'il existe, au moins en ce qui concerne les masses d'eau superficielles, un mouvement de dérive générale vers l'est, au large du Cotentin.

4.2 Niveaux de radioactivité

Les travaux précédemment mentionnés et effectués entre 1975 et 1978 ont comporté, en ce qui concerne les émetteurs alpha et gamma, un nombre considérable de mesures de radioactivité sur les espèces et les sédiments prélevés dans le secteur des rejets et à distance sur l'ensemble des côtes françaises de la Manche.

D'une façon générale il apparait que, parmi les radionucléides rejetés par l'usine de La Hague, ce sont le ruthénium 106 et le cérium 144 qui constituent la part la plus importante de la radioactivité artificielle gamma observée dans les espèces et les sédiments.

Nous nous bornerons à exposer les résultats concernant la zone proche du point de rejet (station de Goury) - c'est-à-dire celle où risque, en principe, de se produire l'accumulation radioactive la plus notable.

4.2.1 Espèces.

On trouvera dans le tableau IV les valeurs moyennes de radioactivité chez quatre espèces choisies en raison de leur abondance et de leur pouvoir d'accumulation (deux algues : Fucus serratus et Fucus vesiculosus et deux mollusques : Patella sp. et Gibbula sp.), d'après les mesures effectuées mois par mois pendant deux ans, entre 1976 et 1978, sur du matériel prélevé à la station de Goury. Ces mesures se rapportent au ruthénium 106 et au cérium 144 - radionucléides dont les teneurs sont de beaucoup les plus élevées comme nous l'avons indiqué - ainsi qu'au césium 137. La valeur moyenne de la radioactivité due au potassium 40 est fournie à titre d'indication.

D'après ces données dont il importe de souligner la valeur représentative en raison de la fréquence de prélèvements s'échelonnant sur une longue période de temps, la radioactivité artificielle (émetteurs gamma) chez les espèces mentionnées était de l'ordre de 3.000 à 20.000 pCi/kg frais.(A comparer aux niveaux observés, tout au moins en ce qui concerne les algues, avant le fonctionnement de l'usine).

D'autres mesures également effectuées sur du matériel prélevé à la station de Goury ont porté sur un ensemble d'espèces

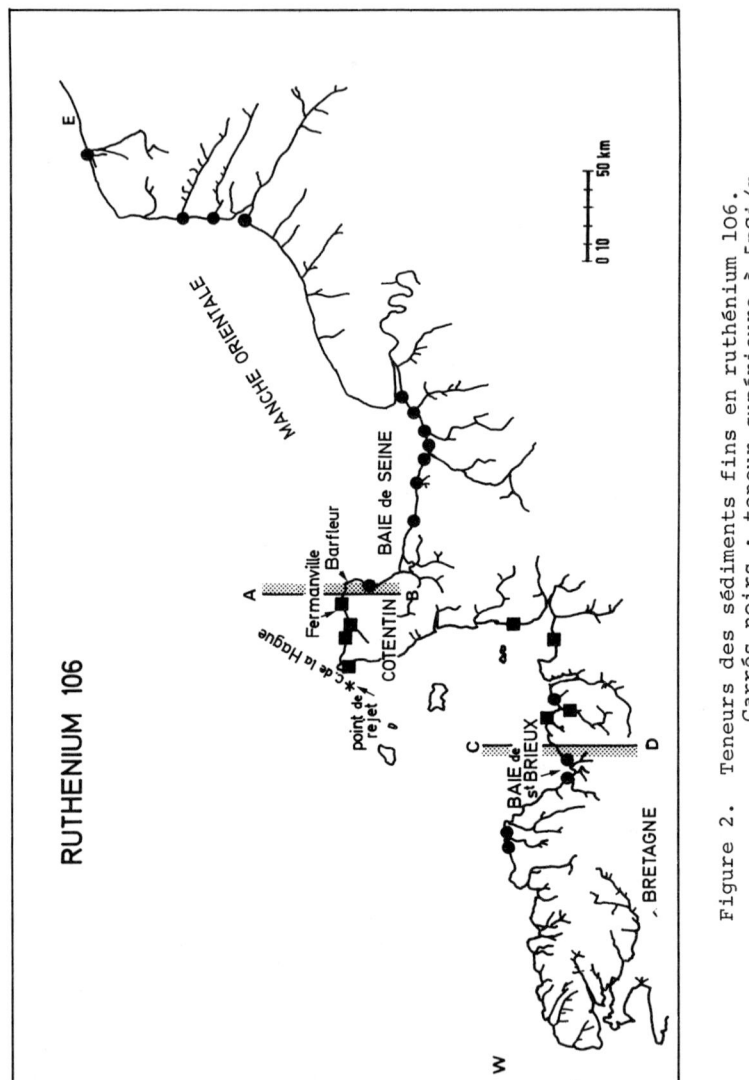

Figure 2. Teneurs des sédiments fins en ruthénium 106.
Carrés noirs : teneur supérieure à 5pCi/g.
Ronds noirs : teneur inférieure à 5pCi/g.
(D'après /5/).

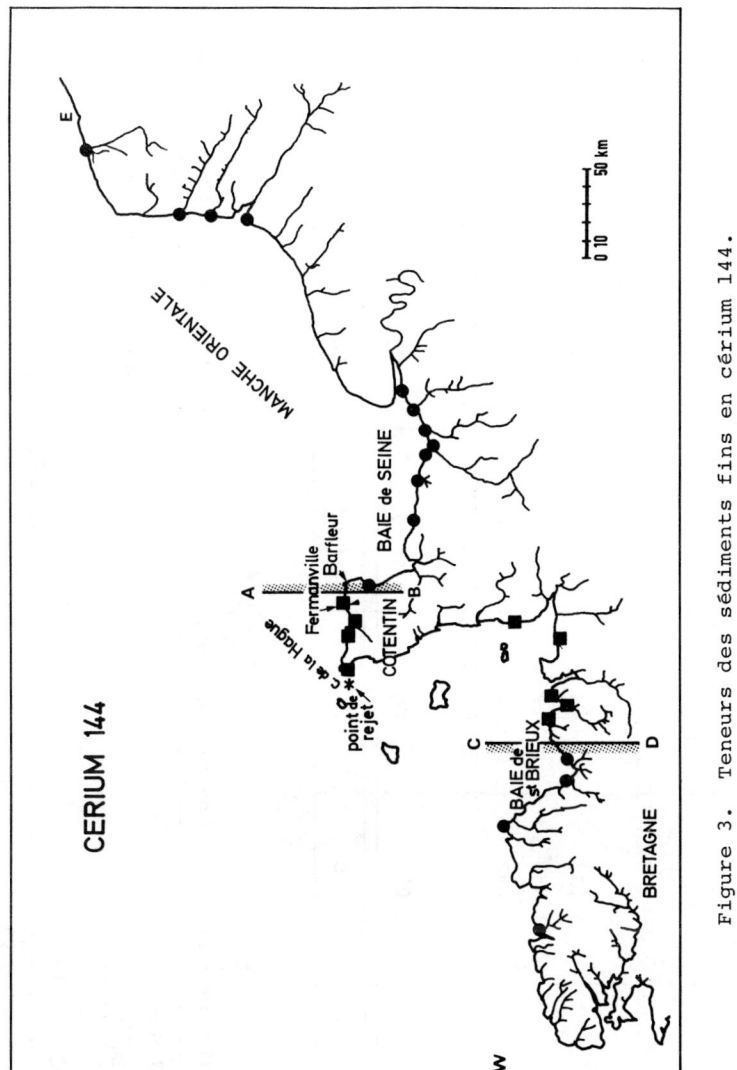

Figure 3. Teneurs des sédiments fins en cérium 144.
Carrés noirs : teneur supérieure à 2pCi/g.
Ronds noirs : teneur inférieure à 2pCi/g.
(D'après [5]).

TABLEAU V
(d'après [4])
Radioactivité de diverses espèces marines comestibles, prélevées à la station de Goury
(secteur du point de rejet), entre 1976 et 1978 (valeurs maximales en pCi/kg frais)

	^{144}Ce	^{125}Sb	^{106}Ru	^{137}Cs	^{110m}Ag	^{65}Zn
CRUSTACES (global)						
Cancer pagurus	1 173	6	3 110	25	192	715
Carcinus mœnas	1 350	19	3 729	32	86	298
Homarus vulgaris	443	-	1 403	42	133	217
MOLLUSQUE (chair)						
Patella sp.	2 011	96	6 434	78	180	298
POISSONS (organes divers)	210	6	1 201	346	13	72

TABLEAU VI
(d'après [5])
Radioactivité de sédiments fins prélevés à la station de Goury (secteur du point de rejet), en 1977 et 1978. (valeurs moyennes)

Ruthénium 106	8,6 pCi/g
Cérium 144	4,7 "
Césium 137	0,31 "
Zirconium 95	0,10 "

comestibles. Les valeurs maximales observées (et retenues de préférence aux valeurs moyennes, comme répondant mieux à un objectif de protection) sont reportées dans le tableau V.

D'après ces données les ordres de grandeur de la radioactivité artificielle maximale due aux émetteurs gamma seraient les suivants pour les trois groupes d'espèces comestibles :

. Crustacés : 5.000 pCi/kg/frais
. Mollusques : 10.000 pCi/kg/frais
. Poissons : 2.000 pCi/kg/frais

4.2.2 Sédiment.

Divers résultats de mesures effectuées sur les sédiments fins prélevés à cette même station de Goury entre novembre 1977 et mai 1978, figurent dans le Tableau VI (valeurs moyennes). Compte tenu de l'ensemble des données publiées $\int 5\int$ on peut estimer que la radioactivité artificielle (émetteurs gamma) de ces sédiments était de l'ordre de 15.000 à 20.000 pCi/kg.

4.3. Incidences sur le plan sanitaire

On est appelé à se poser la question de savoir quelle est la signification de ces résultats sur le plan sanitaire.

Au point de vue de l'irradiation interne due à la consommation alimentaire, en tenant compte de la consommation moyenne journalière retenue à la suite de l'étude prévisionnelle pour le groupe le plus exposé (crustacés : 19 g., mollusques : 12 g., poissons : 48 g.) et de la radioactivité maximale observée pour les crustacés (5 pCi/g), les mollusques (10 pCi/g) et les poissons (2 pCi/g), on constate que cette consommation entrainerait pour une personne l'ingestion de 311 pCi par jour. Pour une CMA de 10^{-4} µCi/cm^3 - qui est celle du ruthénium 106 et du cérium 144 - ceci correspondrait à une irradiation représentant $1,43.10^{-2}$ de la dose admissible (1) : chiffre qui se trouve être légèrement en dessous de celui mentionné dans l'étude prévisionnelle ($1,52.10^{-2}$) pour un rejet de 45.000 curies par an. De plus, il comporte très certainement une appréciable marge de sécurité puisqu'il implique que les produits consommés proviennent exclusivement de la zone proche des rejets et puisqu'il repose sur un ensemble de valeurs maximales de la radioactivité constatée chez diverses espèces comestibles.

Au point de vue de l'irradiation externe due aux sédiments radioactifs, les évaluations sont beaucoup plus délicates en l'absence de relation bien établie entre la radioactivité massique du sédiment et l'irradiation résultante au contact.

---

(1) L'équivalent de dose maximal pour "les personnes du public" (1,5 rem par an, tractus gastro-intestinal) implique que, pour une CMA de 10^{-4} µCi/cm^3, l'absorption maximale quotidienne par un individu ne doit pas dépasser :

$$\left[10^{-4} \text{ µCi/cm}^3 \times \frac{1}{10} \begin{pmatrix} \text{personne} \\ \text{du public} \end{pmatrix} \times 2.200 \text{ cm}^3 \begin{pmatrix} \text{consomm.} \\ \text{journal.} \end{pmatrix} \right] \text{ µCi/jour,}$$

soit 22.000 pCi/jour. L'ingestion de 311 pCi correspond à $\frac{311}{22.000} = 1,43.10^{-2}$ de la dose admissible.

Selon les données auxquelles il est fait référence dans l'étude prévisionnelle, un lit de sable ayant en épaisseur une activité de 25 pCi/g entrainerait (émission gamma et bêta) à son contact une dose d'environ 15.10^{-3} millirads/heure. Sur cette base une activité de 15-20 pCi/g entrainerait, pour un pêcheur travaillant 2.000 heures en mer par an, une dose annuelle légèrement inférieure à 1,5 millirads, soit 3.10^{-3} de la dose admissible (2) - chiffre correspondant à celui établi dans les prévisions et vraisemblablement assorti, là encore, d'une notable marge de sécurité, puisqu'il résulte de mesures faites dans la zone proche des rejets et sur des sédiments fins qui ont un pouvoir de rétention des radionucléides sensiblement plus élevé que les sédiments moyens ou grossiers, les plus répandus.

+ +

+

En conclusion, il se confirmerait, d'après les mesures portant sur de multiples échantillons prélevés dans le secteur proche des rejets de l'usine de La Hague au cours des deux ou trois dernières années, que les doses d'irradiation liées à l'utilisation du milieu marin et délivrées aux personnes faisant partie du groupe le plus exposé, n'ont pas dépassé, en ce qui concerne les émetteurs gamma, le 1/100e de la dose admissible et, en tout état de cause, ont dû se situer assez nettement au dessous de cette valeur compte tenu de l'importance des marges de sécurité intervenant dans les calculs.

REFERENCES BIBLIOGRAPHIQUES

/1/ ANCELLIN J., GUEGUENIAT P., GERMAIN P. Radioécologie marine. Etude du devenir des radionucléides rejetés en milieu marin et applications à la radioprotection. 256 p. Paris : Eyrolles (1979).

/2/ SCHEIDHAUER J., AUSSET R., PLANET J., COULON R. Programme de surveillance de l'environnement marin du Centre de La Hague. In : Population dose evaluation and standards for man and his environment. Proceedings of a seminar, Portoroz, May 1974, pp. 347-366. Vienne : I.A.E.A. (1974).

/3/ FRAIZIER A., GUARY J.C. Diffusion du plutonium en milieu marin: étude quantitative effectuée sur des espèces marines du littoral de la Manche, de Brest (Pointe Saint Mathieu) à Honfleur. Rapport C.E.A.-R-4822, 16 p. SACLAY : Serv.Docum.Commiss. Energie Atom. (1977).

---

(2) $2.000 \text{ h.} \times 15.10^{-3} \times \frac{1}{20}$ (manip.engins de pêche) = 1,5 millirads/an,

ce qui équivaut à 3.10^{-3} de la dose admissible (0,5 rem/an, organisme entier, pour les "personnes du public").

[4] GERMAIN P., MASSON M., BARON Y. Etude de la répartition de radionucléides émetteurs gamma chez des indicateurs biologiques littoraux des côtes de la Manche et de la Mer du Nord, de février 1976 à février 1978. Rapport Commissariat à l'Energie Atomique (sous presse).

[5] GUEGUENIAT P., AUFFRET J.P., BARON Y. Evolution de la radioactivité artificielle gamma dans des sédiments littoraux de la Manche pendant les années 1976, 1977, 1978. Oceanologia Acta. Vol. 2, n° 2, pp. 165-180 (1979).

[6] PRESTON A. Radioactivity in the marine environment. In : Sea fisheries research, pp. 305-329. Jones H., Ed. Londres : Elek Science (1974).

[7] HETHERINGTON J., JEFFERIES D., LOVETT M. Some investigations into the behaviour of plutonium in the marine environment. In : Impact of nuclear releases into the aquatic environment. Proceedings of a Symp., Otaniemi, July 1975, pp. 193-212. Vienne : I.A.E.A. (1975).

Discussion

K. SARUHASHI, Japan

Will you telle me the level of Cs-content in surface water outside the area where the Cs-137 content is less than 0.01 pCi/l per Ci discharged per day in the figure ? Is it the same as the fallout level, that is 0.2-0.3 pCi/l ?

J. ANCELLIN, France

Les teneurs de l'eau de mer en césium 137 au-delà de la zone des isoconcentrations 0,01-0,03 pCi/l déterminée pour un rejet de 1 curie/jour sur le site de La Hague (Figure 1) sont plus élevées, en général, que celles qui seraient dues aux seules retombées. (Le Dr. Kautsky donne des précisions à ce sujet, cf. infra.)

H. KAUTSKY, Federal Republic of Germany

The Cs-137 content of the water in the English Channel East of La Hague and in the Southern North Sea is for the time being about 1 pCi/l, un some cases up to 3 pCi/l, sometimes even lower than 1 pCi/l.

THE BEHAVIOUR OF TRANSURANIC AND OTHER LONG-LIVED RADIONUCLIDES
IN THE IRISH SEA AND ITS RELEVANCE TO THE DEEP SEA DISPOSAL
OF RADIOACTIVE WASTES

R. J. Pentreath, D. F. Jefferies, M. B. Lovett and D. M. Nelson*
Ministry of Agriculture, Fisheries and Food
Directorate of Fisheries Research
Fisheries Radiobiological Laboratory
Hamilton Dock
Lowestoft
Suffolk
England

*Argonne National Laboratory
9700 South Cass Avenue
Argonne
Illinois 60439
USA

ABSTRACT

The discharge of transuranic nuclides as part of the authorised, low-level releases from the BNFL Windscale site into the Irish Sea provides a unique opportunity to study the behaviour of these elements in the marine environment. Other long-lived radionuclides, such as ^{99}Tc, are also discharged. Many of the questions which need to be addressed with regard to the introduction of longer-lived nuclides into coastal waters are very similar to those which arise in connection with deep sea disposal, although not necessarily with the same priority. Environmental studies related to the Windscale discharges have centred on several areas of mutual interest: chemical speciation in sea water, adsorption on to sedimentary materials, the permanence of radionuclide incorporation into sediments and, of particular interest, the relative biological availabilities of these elements - not only relative to each other but relative to the quantities discharged and to their rates of introduction. Many of these data will prove valuable in providing an input to numerical models, and in highlighting those areas where research is required to fill the substantial gaps in our present knowledge.

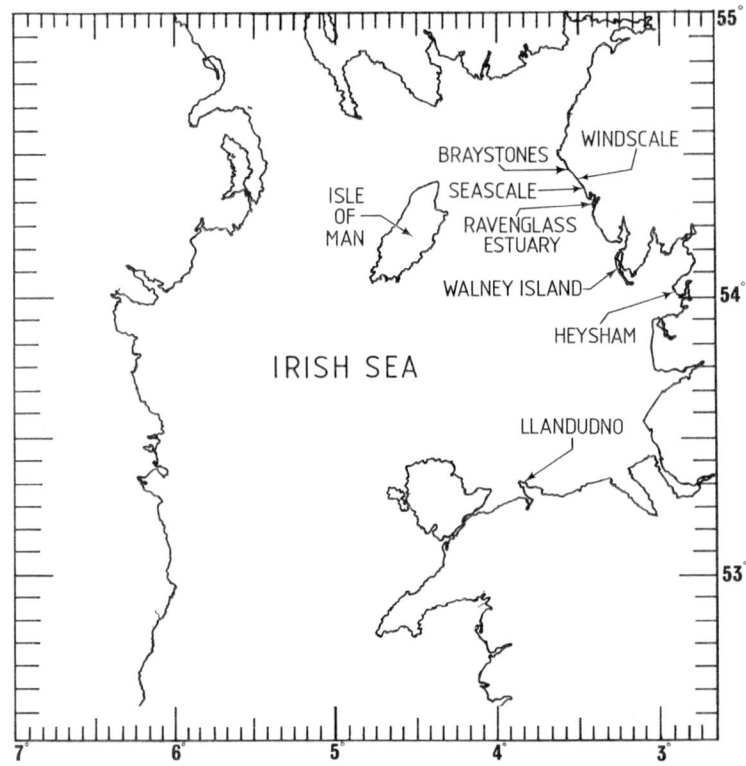

Figure 1

Map of the Irish Sea showing sample locations mentioned in the text.

INTRODUCTION

In order to assess the feasibility of the permanent disposal of high-level radioactive wastes in or on the sea bed it is necessary to extend considerably certain aspects of our knowledge of the world's oceans. The purpose of such feasibility assessments is to permit realistic estimates to be made of the likely dose equivalent commitment to man from the possible short-term and the eventual long-term release of radionuclides. To this end a number of mathematical models have been developed and these have highlighted several areas in which our data are substantially lacking. For example there is a need to obtain a better understanding of deep-water dynamics and mixing processes, to learn more of the geology and stability of the ocean floor, and to improve our assessments of potential biological pathways back to man.

Of equal importance, however, is a knowledge of the potential behaviour in the marine environment of those long-lived radionuclides which are likely to predominate in the high-level wastes. The list of such important radionuclides is a long one but of particular interest are the transuranium nuclides; not only because of the quantities involved but also because they are nuclides of elements which are, to all intents and purposes, entirely man-made. Also within this category is the long-lived nuclide ^{99}Tc.

As a result of the atmospheric testing of nuclear weapons some of these radionuclides are widely dispersed - notably $^{239/240}$Pu and ^{241}Am - although the concentrations are too low to permit an adequate detailed study of their behaviour. The other radionuclides of interest, however, such as those of curium, neptunium and technetium, are virtually absent throughout the world. It is therefore of particular interest that, as a result of the authorised low-level aqueous discharges from the British Nuclear Fuels plant at Windscale in Cumbria (Figure 1), the Irish Sea provides a unique opportunity to study the behaviour of some of these elements in a natural marine ecosystem. The bulk of the Windscale discharges consists principally of fairly short-lived fission products such as ^{95}Zr/^{95}Nb, ^{106}Ru, ^{144}Ce and ^{137}Cs which have been studied in some detail over a number of years. Nevertheless, the quantities of the longer-lived nuclides which are discharged under authorisation can be substantial: for example, during 1978 over 1 kCi (37 TBq) of $^{239/240}$Pu, about 200 Ci (7.4 TBq) of ^{241}Am and about 5 kCi (185 TBq) of ^{99}Tc were discharged, plus a few Ci of ^{242}Cm, ^{244}Cm and ^{237}Np.

It is of course appreciated that the deep sea environment is a very different one from that of such coastal waters as the Irish Sea. There are differences in temperature, pressure, solute concentration and so on which are known to have effects on chemical behaviour; they may therefore considerably modify those reactions which take place in surface waters. But it is only from studying such areas as the Irish Sea that any basic surface water data can be obtained.

METHODS

Samples obtained from the Irish Sea are analysed in a number of different ways. The concentrations of gamma-emitting fission product nuclides are determined by gamma-spectrometric techniques using NaI(Tl) or Ge(Li) crystals. Alpha-emitting nuclides are determined by alpha-spectrometry using silicon surface barrier detectors with ^{236}Pu and ^{243}Am as yield tracers. When measurements of plutonium oxidation states are made, ^{236}Pu and ^{242}Pu are used as yield tracers (1). The pure beta-emitting ^{99}Tc is separated from most other radionuclides by distillation of the Tc_2O_7 and, after decontamination from ruthenium nuclides, electrodeposited on to stainless steel discs for subsequent thin end-window beta-counting.

SEA WATER AND SEDIMENT STUDIES

The value of studying the higher concentrations of radionuclides in the Irish Sea is readily demonstrated by a discussion of the chemical behaviour of plutonium in sea water. Theoretical studies had indicated that plutonium could exist in a number of oxidation states (2) but their presence in sea water had not been confirmed. Initial observations in the Irish Sea had demonstrated that the majority of $^{239/240}$Pu discharged from Windscale was apparently associated with

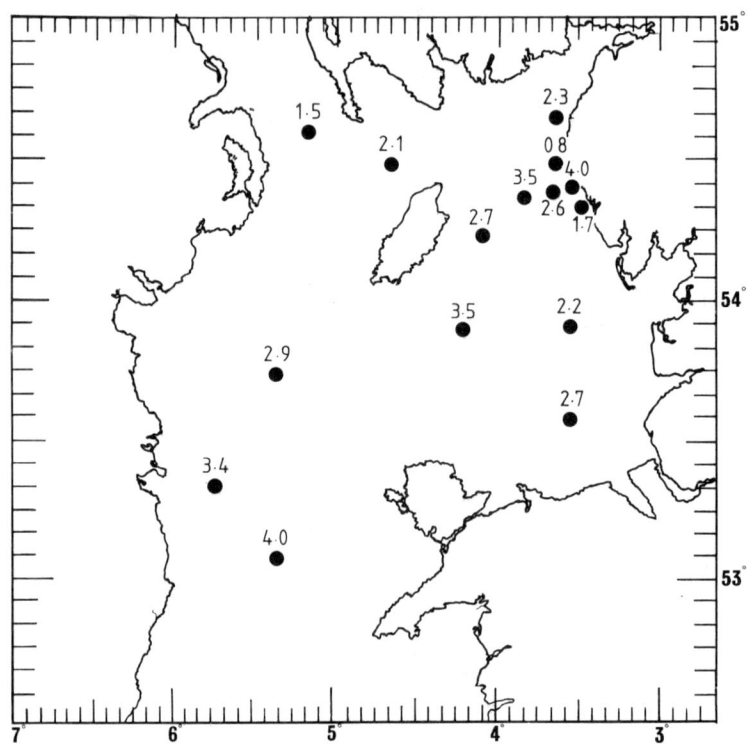

Figure 2

Relationship between ^{137}Cs and $^{239/240}$Pu in filtered surface sea water taken in the Irish Sea during September 1977. Values are pCi.l^{-1} of ^{137}Cs x 10^{-3}/pCi.l^{-1} of $^{239/240}$Pu.

particulate material and thus rapidly removed from the water column (3). It was also observed, however, that some plutonium did remain in filtered (< 0.22 μm) sea water, and that this fraction maintained a fairly constant ratio to ^{137}Cs - a nuclide considered to be relatively conservative to sea water - throughout the whole of the Irish Sea (Figure 2). A more detailed analysis of sea water taken close to the coast revealed that this fraction is predominantly Pu (V + VI) whereas the fraction removed by filtration is principally Pu (III + IV) (1). Subsequent analyses have shown that this conclusion holds good throughout the whole of the Irish Sea.

The oxidation state of plutonium in the Windscale discharges is not measured routinely. However, one series of samples has been taken and, as expected from the other data available, it was found that the plutonium existed virtually entirely as Pu (III + IV), the inference being, therefore, that subsequent to release some plutonium has been raised to a higher oxidation state. This interesting possibility is now being studied under controlled laboratory conditions using ^{237}Pu. A conversion of ^{237}Pu (III + IV) to ^{237}Pu (V + VI) can be demonstrated. The reverse reaction also takes place, however, and thus it is not as yet clear to what extent these initial observations are valid for the Windscale environment or are merely a consequence of the laboratory conditions used.

There is no environmental evidence of the oxidation states of the other transuranium nuclides discharged by Windscale. Americium is thought to exist as Am (III) but this has not yet been confirmed. Curium is expected to be similar to americium, but neptunium is more likely to exist as Np (V).

Technetium is another element which can potentially exist in a number of oxidation states. Again as yet we have no direct observations on the state of technetium in sea water but there is indirect evidence which suggests that the ^{99}Tc discharged remains waterborne to a greater extent than ^{137}Cs. This evidence comes from a comparison of the ratios of ^{99}Tc to ^{137}Cs in the laminae of one species of alga, *Fucus vesiculosus*, collected along the shore-line of the Irish Sea in November 1978. The sampling locations are indicated in Table I and Figure 1. If one makes the reasonable assumption that the concentration factors for both elements by this species remain constant at different locations - there is no temperature difference or difference in any other environmental parameter to suggest otherwise - then the quotients determined in Table I suggest that ^{99}Tc remains available for accumulation over a greater distance than does ^{137}Cs. All of the samples were collected over a two-week period and the average $^{99}Tc/^{137}Cs$ value in the discharge over the preceding three months was 0.01; in fact throughout the whole of 1978 prior to sample collection the monthly quotient varied only from 0.01 to 0.2, with an average value of 0.06.

The value of determining the chemical states of these long-lived radionuclides in sea water lies in the subsequent possibility of predicting their gross distribution and biological availability. The degree to which radionuclides are adsorbed to sedimentary materials, and the permanence of such an association, is clearly of interest with regard to both the Windscale discharges and to the deep sea disposal option. There are two ways in which the adsorption coefficient (K_d) for sedimentary particles can be derived from environmental samples: by obtaining the value for material in suspension, and by obtaining the value for settled sediment on the sea floor. In order to obtain a K_d for material in suspension, pre-weighed filters (0.22 μm) are used and the quantity of radionuclide per unit dry weight of particulate is divided by the quantity of radionuclide in the same weight of filtrate. A large number of determinations have been made, using this method, on surface water samples collected throughout the Irish Sea on cruises made in September 1977 and May 1978. The suspended loads varied from ~ 1 mg.l^{-1} to ~ 10 mg.l^{-1}. The results are given in Table II.

A number of features are of immediate interest: first of all in a comparison of the data for americium and curium. The ^{242}Cm K_d values are not significantly different inside and outside the Irish Sea, and neither are the ^{244}Cm values (P > 0.25). Taking all of the curium data collectively, therefore, the two isotopes appear to behave the same. (In fact, the ^{244}Cm measurements also include any ^{243}Cm present in the sample.) Similarly, the ^{241}Am K_d values are not significantly different inside and outside the Irish Sea (P > 0.25). Taking all of the americium and curium data collectively, therefore, it can be shown that the K_d

TABLE I WET WEIGHT CONCENTRATIONS ($\pm$ 2σ PROPAGATED COUNTING ERRORS) OF ^{99}Tc AND ^{137}Cs IN *Fucus vesiculosus* COLLECTED AROUND THE COASTLINE OF THE IRISH SEA

Location	Concentration						$\dfrac{^{99}\text{Tc}}{^{137}\text{Cs}}$*
	^{99}Tc			^{137}Cs			
	pCi.g^{-1}		(mBq.g^{-1})	pCi.g^{-1}		(mBq.g^{-1})	
Braystones	164 $\pm$ 7	($\pm$	6068) 259)	43 $\pm$ 1.4	($\pm$	1591) 52)	3.81 $\pm$ 0.20
Seascale	411 $\pm$ 8	($\pm$	15207) 296)	47 $\pm$ 0.7	($\pm$	1739) 26)	8.74 $\pm$ 0.21
Walney Island	575 $\pm$ 12	($\pm$	21275) 444)	22 $\pm$ 0.3	($\pm$	814) 11)	26.1 $\pm$ 0.65
Heysham	267 $\pm$ 5	($\pm$	9879) 185)	15 $\pm$ 0.2	($\pm$	555) 7)	17.8 $\pm$ 0.41
Llandudno	123 $\pm$ 5	($\pm$	4551) 185)	3.5 $\pm$ 0.1	($\pm$	130) 4)	35.1 $\pm$ 1.75

*The average ^{99}Tc/^{137}Cs quotient in the Windscale discharge over the three months prior to sample collection was 0.01.

TABLE II K_d VALUES (fCi.kg^{-1} OR Bq.kg^{-1} DRY WEIGHT DIVIDED BY fCi.l^{-1} OR Bq.l^{-1}) FOR PARTICULATE MATERIAL IN SURFACE WATER SAMPLES COLLECTED IN UK WATERS IN 1977 AND 1978. ($\bar{x}$ = mean; $s_{\bar{x}}$ = standard error of the mean; n = number of samples)

Location		Radionuclide			
		$^{239/240}$Pu	^{241}Am	^{242}Cm	^{244}Cm
Irish Sea * (September 1977)	$\bar{x}$	3.1×10^5	2.4×10^6	-	-
	$s_{\bar{x}}$	0.8×10^5	0.4×10^6	-	-
	n	15	15	-	-
(May 1978)	$\bar{x}$	2.8×10^5	2.2×10^6	1.2×10^6	1.5×10^6
	$s_{\bar{x}}$	0.4×10^5	0.3×10^6	0.3×10^6	0.5×10^6
	n	21	21	6	6
Outside Irish Sea (September 1977)	$\bar{x}$	6.3×10^5	2.8×10^6	-	-
	$s_{\bar{x}}$	2.7×10^5	1.5×10^6	-	-
	n	5	6	-	-
(May 1978)	$\bar{x}$	7.5×10^5	1.9×10^6	1.5×10^6	9.4×10^5
	$s_{\bar{x}}$	1.2×10^5	0.2×10^6	0.4×10^6	4.0×10^5
	n	31	31	6	6

*The relative quantities of $^{239/240}$Pu and ^{241}Am discharged by Windscale have changed considerably over recent years. The values in this table are derived from two cruises when the ^{241}Am discharges were very low; the extent to which they may be influenced by a long retention time of americium on sedimentary particles is not known.

TABLE III K_d VALUES (fCi.kg^{-1} OR Bq.kg^{-1} DRY WEIGHT DIVIDED BY fCi.l^{-1} OR Bq.l^{-1} FILTERED SEA WATER) FOR SEDIMENT FROM THE SURFACE OF MUDDY AREAS OF THE IRISH SEA. ($\bar{x}$ = mean; $s_{\bar{x}}$ = standard error of the mean; n = number of samples)

Time of collection*		Radionuclide	
		$^{239/240}$Pu	^{241}Am
January 1976	$\bar{x}$	1.4×10^5	1.3×10^6
	$s_{\bar{x}}$	0.4×10^5	0.2×10^6
	n	7	6
September 1977	$\bar{x}$	1.3×10^5	1.4×10^6
	$s_{\bar{x}}$	0.6×10^5	0.5×10^6
	n	5	5

*The relative quantities of $^{239/240}$Pu and ^{241}Am discharged by Windscale have changed considerably over recent years. The values in this table are derived from two cruises when the ^{241}Am discharges were very low; the extent to which they may be influenced by a long retention time of americium on sedimentary particles is not known.

TABLE IV CONCENTRATIONS (WET WEIGHT) OF $^{239/240}$Pu AND ^{241}Am IN THE STARFISH *Asterias rubens* AND THE ECHINOID *Echinus esculentus* COLLECTED CLOSE TO WINDSCALE IN 1978 (ERRORS ARE $\pm 2\sigma$ PROPAGATED COUNTING ERRORS)

Species	Radionuclide			
	^{241}Am		$^{239/240}$Pu*	
	fCi.g^{-1}	(mBq.g^{-1})	fCi.g^{-1}	(mBq.g^{-1})
Asterias rubens				
Aboral body wall	1024 $\pm$ 32	(37.9) ($\pm$ 1.2)	946 $\pm$ 32	(35.0) ($\pm$ 1.2)
Digestive gland	468 $\pm$ 16	(17.3) ($\pm$ 0.6)	719 $\pm$ 22	(26.6) ($\pm$ 0.8)
Gonad	90 $\pm$ 8	(3.32) ($\pm$ 0.28)	92 $\pm$ 6	(3.40) ($\pm$ 0.22)
Echinus esculentus				
Aboral test	505 $\pm$ 22	(18.7) ($\pm$ 0.8)	497 $\pm$ 27	(18.4) ($\pm$ 1.0)
Gonad	362 $\pm$ 11	(13.4) ($\pm$ 0.4)	122 $\pm$ 12	(4.50) ($\pm$ 0.46)

*The average $^{239/240}$Pu/^{238}Pu value is 4.06 $\pm$ 0.22.

values for ^{241}Am are significantly higher (P < 0.1) than those for the curium isotopes on the basis of the samples analysed to date. It is clearly necessary to obtain more data, however, particularly with regard to curium for which there is a large spread in the observations made so far, as can be seen from the standard errors in Table II. For statistical purposes the data have been treated assuming a normal distribution.

With regard to plutonium, the K_d values for $^{239/240}$Pu inside the Irish Sea are significantly lower than those of curium (P < 0.001); and the same holds true outside the Irish Sea (P < 0.1). Thus the overall order of K_d values is ^{241}Am > ^{242}Cm and ^{244}Cm > $^{239/240}$Pu. Unlike curium and americium, however, the K_d values for $^{239/240}$Pu outside the Irish Sea are significantly higher (P < 0.01) than the values inside the Irish Sea. One explanation for this difference could lie in the nature of the material in suspension, for example whether organic or inorganic; if so, the difference applies only to the adsorption of plutonium. An alternative explanation is that such differences are a reflection of the plutonium oxidation states. A detailed examination of samples collected close to Windscale has shown that the K_d value of Pu (III + IV) exceeds the value for Pu (V + VI) by some two orders of magnitude (1). Our knowledge of plutonium oxidation states in filtered sea water outside the Irish Sea is less complete because whereas the total plutonium concentration in sea water close to Windscale is of the order of 1 pCi.l^{-1} (37 mBq.l^{-1}) it falls to between about 1 and 10 fCi.l^{-1} (0.037 to 0.37 mBq.l^{-1}) outside the Irish Sea. Nevertheless, some initial data have been obtained and these have demonstrated a decreased proportion of the higher oxidation state in sea water outside the Irish Sea. There is, however, obviously still much to be learned with regard to the chemistry of these nuclides in sea water and the effects which materials in suspension have on them.

Estimates of K_d values for surface sediment samples will clearly depend upon the nature of the sediment. It is to be expected that such values would be lower than those derived from surface water particulates because only the finer particles are likely to be held in suspension; and it is reasonable to assume that such finer particles would adsorb more of the radionuclides per unit weight. In fact there is not a large difference between K_d values calculated for the surface of muddy areas in the Irish Sea (Table III) and those given in Table II. Not all of the Irish Sea bed is covered by mud, however, and there are large areas of sand, clay and gravel. Thus K_d values for the Irish Sea bed as a whole are considerably lower; for example, about 4.0×10^4 ($\pm 1.3 \times 10^4$) for $^{239/240}$Pu, based on the analyses of 27 samples collected on both muddy and sandy areas.

From the deep sea disposal aspect, many more data are required. Of particular interest would be a more detailed understanding of the relationship between sediment mineralogy and K_d, as well as data on more obvious aspects such as the effect of particle size. A start has been made in these directions and sediment cores from the middle of the North Atlantic have been obtained for laboratory analysis and experimentation in order to see to what extent comparisons can be made with coastal sediment observations.

One subject of particular interest is the permanence of the radionuclide-to-sediment bond. In this respect there is clearly a difference between suspended sediment and sediment which has been consolidated into the sea floor. If a few mg of surface sediment from the Windscale area are shaken up with clean sea water for a short period of time, sufficient $^{239/240}$Pu can be removed to establish an equilibrium value that would be expected from the K_d values of Table II. In other words the sea water is found to contain $^{239/240}$Pu at a concentration of about 10^{-5} of that of the sediment. Of greater interest, however, is the possible remobilisation of the transuranium nuclides from materials which have become incorporated into the sea bed. One approach to this interesting question is to examine core profiles, both for radionuclide concentration and for isotopic ratios. There are areas offshore from Windscale where all of the evidence suggests a steady sedimentation process in which the $^{239/240}$Pu concentrations decrease steadily with depth, and the $^{239/240}$Pu/^{238}Pu values are consistent with what is known of these ratios in the discharges over a number of years. A core profile from such an area is shown in Figure 3. This core was taken in 1977, at which time the $^{239/240}$Pu/^{238}Pu quotient in the discharge was approximately 4.0. This value has gradually decreased, as evidenced from surface sediment samples taken in an estuary close to Windscale, from ∼19 in 1966, to ∼10 in 1969, down to ∼5 in

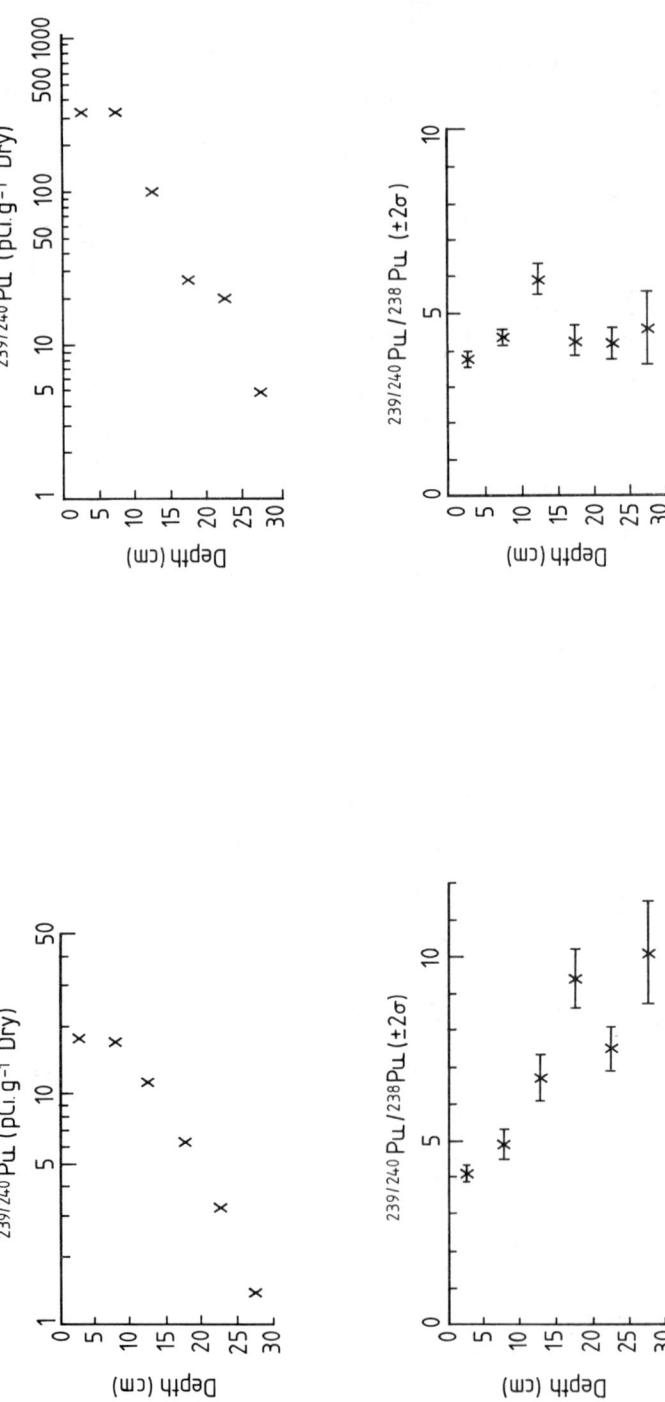

Figure 3

Plutonium profiles from a core taken in 1977 approximately 15 km from the Windscale pipeline.

Figure 4

Plutonium profiles from a core taken in 1977 approximately 1 km from the Windscale pipeline.

1973 (4,5). The $^{239/240}Pu/^{238}Pu$ quotients in Figure 3 agree well with this known history, and it is reasonable to assume that the sediment at 25-30 cm depth had been in direct contact with the surface water some eight years before collection.

Not all of the cores are so nicely labelled. This is especially true of samples taken closer to the discharge-pipe, such as the sample for which a profile is shown in Figure 4. This profile differs from that of Figure 3 in two respects: the concentration decreases much more rapidly with depth, and the $^{239/240}Pu/^{238}Pu$ value is relatively constant throughout. The profile is therefore far more difficult to interpret and a number of possibilities present themselves including, it could be argued, a movement of the plutonium, although this seems the least likely.

An alternative approach to mere speculation is to seek a possible mechanism by which $^{239/240}Pu$ or other long-lived radionuclides could return to the sediment surface. If plutonium is to move out of consolidated sediment then such movement is most likely to occur via the interstitial water. A knowledge of radionuclide speciation is thus again of importance, and as yet we have very few data. A profile of $^{239/240}Pu$ oxidation states in interstitial waters extracted from a core taken close to Windscale, together with the $^{239/240}Pu$ concentrations of the dried sediment, is shown in Figure 5. Whereas the plutonium in the sea water overlying the sediment is predominantly present as Pu (V + VI), that below 5 cm is predominantly Pu (III + IV), the least mobile of the two sets. There are, of course, other possibilities for mobilisation - such as complexation with organic ligands - and these, too, are being sought. A number of core profiles have now been examined with respect to Eh, pH, ferrous and ferric iron, manganese, nitrate and nitrite, and it is hoped by these and other observations, including those on organic substances, to improve our understanding of the potential for radionuclide mobility in sediments. An entirely different possibility is that of bioturbation, and this again is a research area which warrants more detailed examination in order to establish its relative importance.

BIOLOGICAL AVAILABILITY

The biological availability of long-lived radionuclides is, in fact, the most important aspect in relation to assessment of potential biological pathways back to man, for were such nuclides not accumulated by the biotic component of the marine environment their potential hazard to man would be greatly diminished. In the first instance it is instructive to relate the relative concentrations of transuranic nuclides in biota to what is known of their relative concentrations in the Windscale discharges. Unfortunately the relative quantities of the two principal transuranic elements discharged - plutonium and americium - vary considerably from month to month and from year to year. The mean, monthly, quantities of ^{238}Pu, $^{239/240}Pu$ and ^{241}Am discharged each year from 1972 to 1978 are shown in Figure 6. The picture is even more complicated, however, because in addition to discharging ^{241}Am directly Windscale also discharges ^{241}Pu which decays by beta-emission to ^{241}Am. Thus, for example, during 1978 some 50 kCi (1.85 PBq) of ^{241}Pu were discharged. Such an annual rate of discharge would give rise to some 41 Ci (1.52 TBq) of ^{241}Am after one year, 157 Ci (5.8 TBq) after two years, 353 Ci (13.1 TBq) after three years, and so on. For the period 1976 to 1978 therefore, when the direct ^{241}Am discharges were relatively low, the total quantity of ^{241}Am discharged would be about double the value shown in the figure - although still considerably less than $^{239/240}Pu$. In fact the 1976 and 1977 ^{241}Pu discharges were substantially lower than those of 1978.

The biological samples from the Irish Sea which have been studied in greatest detail, with regard to their content of alpha-emitting nuclides, are fish species of commercial importance (6,7). There is, however, a large variety of biological material available for study and, as an example, Tables IV to VI contain data derived from two echinoderm species, the starfish *Asterias rubens* and the echinoid *Echinus esculentus*. Echinoderms are of particular interest because of their ubiquity in the marine environment - especially in the deep sea. Both americium and plutonium are readily detected in samples collected close to Windscale. The data in Table IV were derived from bulked samples, not individual animals, and the two species were collected on separate occasions. In order to examine the relative accumulation of the two elements, therefore, the $^{241}Am/^{238}Pu + ^{239/240}Pu$ quotients obtained from these data (Table V) have been compared with the quotients

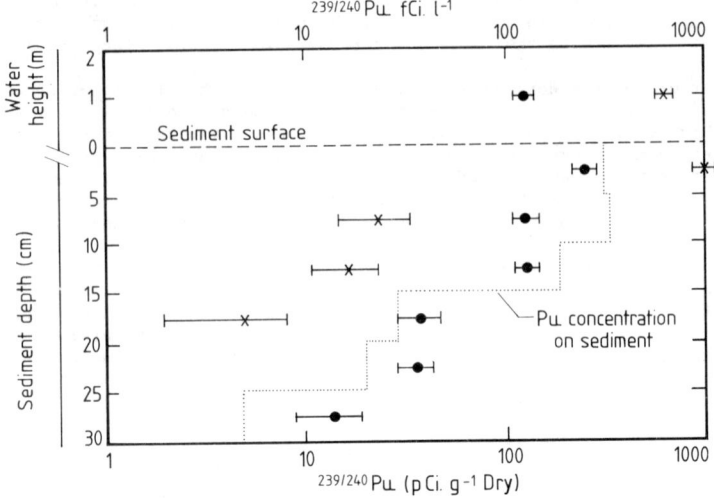

Figure 5

Oxidation states of $^{239/240}$Pu in interstitial water removed from a sediment core taken close to Windscale, and in the immediate overlying water. The $^{239/240}$Pu concentration (dry weight) of the sediment down the core is also shown. (x Pu (V + VI) in water; • Pu (III + IV) in water).

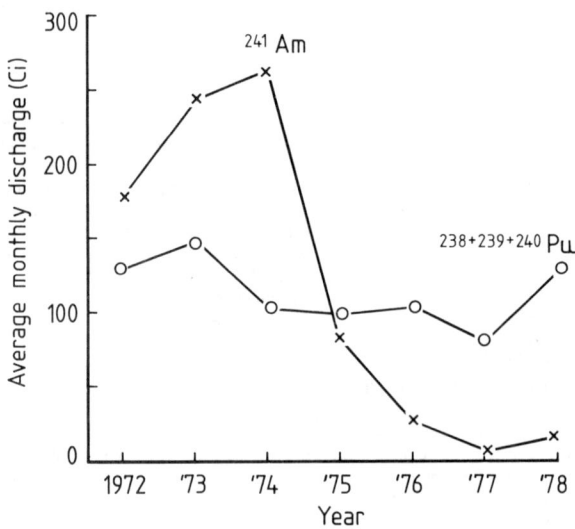

Figure 6

Mean, monthly, quantities of ^{238}Pu plus $^{239/240}$Pu, and of ^{241}Am discharged from Windscale for the period 1972 to 1978. (From returns made by British Nuclear Fuels Limited).

TABLE V $^{241}Am/^{238}Pu + ^{239/240}Pu$ VALUES FOR ECHINODERM AND SEA WATER SAMPLES AT TIME OF COLLECTION IN 1978 (ERRORS ARE BASED ON 2σ PROPAGATED COUNTING ERRORS)

Sample	$\dfrac{^{241}Am}{^{238}Pu + ^{239/240}Pu}$
Asterias rubens	
Aboral body wall	0.87 ± 0.04
Digestive gland	0.53 ± 0.02
Gonad	0.78 ± 0.08
Shore-line sea water	
Filtrate (< 0.22 µm)	0.08 ± 0.01
Particulate (> 0.22 µm)	0.53 ± 0.04
Discharge	
Previous 3 months	0.20
Previous 12 months	0.17
Echinus esculentus	
Aboral test	0.81 ± 0.05
Gonad	2.38 ± 0.22
Shore-line sea water	
Filtrate (< 0.22 µm)	0.13 ± 0.01
Particulate (> 0.22 µm)	0.42 ± 0.02
Discharge	
Previous 3 months	0.27
Previous 12 months	0.15

TABLE VI Pu/Cm AND Am/Cm VALUES FOR ECHINODERM AND SEA WATER SAMPLES AT THE TIME OF COLLECTION IN 1978 (ERRORS ARE BASED ON 2σ PROPAGATED COUNTING ERRORS)

Sample	$\dfrac{^{238}Pu + {}^{239/240}Pu}{^{244}Cm}$	$\dfrac{^{241}Am}{^{244}Cm}$
Asterias rubens		
Aboral body wall	273 ± 103	237 ± 89
Digestive gland	366 ± 163	192 ± 86
Shore-line sea water		
Particulate (> 0.22 μm)	198 ± 132	105 ± 70
Discharge		
Previous 3 months	235	46
Echinus esculentus		
Aboral test	383 ± 256	312 ± 208
Gonad	43 ± 14	103 ± 32
Shore-line sea water		
Particulate (> 0.22 μm)	195 ± 56	83 ± 24
Discharge		
Previous 3 months	205	54

obtained from shore-line sea water samples, both filtrate and particulate, collected close to Windscale during the same months as the biological samples, and with the ratio in the Windscale discharges for both the three months and twelve months prior to sample collection. It is immediately apparent that, in comparison with the discharges, the filtrate fraction of the shore-line sea water is depleted in ^{241}Am relative to plutonium, and that the particulate fraction is enhanced, as would be expected from their K_d values; although the particulate fraction here is of an entirely different nature, being predominantly of sand. There is also a high sediment load in these samples. The echinoderm organs, however, are even more enhanced in ^{241}Am relative to plutonium. The lowest value, that for starfish digestive gland, is the same as that for the particulate material but all of the other quotients are considerably higher, especially that of the echinoid gonad. It seems quite clear, therefore, that ^{241}Am is more biologically accumulated than plutonium, even allowing for grow-in from ^{241}Pu.

A comparison of these two elements with curium is more difficult because of the very low concentrations of the latter, resulting in large counting errors (Table VI). Nevertheless, some interesting trends are apparent. In comparing the quotients with the previous three months' discharge, plutonium appears to be slightly enhanced relative to curium in all of the organs except for the echinoid gonad which appears to be significantly different; americium, too, is clearly enhanced relative to curium.

A more interesting comparison, although unfortunately not with echinoderms, is that between technetium and caesium (Table VII). Caesium is removed to sediment more than technetium, as evidenced by silt samples from an estuary close to Windscale. In biological materials, however, whereas the concentrations of ^{137}Cs in the claw muscle of the crab, the digestive gland of the crab, and the total soft parts of the mussel are all rather similar - within a factor of three - the concentrations of ^{99}Tc are quite different. The data in Table VII compare the ^{99}Tc/^{137}Cs quotients in these samples with those of the discharges for the three months prior to sample collection. From them it can be inferred that ^{137}Cs is preferentially accumulated in the claw muscle of the edible crab, by a factor of about ten. In marked contrast, the digestive gland of the crab and the total soft parts of the mussel preferentially accumulate ^{99}Tc, by a factor of about three. It is possible that such differences are a reflection of changes in oxidation state depending upon the route of uptake into the animal, for example, via the gills or the gut of the crab, in which case the data on the alga *Fucus vesiculosus* are of particular interest because this species appears to preferentially accumulate ^{99}Tc over ^{137}Cs two orders of magnitude greater than the animal material analysed to date. Of course there are many other factors to be considered, such as the relative rates of intake and loss of the different elements, but these few observations serve to demonstrate our lack of knowledge regarding their general biological availability.

From the mathematical modelling point of view data are usually required in the form of concentration factors - the biological equivalent of a K_d for sediments, only expressed per unit wet weight. There are many parameters which can affect the calculation of a concentration factor, and these are usually thought of in terms of their biological effect. In an area such as the Irish Sea, however, the difficulty is usually one of selecting the most suitable sea water value to use as a denominator - total sea water, filtrate only, surface water, bottom water, and so on. The difficulty is even more acute when the samples are obtained close to Windscale because here the water concentrations fluctuate markedly as a result of the day to day content of the discharges. In order to obtain more reliable data, therefore, it will be necessary to obtain samples at a greater distance from Windscale where the water concentrations are less susceptible to short-term variations. The disadvantage in this approach is that the concentrations are much lower and thus a considerably greater analytical effort is required. From the limited data available from samples taken close to Windscale the concentration factors over filtrate sea water for plutonium, americium and curium are generally of the order of 10^3 for invertebrate samples; values for americium are generally the highest. Values for fish are much lower, and vary considerably from one organ to another [6].

An equally valuable approach is to relate the concentrations of radionuclides in environmental samples to their rates of input, for example as $fCi.g^{-1}$

TABLE VII ^{99}Tc/^{137}Cs VALUES FOR SAMPLES COLLECTED IN THE WINDSCALE AREA (ERRORS ARE $\pm 2\sigma$, BASED ON PROPAGATED COUNTING ERRORS)

Sample	^{137}Cs g^{-1} wet pCi (mBq)		$\dfrac{^{99}\text{Tc}}{^{137}\text{Cs}}$ in sample	$\dfrac{^{99}\text{Tc}}{^{137}\text{Cs}}$ in discharge over previous 3 months
Ravenglass silt	177 $\pm$ 1.0	(6549) ($\pm$ 37)	0.0014 $\pm$ 0.0008	0.028
Cancer pagurus (edible crab)				
Claw muscle	13 $\pm$ 0.3	(481) ($\pm$ 11)	0.010 $\pm$ 0.002	0.098
Digestive gland	20 $\pm$ 1.0	(740) ($\pm$ 37)	0.33 $\pm$ 0.02	0.098
Mytilus edulis (edible mussel)				
Total soft parts	7.5 $\pm$ 0.1	(278) ($\pm$ 4)	0.36 $\pm$ 0.09	0.098
Fucus vesiculosus (brown alga)	47 $\pm$ 0.7	(1739) ($\pm$ 26)	8.7 $\pm$ 0.2	0.013

TABLE VIII RELATIONSHIP BETWEEN CONCENTRATIONS OF ^{238}Pu AND $^{239/240}$Pu, AND ^{241}Am IN LIVER AND MUSCLE OF THE PLAICE (*Pleuronectes platessa*) AND THE QUANTITIES DISCHARGED, EXPRESSED AS fCi.g^{-1} PER Ci.day^{-1}. (THE Ci.day^{-1} VALUES ARE DERIVED FROM THE AVERAGE FOR THE PRECEDING THREE MONTHS)

Month of collection	Liver			Muscle		
	^{238}Pu + $^{239/240}$Pu		^{241}Am	^{238}Pu + $^{239/240}$Pu		^{241}Am
1975						
February	12		61	0.06		0.24
May	2		31	0.09		0.24
August	2		13	0.18		0.33
November	10		49	0.12		0.27
1976						
February	10		148	0.09		0.60
May	16		153	0.36		1.35
August	3		19	0.09		0.21
November	2		54	0.03		0.63
1977						
February	3		79	0.03		0.63

wet of sample per Ci discharged day^{-1}. This method is used with considerable effect in the radiological control of fission product discharges from Windscale. Such data have also been derived over a short period of time for plutonium and americium in the commercial flatfish the plaice (*Pleuronectes platessa*) caught at regular intervals in an area some 5 km from the pipeline [6]. The three-monthly values (Table VIII) are remarkably constant. Incidentally, these data also indicate that ^{241}Am appears to be more available to the fish than $^{239/240}$Pu, per unit discharge, without our having to resort to concentration factor calculations. The contamination of plaice and other fish species in the Windscale area is such that the local fish eaters still only sustain a plutonium and americium intake which is less than 0.1% of the ICRP-recommended intake [7], even though Windscale regularly discharges some 100 Ci (3.7 TBq) per month of plutonium into the Irish Sea. This is an interesting observation when it is considered that the Irish Sea has a relatively high biomass to water ratio, the fish live close to the shore where the discharge is made, and the fish are directly eaten by man.

DISCUSSION

It is a truism to state that many more data are required on the behaviour of long-lived radionuclides in the marine environment. It is also very clear that the Irish Sea is unique with regard to some of the radionuclides which are present in it. One element in particular has not been discussed, and that is ^{237}Np. This nuclide is present in the Windscale discharges in very small amounts and it has been detected in environmental samples. As yet, however, only qualitative data are available; these are very interesting. (Quantitative data are currently being obtained using ^{239}Np as a yield tracer.) The qualitative data indicate that at least some of the neptunium appears to behave in a manner similar to that of the higher oxidation states of plutonium, and thus should provide some interesting comparisons with the other transuranium nuclides which have been studied.

The importance of such data with regard to discharges of low-level radioactivity into coastal waters is very obvious. Their relevance to the deep sea disposal option, however, may at first appear to be somewhat tenuous, but there would appear to be two worthwhile approaches which can be made. The first is that contaminated coastal waters provide the only means by which theoretical considerations of the chemical behaviour of the long-lived radionuclides can be tested. It is clearly highly desirable to be able to characterise the chemical behaviour of these radionuclides at low gravimetric concentrations and thus develop more accurate thermo-dynamic models: given such models it should be perfectly feasible to extrapolate to the conditions of the deep sea. Even this approach, however, will have considerable pitfalls, and it is therefore also necessary to make a different, but complementary one. It is often stated that uranium and thorium should provide useful analogues to the higher and lower oxidation states of plutonium [2]. Unfortunately our knowledge of these two elements in the marine environment is extremely poor, particularly with regard to their biological availability and long-term cycling [8,9]. Nevertheless, the general statement is a valid one and a more detailed study of the naturally-occurring actinides in the marine environment is long overdue. It is necessary first of all to obtain a comparison between the behaviour of individual transuranium elements and naturally-occurring actinides in coastal waters; and secondly, to compare the behaviour of these naturally-occurring actinides in coastal waters with their behaviour in the deep sea. Biological availability is obviously one area which can immediately be studied by comparative analyses. There are, in fact, very few data available on uranium or thorium in coastal water organisms; and even less on deep-water species. Preliminary analyses which have been made on the plaice, using spark-source mass spectrometric techniques, indicate that the concentration factors for uranium in bone, liver and muscle are all lower than those of plutonium and americium, and the same appears to be true for the cod (*Gadus morhua*). Thorium (^{232}Th) is very low in all of the samples analysed to date by mass spectrometry, typically being below the limits of detection, which are 0.5 pCi (18.5 mBq).kg^{-1} wet in bone and 0.1 pCi (3.7 mBq).kg^{-1} wet in muscle. Thus there is clearly much to be learned; opportunities to do so do exist, and advantage should be taken of them.

REFERENCES

(1) Nelson, D. M. and Lovett, M. B.: "Oxidation state of plutonium in the Irish Sea", Nature, London 276, 599-601 (1978).

(2) Schell, W. R. and Watters, R. L.: "Plutonium in aqueous systems", Health Physics 29, 589-597 (1975).

(3) Hetherington, J. A., Jefferies, D. F. and Lovett, M. B.: "Some investigations into the behaviour of plutonium in the marine environment", Impacts of Nuclear Releases into the Aquatic Environment, pp. 193-212, IAEA, Vienna 1975.

(4) Hetherington, J. A.: "The behaviour of plutonium nuclides in the Irish Sea", Environmental Toxicity of Aquatic Radionuclides: Models and Mechanisms, edited by M. W. Miller and J. N. Stannard, pp. 81-106, Ann Arbor Science Publ. Inc., Michigan 1976.

(5) Hetherington, J. A.: "The uptake of plutonium nuclides by marine sediments", Marine Science Communications 4, 239-274 (1978).

(6) Pentreath, R. J. and Lovett, M. B.: "Transuranic nuclides in plaice (*Pleuronectes platessa*) from the north-eastern Irish Sea", Marine Biology 48, 19-26 (1978).

(7) Pentreath, R. J., Lovett, M. B., Harvey, B. R. and Ibbett, R. D.: "Alpha-emitting nuclides in commercial fish species caught in the vicinity of Windscale, United Kingdom, and their radiological significance to man", Biological Implications of Radionuclides Released from Nuclear Industries, IAEA, Vienna (in press).

(8) Cherry, R. D. and Shannon, L. V.: "The alpha radioactivity of marine organisms", Atomic Energy Review 12, 3-45 (1974).

(9) Pentreath, R. J.: "Radionuclides in marine fish", Oceanography and Marine Biology Annual Review 15, 365-460 (1977).

Discussion

<u>B.R.R. PERSSON</u>, Sweden

Do you have any data on Np, or do you have any opinion about the K_d values, or concentration factors, relative to Pu and Am ?

<u>R.J. PENTREATH</u>, United Kingdom

As I indicated in the paper, our data on Np so far are qualitative rather than quantitative. These data indicate that at least some of the Np behaves in a manner similar to that of the higher oxidation state of Pu. We are currently obtaining quantitative data and should soon be able to provide the answers to your question with a much greater degree of confidence.

<u>A. AARKROG</u>, Denmark

Do you consider Np-237 radiologically important as compared to Pu in the Windscale releases.

<u>R.J. PENTREATH</u>, United Kingdom

No, on a per caput and collective dose basis, I do not - at the current low rate of discharge. Our main interest in Np-237 is to assess its importance in termes of dose commitment, and to provide data which would be useful in the modelling of potential pathway back to mean when considering the marine disposal option for the disposal of high-level wastes.

Session 5

Chairman - Président
Dr. R.J. PENTREATH
(United Kingdom)

Séance 5

BIOTURBATION OF SURFICIAL SEDIMENTS ON THE CONTINENTAL SLOPE, EAST OF NEWFOUNDLAND

J.N. Smith
Atlantic Oceanographic Laboratory

and

C.T. Schafer
Atlantic Geoscience Centre

Bedford Institute of Oceanography, Dartmouth, N.S., Canada B2Y 4A2

Abstract

Data on regional sedimentary processes, in conjunction with measurements of Pb-210 activity distributions in short cores, facilitate the differentiation of local bioturbation regimes on the continental slope, east of Newfoundland, and provide insight into the factors controlling sediment mixing in these and comparable environments. Excess Pb-210 is confined to the upper 2 cm of sediments underlying the southerly-flowing, Western Boundary Undercurrent (WBU) at 2600 m indicating that minimal contemporary biological reworking of material has occurred at this location. This regime represents an inhospitable environment for the establishment of bioturbating communities because: (1) deposition of fine-grained, organic-rich material which serves as a food source for benthic infauna is inhibited by bottom currents which attain speeds of at least 20 cm s^{-1}, and (2) the variable current regime produces an unstable bottom substrate in which smaller organisms (e.g. meiofauna) are frequently buried and their burrows are rapidly infilled. In contrast, the higher flux of fine-grained, organic-rich material to the middle slope (700 to 2000 m), and the more stable sedimentary conditions which prevail in this low energy regime, are conducive to active colonization of this substrate by bioturbating macrofauna. Maxima in Pb-210 activity profiles at depths of 7 cm to 11 cm in sediment cores are suggestive of heterogeneous, downward transport of surface material, possibly through infilling of macrofaunal burrows.

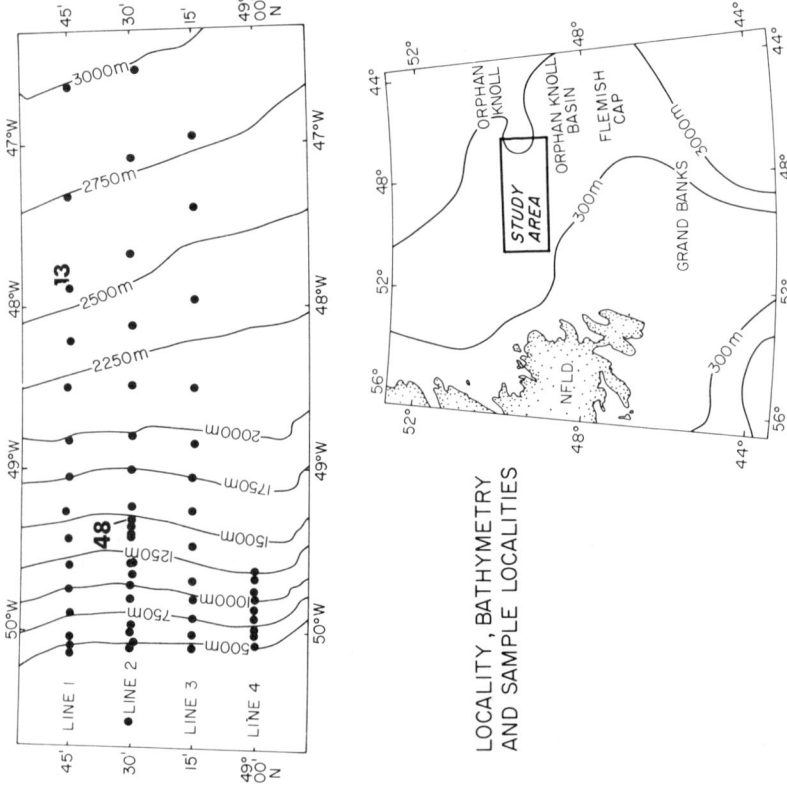

Figure 1: Location of study area, sampling stations, and sites for collection of box core 48 and Lehigh core 13-A.

1. INTRODUCTION

Benthic macrofauna can have a pronounced effect on the physical and chemical properties of marine sediments as a consequence of their burrowing, feeding and physiological activities.[1-4] Vertical redistribution of sediment by benthic infauna can lead to perturbations in stratigraphies, geochemical profiles, and geotechnical properties of the sediments.[5-7] These perturbations can produce secondary feedback effects which may alter the original structure and diversity of the benthic communities. [8] Burrow construction facilitates the downward transport of oxygenated and nutrient-rich bottom water into the deeper sediment strata and can increase the surface area of the transition zone between aerobic and anerobic conditions.[9] Changes in the biogeochemical regime, induced by bioturbation, may promote the remobilization of some radionuclides (e.g. Pu-239, 240) at depth in the sediment by means of organically-mediated reactions.[10, 11] The effect of bioturbation on the stability and composition of marine sediments influences their role as sinks or sources of radionuclides, and, in this sense, may have important implications with respect to the global cycling of radionuclides.

Pb-210 is a useful sediment tracer because of its limited chemical mobility in sediments both under oxidizing and reducing conditions. [12] In addition, its half life ($t_{\frac{1}{2}}$ = 22.3 yr) provides a convenient time scale for the resolution of recent sediment mixing phenonema. Tracer applications of Pb-210 to studies of bioturbation in nearshore sediments are complicated by difficulties associated with the separation of the effects of biological mixing from those of sediment accumulation.[13]

However, on 'constructional' continental slopes [14] the low rate of sediment accumulation is sufficiently dominated by bioturbation so as to permit the use of short-lived particle tracers to study sediment mixing processes. This paper describes the initial results of a study on the use of Pb-210 to measure mixing rates and identify mixing processes operative in continental slope sediments of the northwest Atlantic Ocean.

2. FIELD AND LABORATORY METHODS

In 1977, the CSS Dawson (BIO Cruise 77-034) occupied a total of 59 stations along four transects that extended from the continental shelf break, at 300 to 350 m depth, down the slope and rise to the 3000 m isobath (Figure 1). Samples collected included water samples from 4 m and 100 m above the bottom, seabed photographs, Van Veen grab samples, several box cores, and Lehigh gravity cores from the middle slope and upper continental rise. The box cores were subsampled using 10 cm diameter PVC pipe which also served as the combination core barrel and liner for Lehigh gravity cores. All cores were stored and transported vertically at 4°C and subsequently subsampled at 1 cm intervals in the laboratory. Pb-210 and Ra-226 activities were measured on sediment samples using the techniques outlined by Smith and Walton (1980).[15] Cores were x-radiographed using a Philips K-200 industrial x-ray system.

3. ENVIRONMENTAL SETTING

The uppermost slope at latitude 49°30'N is bounded by the shelf break which lies between the 300 and 350 m isobaths. From here the slope dips gently seaward at an angle of 0.8 degrees to the lower slope (2000 to 2500 m) where the dip decreases to less than 0.5 degrees. The lower slope and the upper continental rise (2500 to 3000 m) are swept by the Western Boundary Undercurrent (WBU) which moves along the Labrador continental margin, through the study area, and then to the south around the Tail of the Grand Bank. Recent bottom current meter measurements indicate that the WBU has a high speed core, characterized by transient speeds of at least 20 cm s^{-1}, that is situated at a depth of about 2800 m to 3000 m (Schafer and Carter, in prep.). At this depth the sediments are relatively coarse silty-sands and gravels, and display maximum development of current structures in bottom photographs. Sediments on both sides of the core are finer grained and suggest waning current speeds.[16]

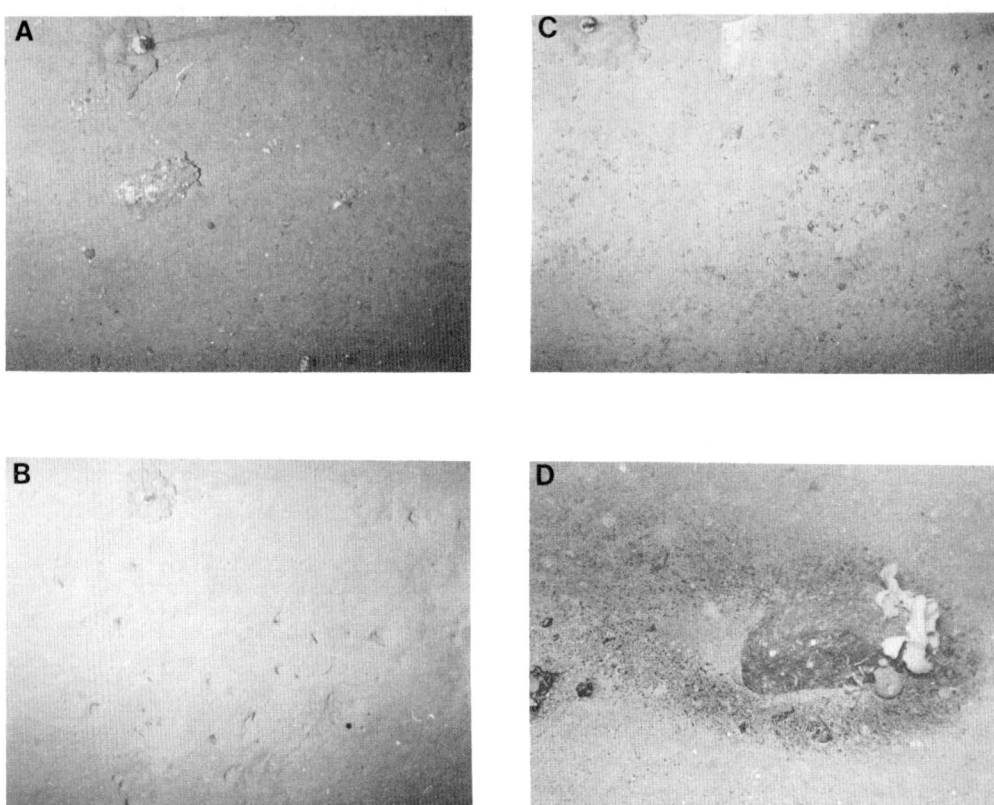

Figure 2: Bottm photographs show: (A) **partly buried ice-rafted boulder** on the upper slope. Diameter of the trigger weight near the upper edge of photograph is 7 cm, 600 m; 49°15.0'N, 50°01.6'W. (B) bioturbated muds on middle slope, 1400 m; 49°29.8'N, 49°24.6'W. (C) muddy reworked foraminiferal sands strewn with manganese-coated pebbles, 2600 m; 49°14.6'Y, 47°23.8'W". (D) oblique image of manganese-coated boulder and pebbles on the upper continental rise. General southwesterly flow of the WBU is indicated by the configuraiton of the pebble halo around the boulder and by the attached epibenthos on the upstream side of the boulder, 2758 m; 49°29.8'N, 49°00.8'W.

The fine scale character of the seabed, as recorded in underwater photographs, changes progressively with depth. The upper slope, down to about 700 m, is mantled by sandy sediments strewn with subangular to subround cobbles and boulders (Figure 2a). Evidence of current activity is restricted to the isolated occurrence of scour moats at the base of boulders and cobbles, infilled burrows, and a general alignment of linear biogenic remains such as worm tubes. The middle slope between 700 and 1400 m is heavily populated by burrowing macroinvertebrates. Down to 1000 m the prominent bioturbation marks are shallow craters (4 to 8 cm in diameter) produced by browsing echinoids, small rimless burrows (~ 1 cm diameter), and a few short, linear trails (Figures 2b). From 1000 to 1400 m, burrows become more numerous and more diverse in form and size. Conspicuous forms are cones, small (1 to 2 cm diametr) symmetric, rimless burrows, and large (5 cm diameter) irregular burrows. The excellent preservation of the bioturbation marks precludes strong current movement near the seabed. At 2000 m, on the lower slope, sediment-coated pebbles and a few cobbles are again evident. Bioturbation marks are common and well preserved. At 2600 m on the upper rise the seabed is strewn with manganese-coated gravel which, together with scour moats and alignment of linear fragments, attest to the power of local bottom currents (Figures 2c and 2d). Bioturbation structures are few and consist of burrows and short simple trails.

Upper slope sediments are predominantly terrigenous sands and muds (Figure 3) containing a total of about 1 to 100 benthonic foraminifera per cubic centimetre of wet sediment and $>0.4\%$ organic carbon. At the other end of the spectrum, lower slope and upper rise sediments may contain 10,000 plankton foraminifera and 100 to 1000 benthonic foraminifera per cubic centimetre of wet sediment and $\lesssim 0.2\%$ organic carbon (Figure 4).

Suspended particulate matter (SPM) levels within the bottom nepheloid layer at levels 4 and 100 m above the seabed display trends that are related to proximity to the shelf, contour current activity, and sediment distribution patterns (Figure 5). Lowest values (0.09 to 0.11 mg l^{-1}) occur in the vicinity of the shelf break beyond which values increase sharply to about 0.25 mg l^{-1} in water depths between 800 and 1000 m. Thereafter, SPM at the 100 m level remains stable but at the 4 m level it tends to show a 0.14 mg l^{-1} variation. [16] The largest fluctuations at the 4 m level occur over the middle slope (1200 to 1800 m) where bioturbation appears to be intense, and over the upper rise (2500 to 3000 m) where sediment reworking processes are active. Mineral (terrigenous) grains, which make up more than 50% of the SPM, have lowest concentrations near the shelf break and highest concentrations over the upper rise (Figure 5).

In addition to the supply of sediment from the shelf, material is supplied to the middle slope and lower rise as a consequence of the reduction of current speed away from the axis of the WBU. The distribution of SPM, with low concentrations over the shelf edge and upper slope and high concentrations in deeper water, suggest that the WBU is presently the dominant transporting agent of the suspended load; the observed high values of SPM are consistent wih those obtained for the WBU elsewhere.[17] These high SPM concentrations are not source properties of the water mass, but arise from the resuspension and entrainment of particles from sediments previously encountered by the WBU, probably where it is constrained by topographical features. A tentative qualitative model assumes sediment winnowing (and/or non-deposition) beneath the fast-flowing core on the continental rise and deposition under the slow-moving lateral zones of the WBU, especially on the middle slope.

Sources of Pb-210 in the water column include atmospheric inputs and in situ production by radioactive decay from Ra-226. Transport of Pb-210 to the sediments probably involves scavenging by rapidly settling biogenic (e.g. fecal pellets) and inorganic particulate matter in addition to other possible processes such as coprecipitation with Fe and Mn oxides at the sediment-water interface.[18] The higher fluxes of a finer-grained organic-rich component of the SPM load to the middle slope environment is probably the reason for the higher total inventory of excess Pb-210 in these sediments (Profiles 48-A, 48-B, and 48-C, Figure 6) compared to the sediments underlying the WBU (Profile 13-A, Figure 6).

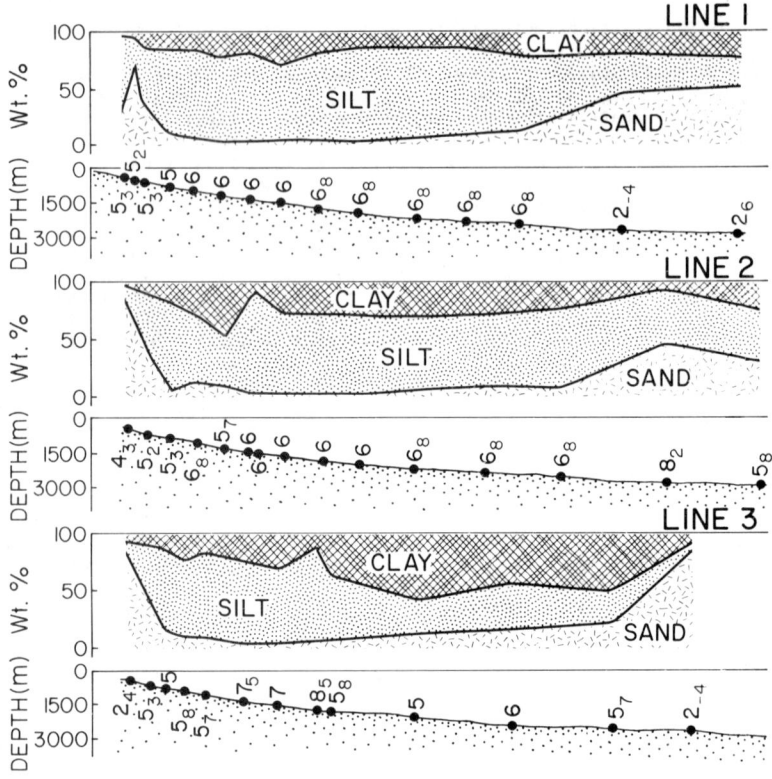

Figure 3: Distribution of sand, silt, and clay-size sediments along three transects on the continental slope.

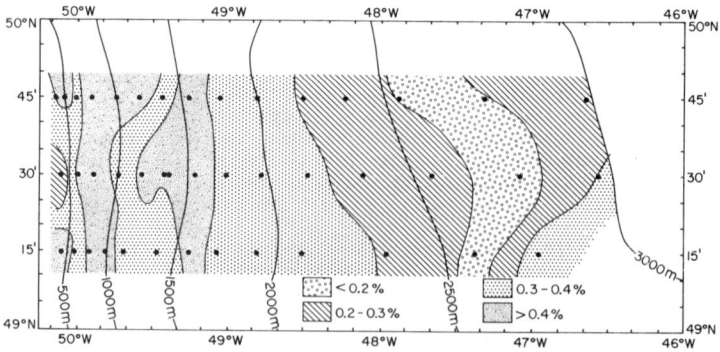

Figure 4: Distribution of total organic carbon (wt %) in bottom sediments on the continental slope.

4. Pb-210 ACTIVITY PROFILES

The sediment in Lehigh core 13-A, collected at a water dpeth of 2600 m on the continental rise (Figure 1), consists of olive-grey to greyish-brown sandy muds (10 to 50% sand) which have been reworked (winnowed) by the WBU to give a gravelly-sand lag deposit that forms the upper 12 cm of the core. Results from box cores collected at Station 13-A in 1979 show that the percent of clay-sized particles increases from 55 to 76% between the 7 cm and 13 cm levels. The good agreement in the depth of the lag deposit between recently collected box cores and Lehigh gravity core 13-A indicates that loss of surface material from the gravity core during the sampling operation was minimal.

The Ra-226 activity measured using the radon gas emanation method [15] for the 21 to 22 cm interval is 1.52 ± 0.18 dpm g^{-1}, which is consistent with the Pb-210 activity (Profile 13-A, Figure 6) of 1.34 ± 0.10 dpm g^{-1} measured for this same sample. This comparison suggests that no significant departure from secular equilibrium has occurred between Ra-226 and supported Pb-210 in the bottom portion of this core. Slightly elevated Pb-210 activities in the 2 to 12 cm interval may reflect an increase in the Ra-226 activity in the winnowed lag deposit. The 13-A, Pb-210 profile exhibits increased activities in the 0 to 2 cm interval (Figure 6). Apparently, below the axis of the WBU most of the excess Pb-210 resides in the upper few centimetres of sediment and there is little evidence of recent, substantial vertical mixing of sediment.

Box core 48, collected at a water depth of 1500 m on the continental slope (Figure 1), contained olive-grey muddy sediments having a finer overall grain size (79% silt plus clay-sized particles) and smaller amounts (20 to 25%) of $CaCO_3$ compared to material from core 13-A. The Pb-210 profile, 48-C (Figure 6), corresponds to material subsampled from the centre of the box core, while profiles 48-A and 48-B are from a sub-core, collected 18 cm from the centre of the box core. The latter two profiles correspond to sediment columns, 1 cm wide, separated by a lateral distance of 3 cm.

Sediment in the vicinity of profiles 48-A and 48-C contained numerous small (1 mm in diameter) burrows in the upper 3 cm and several large burrows (1 to 5 mm in diameter) angling down from the sediment-water interface to depths as great as 12 cm. Profile 48-B sediment had a more mottled texture, a higher porosity, and a greater abundance of burrows of all sizes compared to 48-A and 48-C material.

Pb-210 profile 48-C exhibits elevated activities near the surface and a pronounced maximum in the 6 to 9 cm interval, features which are unlikely to be caused by anomalies in the vertical distribution of Ra-226 or Rn-222. These data, together with x-radiographic evidence of burrows intersecting the 6 to 9 cm interval, indicate that surface sediment containing excess Pb-210 has been recently transported downward either during the excavation of the burrows, during irrigation of the burrows by the organisms, or as a result of infilling of burrows subsequent to the death of the organism or its evacuation from its habitat. The transported surface material has bypassed local Pb-210 activity gradients, resulting in a minimum in the Pb-210 profile in the 2 to 4 cm interval. A reduced Pb-210 maximum is also evident in the 7 to 10 cm interval of Profile 48-A, indicating that heterogeneous downward transport of surface material through the skewed orientation of burrows observed in x-radiographs of the core has also occurred at this location.

Peaks in Cs-137 and Pu-239, 240 activity distributions have been observed below 5 cm in continental slope sediments by Livingston and Bowen.[11] They cite bioturbation as the explanation for the preferential concentration of fallout radionuclides at depth in the sediment column. Benninger et al.[13] have observed excess Pb-210 associated with burrows at depths as great as 70 and 115 cm in estuarine sediments which they ascribed to the infilling of burrows with surface sediment.

Profile 48-B, located only 3 cm from 48-A, has a comparatively constant, negative, Pb-210 activity gradient to a depth of 12 cm. The mottled and reworked appearance of core 48-B sediment, its higher porosity, and increased numbers of

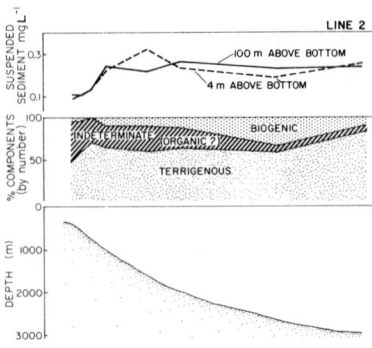

Figure 5: Distribution and composition of suspended particulate matter (SPM) in bottom water during November 1977.

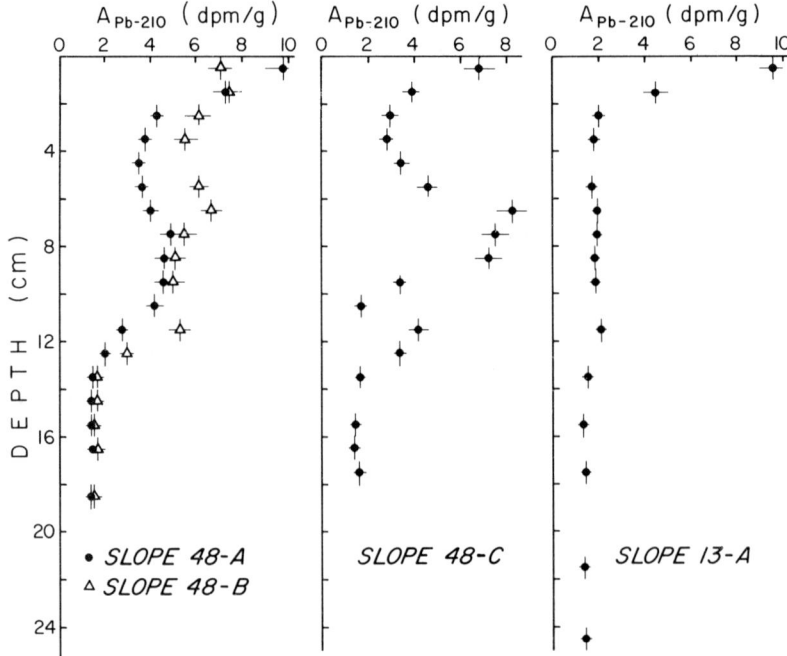

Figure 6: Distribution of total Pb-210 activities with depth in sediments from Lehigh core 13-A (profile 13-A) collected at 2600 m and box core 48 (profiles 48A, 48B, 48C) collected at 1500 m on the continental slope.

burrows compared to 48-A and 48-C, indicate a higher and more uniform rate of mixing through the upper 12 cm. The close correlation of profiles 48-A and 48-B in the 7 to 9 cm interval may reflect enhanced lateral mixing of sediment in this depth interval.

Under certain conditions, biological mixing can be simulated by a diffusion model which implies random, steady-state exchange of adjacent volume elements of sediment by the activities of organisms distributed uniformly through the mixing zone. For these case, the mixing rate can be characterized by a biological mixing coefficient, K_b, which has the same units as a diffusion coefficient. Values for K_b of $1-14 \times 10^{-9}$ cm^2 s^{-1} have been measured for pelagic sediments [19,20] while higher values, in the range of 10^{-6} to 10^{-8} cm^2 s^{-1} have been determined for estuarine sediments [13,20,21] on the basis of uranium series (Pb-210, Th-234) decay profiles.

A diffusion model is not strictly applicable to Pb-210 profiles from core 48 because the mixing process has a non-random, heterogeneous component (on time scales of the order of the Pb-210 half life) as evidenced by the maxima observed in profiles 48-A and 48-C. An increased mixing rate in core 48-B has apparently had an averaging effect on localized maxima, resulting in a Pb-210 profile which is a better approximation to a diffusion profile from 0 to 12 cm. The small decrease in the Pb-210 activity, from 7 dpm g^{-1} at the surface to 5 dpm g^{-1} at the 11 to 12 cm interval of core 48-B, indicates a high degree of biological activity which would be typical of estuarine environments. The absence of excess Pb-210 below 13 cm in all three cores is consistent with x-radiographic evidence for the absence of burrows penetrating below this depth from the sediment-water interface. Evidently, it is the range of the larger burrowing organisms which defines the maximum depth for downward, episodic transport of surface material.

5. SUMMARY

Bottom photography, incipient bedding observed in x-radiographs of cores, and the low numbers of macrofaunal species observed in sediments underlying the WBU all attest to the limited colonization of these bottom sediments by bioturbators. The shallow depths of sediment mixing are consistent with the absence of excess Pb-210 at depths greater than 2 cm below the sediment-water interface. This sedimentary regime constitutes an inhospitable environment for burrowing macrofauna because: (1) bottom currents inhibit both the deposition of fine-grained, organic-rich suspended matter which serves as a food source for benthic infauna, and also promote the winnowing and transport of fine particles from the sediment; and (2) the current regime produces an unstable bottom morphology in which small bioturbators (e.g. epibenthic meiofauna) can be easily buried.

In contrast, the finer-grained, organic-rich, relatively-stable sediments deposited in the quiescent middle slope environment support a richer macrofaunal assemblage which includes actinarians, polychaete worms, echinoids and ophiuroids. The infauna includes a diversity of mollusc species, particularly Nucula sp., which are capable of extensive reworking of muddy sediments.[16] One-dimensional mixing models are inadequate to explain anomalies in Pb-210 profiles that may have been caused by lateral transport of sediment as a result of the activities of a heterogeneous distribution of bioturbators. For this regime, lateral resolution of radionuclide profiles, on the same scale as their vertical resolution, is required to properly assess mixing rates and process-related mechanisms.

REFERENCES

1. Gordon, D.S.: "The Effects of the deposit Feeding Polychaete Pectinaria goulding on the intertidal sediments of Barnstable Harbour", Limnology and Oceanography (11), 327-332 (1966).

2. Peng, T.H., Broecker, W.S., and Kipphut, G.: "The Effect of Bioturbation on the Late Glacial to Holocene Transition as Recorded in Deep sea Cores", Geological Society of America Annual Meeting, Boulder, Colorado, Abstract (8), 1-46 (1976).

3. Peng, T.H. and Broecker, W.S.: "Effect of sediment mixing on calcite dissolution by fossil-fuel CO_2", Transactions of the American Geophysical Union (59), 410 (1978).

4. Clifton, H.E. and Hunter, R.E.: "Bioturbational Rates and Effects in Carbonate Sand, St. John, U.S. Virgin Islands", Journal of Geology (81), 253-268 (1973).

5. Guinasso, N.L., Jr. and Schink, D.R.: "Quantitative Estimates of Biological Mixing Rates in Abyssal Sediments", Journal of Geophysical Research 80, 2032-3043 (1975).

6. Robbins, J.A., McCall, P.L., Fisher, J.B., and Krezoski, J.R.: "Effect of Deposit Feeders on Migration of ^{137}Cs in Lake Sediments", Earth and Planetary Science Letters 42, 277-287 (1979).

7. Gray, J.A.: "Animal-sediment Relationships", Oceanography and Marine Biology Annual Review 12, 223-261 (1974).

8. Rhoads, D.C.: "Organism-sediment Relations on the Muddy Sea Floor", Oceanography and Marine Biology Annual Review 12, 263-300 (1974).

9. Stephens, G.C.: "Uptake of Naturally Occurring Primary Amines by Marine Annelids", Biology Bulletin 149, 297-407 (1975).

10. Bowen, V.T., Livingston, H.D., and Burke, J.C.: "Distribution of Transuranium Nuclides in Sediment and Biota of the North Atlantic Ocean", in Transuranium Nuclides in the Environment (IAEA, Vienna), 107-120 (1976).

11. Livingston, H.D. and Bowen, V.T.: "Pu and ^{137}Cs in Coastal Sediments", Earth and Planetary Science Letters 43, 29-45 (1979).

12. Robbins, J.A.: "Geochemical and Geophysical Applicaions of Radioactive Lead Isotopes", in Biogeochemistry of Lead (ed. J.P. Nriago), 285-393, Elsevier (1978).

13. Benninger, L.K., Aller, R.C., Cochran, J.K., and Turekian, K.K.: "Effects of Biological Sediment Mixing on the ^{210}Pb Chronology and Trace Metal Distribution in a Long Island Sound Sediment Core", Earth and Planetary Science Letters 43, 241-259 (1979).

14. King, L.H. and Young, I.F.: "Paleocontinental Slopes of East Coast Geosyncline (Canadian Atlantic Margin)", Canadian Journal of Earth Sciences 14, 2553-2564 (1977).

15. Smith, J.N. and Walton, A.: "Sediment Accumulation Rates and Geochronologies Measured in the Saguenay Fjord Using the Pb-210 Dating Method", Geochimica et Cosmochimica Acta 44 (in press, 1980).

16. Carter, L., Schafer, C.T., and Rashid, M.A.: "Observations on Depositional Environments and Benthos of the Continental Slope and Rise, East of Newfoundland", Canadian Journal of Earth Sciences 46, 831-846 (1979).

17. McCave, I.N.: "Sediments in the Abyssal Boundary Layer", Oceanus 21, 27-33 (1978).

18. Bacon, M.P., Spencer, D.W., and Brewer, P.B.: "$^{210}Pb/^{226}Ra$ and $^{210}Po/^{210}Pb$ Disequilibria in Seawater and Suspended Particulate Matter", Earth and Planetary Science Letters 34, 167-173 (1976).

19. Nozaki, Y., Cochran, J.K., Turekian, K.K., and Keller, G.: "Radiocarbon and 210-Pb Distribution in Submersible-taken Deep-sea cores from Project Famous", Earth and Planetary Science Letters 34, 167-173 (1977).
20. Turekian, K.K., Cochran, J.K., and Demaster, D.J.: "Bioturbation in Deep-sea Deposits: Rates and Consequences", Oceanus 21 (1), 34-41 (1978).
21. Aller, R.C. and Cochran, K.J.: "$^{234}Th/^{238}U$ Disequilibria in Near-shore Sediment: Particle Reworking and Diagenetic Time Scales", Earth and Planetary Science Letters 29, 37-50 (1976).

Discussion

A.M. ORTINS DE BETTENCOURT, Portugal

In the first part of your paper, when calculating the excess of Pb-210, you admitted that the Ra-226 is in equilibrium with her daughter Pb-210. As radium suffers dynamic processes in rivers, did you confirm this equilibrium ?

J.N. SMITH, Canada

The Ra-226 activity measured in deeper portions of these cores is in good agreement with the measured value for the Pb-210 activity, indicating that secular equilibrium is maintained between Ra-226 and its daughter radionuclides at these sediment depths. Variability in the Ra-226 activity profile through the upper portions of these cores is small, generally less than ± 10 % of the mean value. Although these data do not preclude disequilibria between Ra-226 and supported Pb-210, they do suggest that this phenomenon has had a minimal effect of the excess Pb-210 profile.

J. ANCELLIN, France

Un pic de césium-137 analogue à celui mentionné dans votre travail s'observe également dans les sédiments prélevés par carottage dans le golfe normano-breton. Peut-on estimer que la profondeur variable à laquelle se trouve, d'après nos constatations, la couche où se manifeste le pic des retombées, soit plus particulièrement en relation avec une plus ou moins grande vitesse de sédimentation ?

J.N. SMITH, Canada

In general, it appears that one must be very cautions in the time-stratigraphic interpretation of sedimentary Cs-137 activity profiles in the absence of other geochronological markers because there are a multitude of phenomena such as bioturbation, remobilization of sedimentary Cs-137, and the particle selective nature of Cs-137 deposition at the sediment-water interface, which can produce a divergence of the sedimentary profile from the record of atmospheric deposition resulting from nuclear weapon's testing.

BEHAVIOUR OF NATURAL (Th, U) AND ARTIFICIAL (Pu, Am)
ACTINIDES IN COASTAL WATERS

Elis Holm and Bertil R.R. Persson
Radiation Physics Department
Lasarettet
S-221 85 LUND, Sweden

ABSTRACT

Fucus serratus and F. vesiculosus collected at the Swedish south west coast during 1967-1978 have been analyzed. Uranium was found in highest activity concentrations; 13, 11 and 0.45 Bq/kg dry weight of U-234, U-238 and U-235 respectively, which correspond to dry weight concentration factors of about 700 for uranium isotopes. The activity concentrations of thorium isotopes were 100, 400, and 7 000 mBq/kg dry weight of Th-232, Th-230 and Th-228 respectively. Concentrations of Pu-239+240 in this bioindicator ranged from 100-600 mBq/kg dry weight and of Am-241 from 10-50 mBq/kg dry weight, depending on which year and site of collection. In unfiltered seawater we find in 1978 Pu-239+240 and Am-241 of concentrations in the order of 7-14 µBq/l and 2-3 µBq/l respectively. The dry weight concentration factors were about 12 000 for Pu and 20 000 for Am.

1.   INTRODUCTION

Although seaweed show high concentration factors for actinide elements it has not been used as bioindicator for these element as frequently as mussels. Extremly high concentrations of Pu-239+240 were found in <u>Sargassum fluitans</u> directly after fallout by NOSHKIN et al.(1971) /¯1_7. In many countries different species of algae serve as important food stuff. Using seaweed as a bioindicator for the concentrations of actinides in seawater is very benefical since the concentrations in water are very low and the analytical procedure for seawater is extremly time consuming. We also hope to get a better understanding of transfer mechanisms and biophysical behaviour of actinides. Algae might also play an important role as a link in foodchains.

F. vesicolosus and in same cases F. serratus have been collected at the Swedish south west coast occationally during 1967-1976 /¯2_7. After 1976 a much more frequent and regular collection has taken place. This was mainly done in order to follow fission and activation products released from nuclear power plants in this area. This sample collection was now also used for an actinide investigation .

2.   ANALYSIS AND MEASUREMENTS

About 15-25 g of dry seaweed were ashed at 500 C together with a suitable quantity of yield determinants. For this purpose we used Am-243, Pu-242 and Th-229. In an earlier stage we used Th-230 but in that case we had to perform one analysis without yield determinat also, in order to get the concentration of Th-230 in the sample. Uranium determinations were done separately by the use of U-232 as yield determinant.

For plutonium separation the method described by TALVITIE (1971) was used /¯3_7. In that procedure the sample is poured through an ionexchanger in 9 M HCl and the effluent contains Am and Th isotopes. Thorium was separated by ionexchange from nitric acid medium leaving Am in the effluent and washings. Americium was extracted using a synergistic mixture of HDEHP (di-2-ethyl-hexyl-phosphoric acid) and TBP (tributhyl-phosphate) in toluene at pH 3. Further separation took place on an anionexchange column using nitric acid-acetic acid and nitric acid+methanol mixtures. This method is a slight modification of a procedure previously described by HOLM et. al.(1979) /¯4_7. Thus the actinides were separated sequentially and were then electrodeposited onto stainless steel discs from ammonium sulphate medium at pH 3. The alpha spectra of each element was then measured with silicon surface barrier detectors.

3.   RESULTS AND DISCUSSION

The Pu-239+240 concentration in <u>Fucus</u> has decreased from about 550 to 200 mBq/kg dry weight during 1967-1978. Such a decrease was expected since most plutonium was delivered shortly after the nuclear detonation trials in 1961-1962, and at present a larger fraction in this marine environment is bound to the sediment. The main source for plutonium, in the seawater of this area is delivered from land and resuspension of sediments. The plutonium content found in unfiltered seawater is about 7-14 μBq/l which is 3-4 times lower than in the Atlantic or Mediterranean. Our results are in very good agreement with those reported by MURRAY et al (1978) from the SE part of North Sea and Skagerak $\lfloor 5 \rfloor$.

The concentrations of fallout plutonium and americium in seawater of the northern hemisphere are summarized in Table I $\lfloor 5,6,7,8 \rfloor$. Our results indicate that the transuranium elements are rapidly scavenged out from water by suspended particulate material. The increase of the plutonium concentration along the coast in seaweed further north is not very pronounced. No influence on the plutonium levels in this area from the Windscale reprocessing plant can be observed.

Results concerning plutonium in <u>Fucus</u> reported by different authors are summarized in Table II $\lfloor 1,8,9,10,11 \rfloor$. The concentration factors (i.e. the ratio of activity-concentration per kg dry weight in seaweed to the activity-concentration of water) obtained in the present investigation are in the order of 13 000 for plutonium, which are unusually high values. In a few samples we were also able to measure the Pu-238 concentration, and the activity ratio Pu-238/Pu-239+240 was about 0.04 +- 0.02.

The concentration of fallout Cs-137 has also been measured, and the activity ratio Pu-239+240/Cs-137 in <u>Fucus</u> was found to decrease during this period from 0.09 to 0.02 $\lfloor 12 \rfloor$. This decrease might be explained from the higher solubility of cesium in seawater, which causes decreasing Pu/Cs ratio in that part of activity delivered by run off from land. The results should be compared with the value 0.012 for global fallout. From studies reported by MURRAY et. al. (1978) and AARKROG et. al.(1978) one obtains a Pu-239+240/Cs-137 activity ratio of 0.002-0.003 in seawater $\lfloor 5,13 \rfloor$. The concentration factor for Cs-137 from seawater to seaweed is thus calculated to be in the order of 300.

While plutonium shows a decrease of the concentration in <u>Fucus</u> during 1967-1978, the concentration of americium in <u>Fucus</u> increases. The concentration of Am-241 in seaweed increases from 5-10 to about 55 mBq/kg dry weight. This increase must be connected to the build up of Am-241 (T1/2 433 a) from Pu-241 (T 1/2 14.5 a) in the environment. By using the Pu-241/Pu-239+240 activity ratio from integrated fallout as reported by HOLM and PERSSON(1978) the Am-241 results could be corrected to the date of collection $\lfloor 14 \rfloor$. In samples stored for a long time in the laboratory this correction is rather substantial and increases the uncertainty of the values. Thus all results in Figure 1 are corrected and refer to the date of collection.

The Am-241/Pu-239+240 activity ratio from integrated fallout is today about 0.30 which corresponds very well with the value 0.28 found in seaweed collected in 1978. However the samples from 1967-1975 show lower activity ratio values (~0.05 in 1967) than predicted from integrated fallout. The Am-241 concentration in seawater (1978) is about 2.6 µBq/l which gives a Am-241/Pu-239+240 activity ratio of 0.16. But we can not conclude that the concentration factor of Am-241 in seaweed is less than that of plutonium because the calculations for 1978 gives the contrary results.

Other investigators have also reported results which might indicate lower uptake of americium than plutonium in <u>Fucus</u> $\lfloor 8 \rfloor$. The low Am-241 concentrations found in older samples can partly be explained by that Am-241 has mainly been generated <u>in situ</u> in sediment and in soil on land and the process to reach the seaweed are slow and different to those of plutonium. Am-241 was not present in the water at a predicted level at this time. Observations made by BOWEN et.al. and FUKAI et.al. show that americium is depleted with respect to plutonium in the surface layer in the North Pacific and

Table I

Pu-239+240 and Am-241 in seawater (µBq/l)

	Pu-239+240	Am-241	Reference
N. Pacific (1973)	11-18		(6)
The Mediterranean (1975)	33-44	2-3	(7)
N. Atlantic (1972)	25-32	3-8	(8)
N. North Sea (1978)	7-18	2-3	(5)
Öresund (1978)	7-14	2-3	This work

Table II

Plutonium and americium in _Fucus vesicolosus_
(mBq/kg dry weight)

Place	Year	Pu-239+240	Am-241	Ref.
Cape Cod USA	1971	19-27	1-3	(1)
Mass., USA	1972-74	20-30		(8,9)
Gulf of Finland	1974	430-570		(10)
S. Baltic	1973	93-143		(11)
Öresund	1977	130-400	30-100	This work

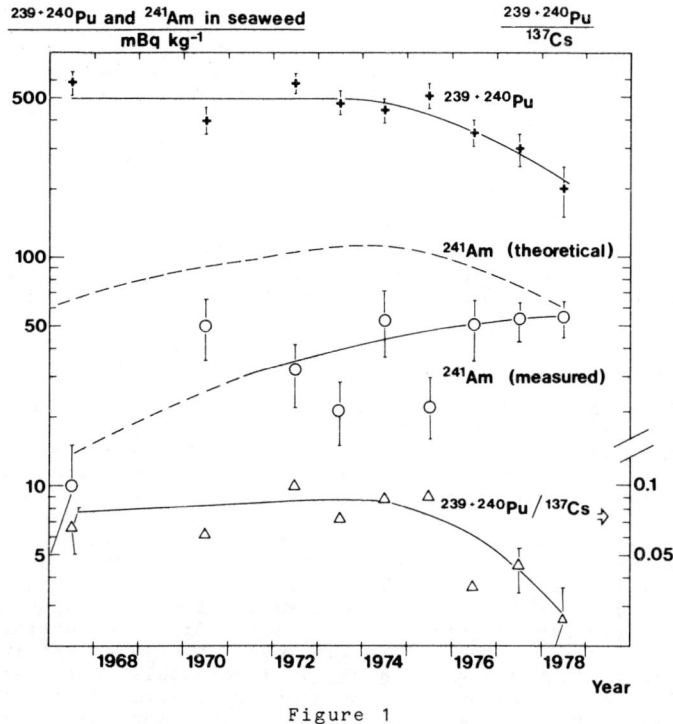

Figure 1

Plutonium-239+240 and americium-241 activity concentrations in seaweed during 1967-1978, and the corresponding activity ratios of Pu-239+240/Cs-137 (lower curve).

Table III

Activity concentrations and concentration factors of actinides in _Fucus_ from Öresund 1978

Actinide	Activity concentration (mBq per kg dry weight)	Concentration Factor (dry)
Th-232	90	750
Th-230	400	950
Th-228	5 000	7 700
U-234	12 700	700
U-235	450	700
U-238	11 100	700
Pu-239+240	190	12 500
Am-241	53	20 600

the Mediterranean $\lfloor$ 15,16 $\rfloor$. This is partly due to higher association of americium with particulate material. At the south west coast of Sweden, where our investigation takes place, we believe that almost all plutonium and americium are associated with particulate matter due to very high resuspension load in these very shallow waters. The high concentration factors we obtain in our investigation might indicate that the sorbtion of actinides to seaweed takes place in particulate form.

The U-234/U-238 activity ratio in the _Fucus_ is about 1.14 i.e. greater than 1.0. This indicates that the sorbtion of uranium takes place in a form which once has gone into solution. But from our results we conclude that we have an organic particulate sorbtion for the actinides. Recently HODGE et. al (1979) showed that uptake of actinides in particulate form by coastal marine organisms was significant $\lfloor$ 17 $\rfloor$.

Thorium in water and seaweed is subject for further investigation at present. Preliminary results show, however, a higher variation from one sample to another than for plutonium and americium. In general the concentrations are fairly high about 100 mBq/kg of Th-232, 400 mBq/kg of Th-230 and 7 000 mBq/kg dry weight of Th-228. These values indicate the importance of radioactivity run off from land and resuspension from sediments in this area. In Table III representative values of concentrations of actinides in _Fucus_ and the concentration-factor(dry) are given. The concentration factors are generally high compared to other areas. Americium, plutonium and Th-228 differ remarkably from others. The high concentrations of Th-228 and high concentration factors can not only be exlplained by build up from precursor Ra-228 _in situ_. A possible explanation is that via Ra-228 a large portion has been transferred into a more bioavailable form (i.e. organic particulate) than the major part of Th-230 and Th-232.

4. REFERENCES

(1) Noshkin,V.E., Bowen,V.T., Wong,K.M. and Burke,JC.,
"Plutonium in North Atlantic Ocean organisms:
Ecological Relationships",
Proc 3rd National Symposium on Radioecology,
Vol 2., pp.681-688 (D.T.Nelson Ed.),CONF-710501-P2,U.S.
U.S. Department of Commerce, Springfield Va. USA (1971).

(2) Mattsson,S
"Long term variation of Cs-137 in _Fucus_ from the Swedish West coast",
To be published

(3) Talvitie, N.A.,
"Radiochemical determination of plutonium in environmental and biological samples by ion-exchange",
Anal. Chem. 43,1827-1830(1971).

(4) Holm,E, Ballestra,S., and Fukai,R.,
"A method for ion-exchange separation of low level americium in environmental materials",
Talanta(in press)

(5) Murray, C.N., Kautsky, H., Hoppenheit, M and Domain, M. :
"Actinide activities in water entering the northern North Sea",
Nature 276,225-230(1978)

(6) Miyake, Y. and Sugimura, Y. :
"The plutonium content of Pacific Ocean waters",
Proc. Symp. on Transuranium Nuclides in The Environment,
pp.91-105, IAEA, Vienna 1976.

(7) Fukai, R., Ballestra, S. and Holm, E. :
"Americium-241 in Mediterranean surface waters"
Nature 264, 739-740(1976).

(8) Livingston, H.D. and Bowen, V.T. :
"Americium in the marine environment-Relationships to plutonium",
In Environmental toxicity of aquatic rdionuclides:
Models and mechanisms, pp. 107-130,
(Ed. M.W. Miller and J.N. Stannard) Ann Arbor Science Publ. Inc. Ann Arbor, USA, 1976.

(9) Livingston, H.D. and Bowen, V.T. :
"Contrast between the marine and freshwater biological interaction of plutonium and americium",
Report COO-3563, COO-3568-7, Wood Hole Oceanographic Institution, 1975.

(10) Miettinen, J.K., Jaakkola, T. and Järvinen, M, :
"Plutonium isotopes in aquatic foodchains in the Baltic Sea",
Proc. Symp. on Impact of Nuclear Releases into
The Aquatic Environment, pp. 147-155
IAEA, Vienna 1975.

(11) Bojanowski, R. and Pempkowiak, J.:
"Accumulation of Sr-90, Cs-137, Ru-106, Ce-144 and Pu-239+240 in Baltic seaweeds"
Oceanologia No. 7, 89-104(1977).in polish

(12) Holm, E, Persson, B.R.R. and Mattsson, S :
"Comparative Studies of Transuranium Elements and Natural Actinides in a Marine Environment",
In Abstracts of the 6th Int. Congr. of Rad. Res.,
p. 154, Tokyo, Japan, 1979.

(13) Aarkrog, A. ,Bötter-Jensen, L., Dahlgaard, H., Hansen, H., Lippert, J., Nielsen, S.P. and Nilsson, K.:
"Environmental Radioactivity in Denmark in 1977",
Risö Report No. 386, Risö, DK-4000 Roskilde, June 1978.

(14) Holm, E. and Persson, R.B.R.,
"Biophysical aspects of Am-241 and Pu-241 in the environment. Rad. and Environm. Biophys. 15,261-276,1978.

(15) Bowen, V.T. :
Personal communication 1979.

(16) Fukai, R., Holm, E. and Ballestra, S :
"A note on vertical distribution of plutonium and americium in the Mediterranean Sea.
Oceanologica Acta 2, in press, 1979.

(17) Hodge, V.F., Koide, M. and Goldberg, E. D. :
"Particulate uranium, plutonium and polonium in the biogeochemistries of the coastal zone".
Nature 277, 206-209, 1979.

PLUTONIUM LEVELS IN THE MARINE ENVIRONMENT AT THULE, GREENLAND

A. Aarkrog
Risø National Laboratory, Health Physics Department
DK-4000 Roskilde, Denmark

ABSTRACT

Since 1968 the marine environment at Thule has contained 25-30 Ci ^{239,240}Pu from an accidental release. The plutonium resides preferentially in the sediments and the benthic fauna. The plutonium inventories in the benthic biota have decreased since 1968, but the decrease was most pronounced in the first years after the accident. A new expedition in August 1979 has included americium in the studies.

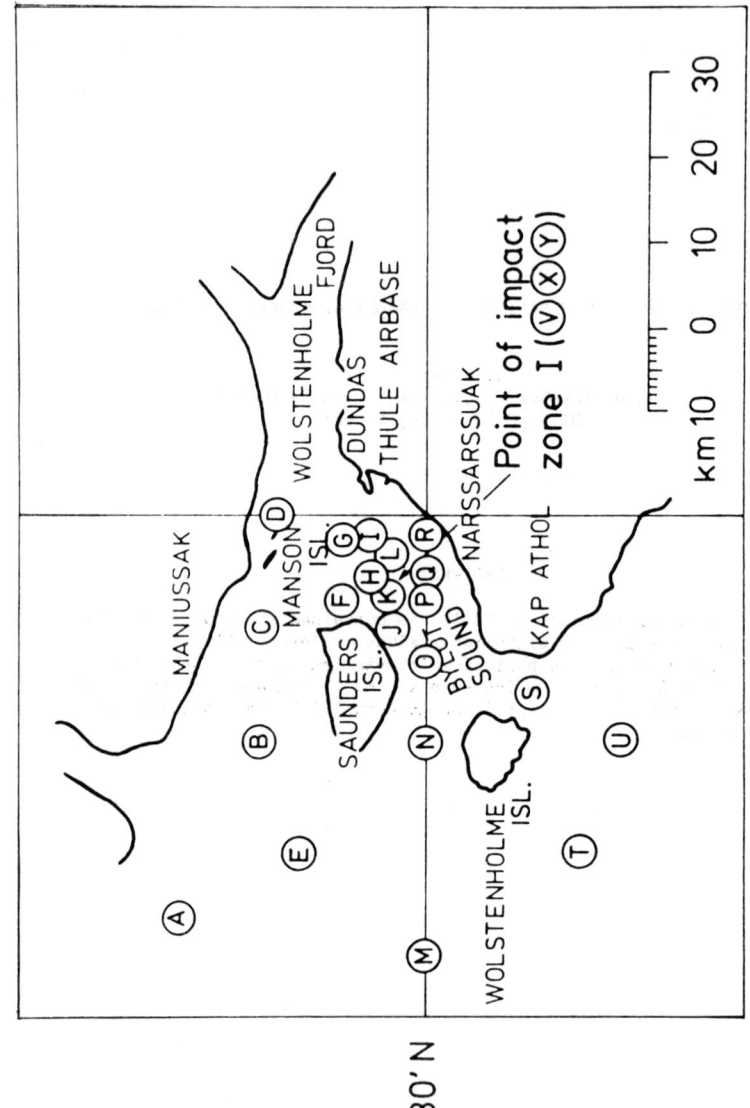

Fig. 1. The sample locations at Thule, Greenland (Location 19 is located in Bylot Sound at 76°27'8 N, 69°37'5 W).

The plutonium contamination at Thule from the B-52 accident in 1968 has been studied with scientific expeditions in 1968, 1970, 1974 and in this year, 1979. The results from the 1979 study are not yet available because the expedition has just been completed.

The following is a short summary of the results obtained of the previous investigations and a noting of the goals for 1979.

The main part of the plutonium currently present at Thule from the accident - which is on the order of a few percent of the total amount released by the accident - resides in sea sediments out to a distance of approximately 30 km from the point of impact. The concentration of plutonium in the sediments decreases approximately exponentially both with distance from the point of impact (cf. Fig. 1) and vertically in the sediments (cf. Fig. 2, Fig. 3). This facilitates an estimate of the Pu-inventory in the sediments at Thule; our estimate is 25-30 Ci ^{239,240}Pu.

The benthic fauna is contaminated with Pu from the sediments. The Pu levels in, e.g., bivalves has changed with time and distance as shown in Fig. 4.

The integrated levels in the biomass (soft parts) of bivalves were 7.7 mCi in 1968, 2.4 mCi in 1970 and 1.9 mCi in 1974.

In Table I the transfer coefficients of ^{239,240}Pu in the benthic environment at Thule is estimated. This estimate is tentative because the environmental decay of plutonium in the various samples is based on observations over a relatively short period (1968-1974).

A main purpose of the 1979 investigations, which has been sponsored by the EEC, is to improve the estimates of the environmental half-lives of plutonium and thus of the transfer coefficients. Another aim is the study of migration (vertical and horizontal) of Pu in the sediments. We intend also to observe higher animals (seabirds, seals and walrus) to check whether these species are still free of plutonium contamination. Table II is a summary of the plutonium levels measured in the various Thule samples since 1968. It is evident that the contamination due to its particulate nature is inhomogenously distributed; hot spots thus occur in most samples.

In conclusion: the plutonium contamination at Thule is not a major radioactive contamination and the health hazards are considered to be negligible. However, the contamination yields a unique opportunity to study the environmental behaviour of plutonium in an arctic benthic environment from a single, point release.

Table I. Estimates of ^{239,240}Pu inventories and transfer-coefficient at Thule

Sample	biomass g m^{-2}	Inventory Ci 1970	1974	Transfer from the release of 1 Ci in Ci·yr
Molluscs				
soft parts	10^2	$2.4 \cdot 10^{-3}$	$1.9 \cdot 10^{-3}$	$1.5 \cdot 10^{-3}$
shells	10^2	$6 \cdot 10^{-3}$	$5 \cdot 10^{-3}$	$4 \cdot 10^{-3}$
Brittle stars	$3 \cdot 10$	$3 \cdot 10^{-3}$	$8 \cdot 10^{-4}$	$6 \cdot 10^{-4}$
Shrimps	$2 \cdot 10$	$1 \cdot 10^{-3}$	$2 \cdot 10^{-4}$	$2 \cdot 10^{-4}$
Worms	10	$5 \cdot 10^{-4}$	$5 \cdot 10^{-4}$	-
Sediments	-	$3 \cdot 10$	$3 \cdot 10$	-

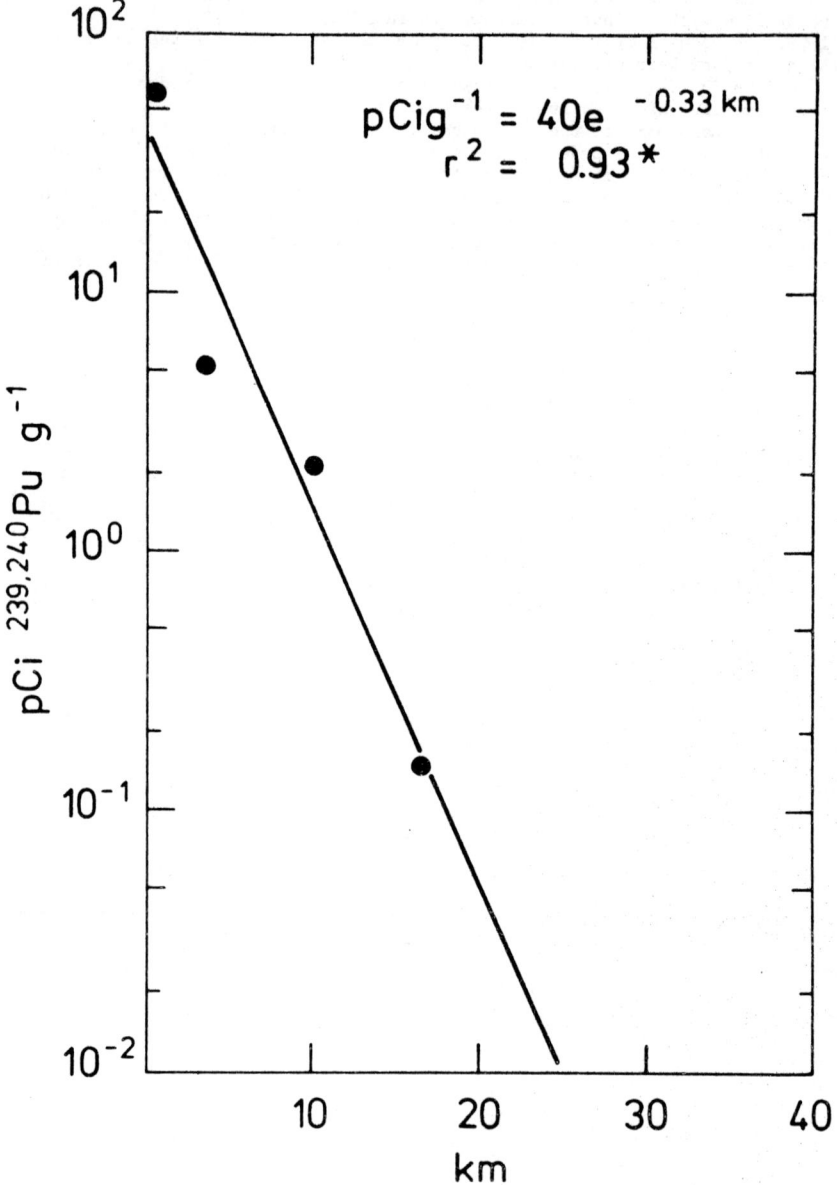

Fig. 2. The ^{239,240}Pu concentration in the sediment surface (0 cm depth) related to the distance (in km) from the point of impact.

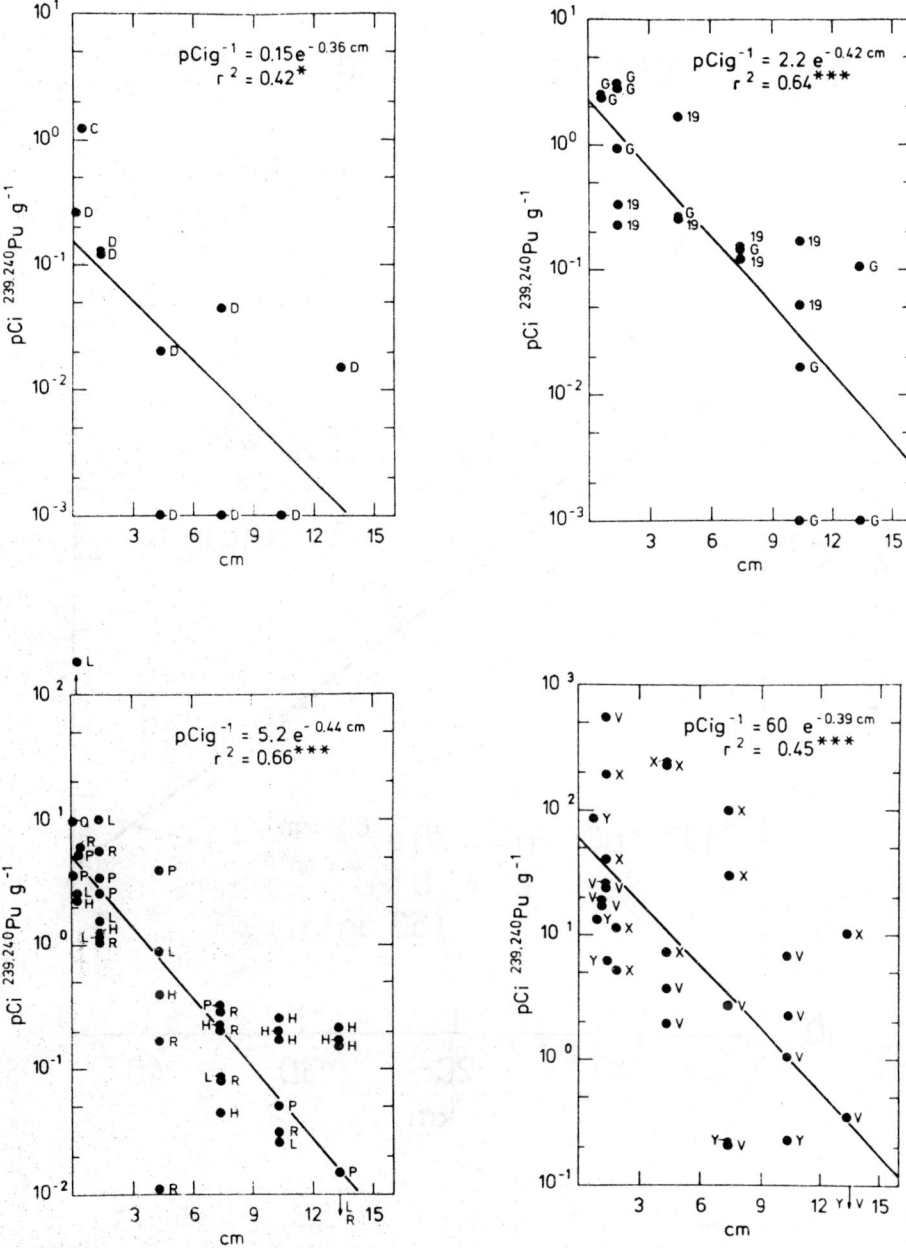

Fig. 3. The vertical distribution of ^{239,240}Pu in the sediments at Thule (cf. fig. 1). The abscisses are the depth in cm of the sediments; r^2 is the correlationscoefficient between observed data and values calculated from the exponential equation (Significance levels: *: 0.05, * *: 0.01, * * *: 0.001).

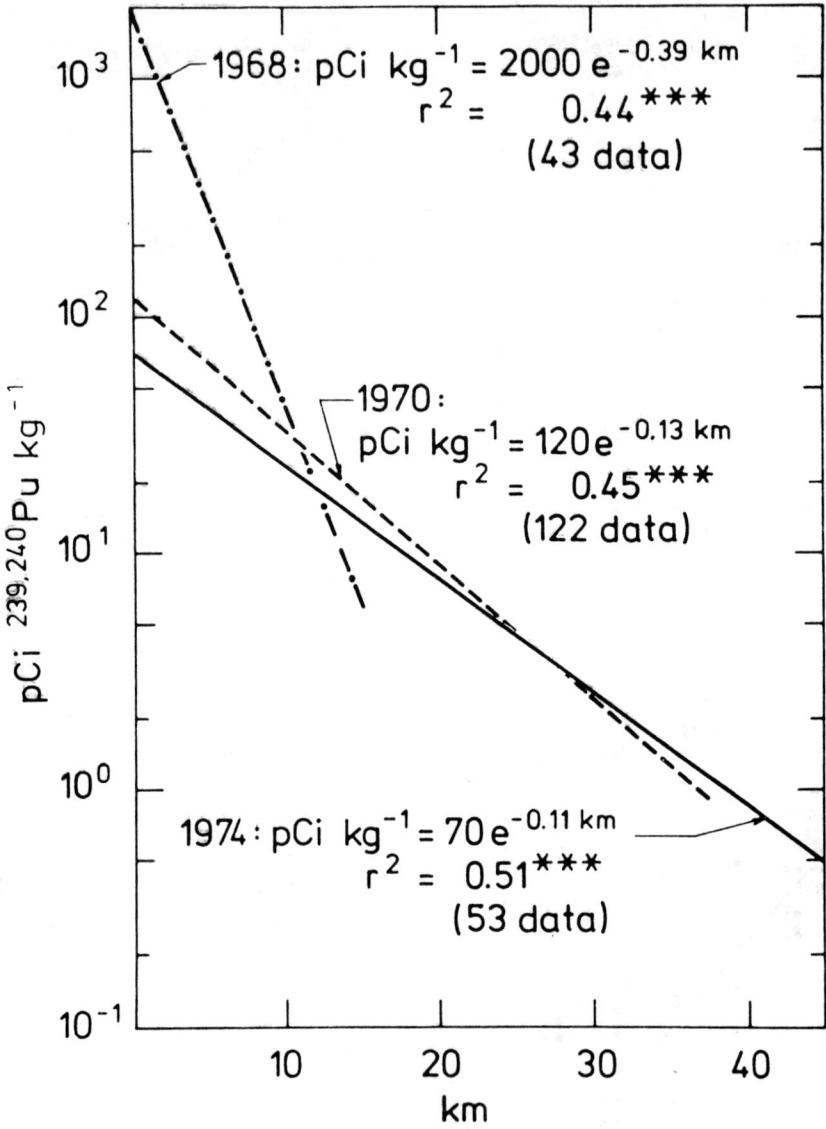

Fig. 4. The ^{239,240}Pu concentration in soft parts of bivalves collected in 1968, 1970 and 1974 as a function of distance from the point of impact; r^2 is the correlationscoefficient between observed data and the values calculated from the exponential equations (Significance level: * * *: 0.001 i.e. highly significant).

Table II. Summary of ^{239,240}Pu data from the sample collections in 1968, 1970, and 1974.

Sample	unit		1968 I 0 - 1 km	1968 II > 1 km	1970 I 0 - 1 km	1970 II > 1 km	1974 I 0 - 1 km	1974 II > 1 km
Seawater (surface: 0-100 m)	fCi l^{-1}	max. min. geometric mean	12 4 6 (4)	67 2 6 (8)	3 2 2 (2)	– 1 (1)	– 1 (1)	– – –
"PK" Sediments (0-1 cm)(ash)	pCi g^{-1}	max. min. geometric mean	130 7 23 (9)	16 0.1 1.0 (7)	86 6 13 (9)	4.7 0.4 1.2 (10)	50 8 17 (3)	78 0.3 5.0 (9)
Sea plants (wet weight)	pCi kg^{-1}	max. min. geometric mean	– – –	74 6 19 (7)	– – –	6 1 2.1 (7)	– – –	8 0.4 2.4 (3)
Worms (ash weight)	pCi g^{-1}	max. min. geometric mean	– 230 (1)	– – –	46 0.3 3.4 (6)	1.1 0.1 0.48 (4)	10.5 2.1 5.7 (4)	6.9 0.06 0.54 (15)
Bivalves (soft parts) (fresh weight)	pCi kg^{-1}	max. min. geometric mean	76000 320 4600 (10)	5400 5 83 (33)	73000 50 390 (15)	13000 1 23 (79)	1900 75 240 (12)	300 2 15 (30)
Brittlestars Seastars (fresh weight)	pCi kg^{-1}	max. min. geometric mean	1120 190 380 (4)	– – –	4400 10 140 (7)	1700 10 44 (7)	250 62 81 (4)	64 1 9 (14)
Shrimps (fresh weight)	pCi kg^{-1}	max. min. geometric mean	– – 41 (1)	12000 22 1130 (4)	76 16 35 (2)	170 1 16 (10)	160 33 72 (2)	64 1 9 (8)
Fish (fresh weight)	pCi kg^{-1}	max. min. geometric mean	– – –	470 1 40 (10)	– – –	400 0.2 4 (11)	– – –	10 < 0.1 1 (4)
Birds (fresh weight entrails)	pCi kg^{-1}	max. min. geometric mean	– – –	7 0.2 2.2 (5)	– – –	130 < 0.1 0.9 (9)	– – –	– – 0.3 (1)
Seal and Walrus (entrails, etc.) (fresh weight)	pCi kg^{-1}	max. min. geometric mean	– – –	4.4 0.1 1.0 (10)	– – –	3.9 < 0.1 0.5 (12)	– – –	5.5 < 0.1 0.5 (15)

(figures in brackets indicate the number of samples).

I: Zone I i.e. less than 1 km from the point of impact

II: Zone II i.e. outside zone I

DIFFUSION OF TRITIATED WATER IN COASTAL AREAS

M. Fukuda, A. Kasai, T. Imai, H. Amano and N. Yanase
Division of Environmental Safety Research
Tokai Research Establishment
Japan Atomic Energy Research Institute
Tokai-mura, Naka-gun, Ibaraki-ken, Japan

   The diffusion of tritiated water discharged by Japan Atomic Energy Research Institute at shore line has been investigated. In continuous discharge, the concentration of tritiated water in samples taken at a point downstream fluctuates largely. To reveal the cause, dye diffusion experiments were made in the coastal area.
   The shapes of dye cloud were photographed by a remote-control camera suspended from a captive balloon as color pictures.
   The movement of dye is so complex that a three-dimensional model must be employed to assess the diffusion in coastal areas.

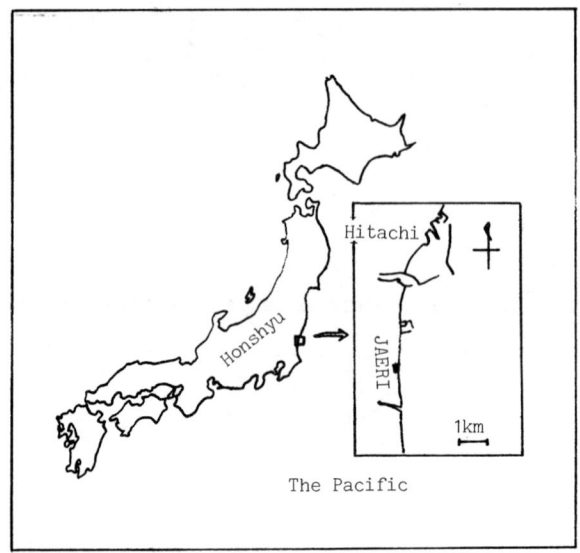

FIG.1. Site of experiment

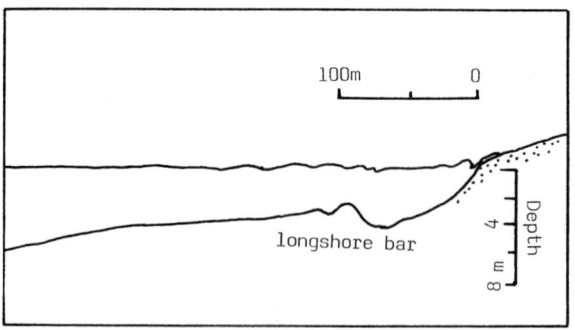

FIG. 2. Profile of sea bottom

- 254 -

1. Introduction

   In the Atomic Energy Research Institute, tritiated water has been discharged more than thirty times between 1972-1978 at shore line of the Pacific. JAERI is situated on a site facing the Pacific Ocean (Fig. 1). JAERI is site on sandy beach running roughly north to south. Profile of the sea bottom is shown in Fig. 2.

   Tritiated water stored in a pond, diluted to the permissible level with fresh water was discharged into the sea. The total volume of tritiated water to discharge was 5∿30 Ci in activity, and the discharge duration was 4∿5 hours. The discharging rate of fresh water for dilution was 0.2∿0.5 m³/sec. Along the beach sampling points were set at 200m intervals downstream, and seawater was sampled when the tritiated water concentration appears to be stationary. In order to measure variation of concentration, sampling was made every 30 minutes 400m downstream.

   Sampled seawater was mixed with emulsinized scintillater and the concentration of tritiated water was measured by a low background scintillation counter. Instrument LSC-LB1, manufactured by Aloka, measures 15 sample bottles of 100ml seawater at a time. The detection limit is down to 20 pCi/l in 500 minutes of measurement. The measured concentration of tritiated water was normalized by dividing with discharge rate and diffusion depth.

   Time variation of concentration and the relationship between the concentration and the distance from the outfall are shown in Figs. 3 and 4 respectively. The fluctuation of concentration with time is more than one figure. In Fig. 4, the concentration is seen to fluctuate largely with measuring day. Because of such large fluctuation, the environmental monitoring must be carried out very carefully. In order to reveal the cause for the fluctuation, dye diffusion experiments were made in the coastal area.

2. Dye diffusion experiment

   Powdered dye, i. e. Rhodamine B or Fluoresceine, was put in a paper case breaking in seawater and it was thrown in the shore. Dye tossed by waves dissolved immediately. Paper cases with dye were also shot to the distance by a line-throwing apparatus using powder.

   Most dye dissolved at sea surface, but part sank to the bottom dissolving in seawater. In this case, dye was distributed in a cylindrical form from sea surface to bottom.

   The shapes and movement of dye were photographed by a remote-control camera suspended from a captive balloon as color picture (Fig. 5). The camera can be raised till 1000m high above sea level, and can taken 30 photographs of 35mm half size each time. The area of sea is smaller than that of land in the picture in sea breeze, so that the dye cloud is often out of the picture. A weak point of this method is that there is a limitation in meteorological condition for the experiment.

3. Analysis of photographs

   Dye cloud in the photograph varies in shape according to sea condition and discharge method. Typical shapes of dye cloud were drawn, and are shown in Figs. 6∿11.

   In Fig. 6, the dye cloud distributes similar to that in a model of homogeneous flow. If dye is discharged continuously, the contour lines of concentration would be in spindle form [1]. A close examination of the figure indicates that the longshore current is dominant and the current decreases rapidly with distance from the shore.

   The shape of dye cloud in Fig. 7 exhibits typical rip current. This shape is often observed in dye experiments. However, because there is a longshore bar further offshore than the rip head, it is not certain that the rip current in Fig. 7 is that rip current illustrated by Shepard and Inman [2]. On edge of the rip head,

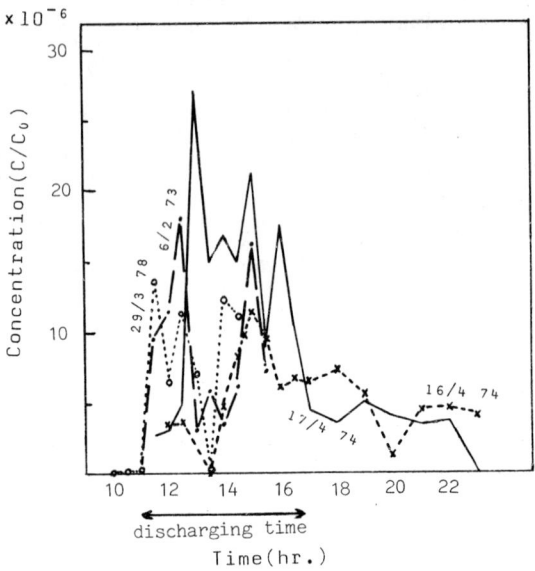

FIG. 3. Variation of concentration with time at 400m downstream

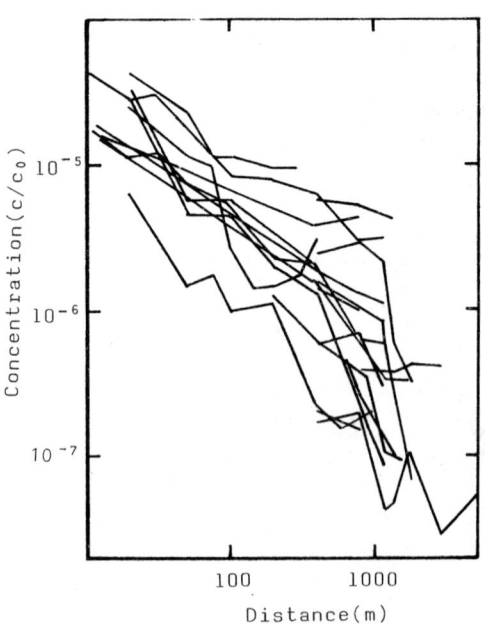

FIG.4. Concentration as a function of distance from out fall

white bubble were often observed. The rip head forms a water mass, which is hardly mixed with external water. Top of the head rotates and reaches the shore occasionally. In this case, the concentration at shore increases suddenly.

In Fig. 8, the direction of longshore current and of offshore current are opposite. Near the shore the horizontal current shear is dominant and the offshore head of dye cloud discharged at shoreline rotates to the shore, as in Fig. 7. As the vertical shear of horizontal current in the offing is small, the apparent horizontal diffusion coefficient is small.

In Fig. 9, the longshore current is small and the surface water in the offing flows off. Structure of this current may be explained by the theory of drift current in the coast.

In Fig. 10, bottom water in the offing flows toward the shore and becomes longshore current. It appears that the bottom current acts as a compensation current to longshore current. Surface water in the offing flows away from shore.

In Fig. 11, it is seen that the area of dye cloud is on decrease. In the beginning, dye diffuses normally. Two different water masses approach to each other, pinching the water with dye, and forms a discontinuous boundary, "Shiome" in Japanese. The dye cloud is drawn into the boundary, and the area of cloud in the picture seems to decrease. The diffusion coefficient becomes a negative value.

4. Discussion

4.1. Currents

In all photographs, we can see strong longshore current. To reveal the cause for this current, the correlation between wave direction and current direction and between wave height and current velocity were obtained. All waves having a north componport of wave progress made the north componport of longshore current, i.e. the longshore current is related with the wave progress. The relationship between wave height and velocity of longshore current is illustrated in Fig. 12. Since the relation between the longshore current and the wind direction is little, waves swashing on the beach may be the cause for the longshore current.

Since the relationship between offshore surface current and wind is considerable, except the case of Fig. 8, the wind is a major cause for offshore current.

The causes for currents in middle and bottom layer are unknown. When white bubble are produced on edge of the rip current, the rip head formes a water mass, and rotates or flows without mixing with external water. Structure of this water mass is not revealed yet. The movement of seawater near the shore is composed of the above discribed currents.

4.2. Diffusion

Usually the diffusion coefficient is used to assess the diffusion of waste water in coastal areas. This parameter is not suitable theoretically, but because of its convenience the parameter is widely used. The diffusion coefficient is estimated easily by measuring the area of dye cloud or the concentration of dye in diffusion experiment.

In Fig. 13, the variation in area of dye cloud measured in the photographs with time is shown. Only in the case on May 28th. 1979, the value of area decreases with time after 1000 seconds from start of the discharge.

Assuming a two-dimensional homogineous horizontal plane, by applying the Fickian equation to an instantaneouse point discharge, the area of dye cloud can be estimated as [3]:

$$S = 4\pi\sqrt{K_x K_y}\, t \ln\{(Q/C) \cdot (1/4\pi\sqrt{K_x K_y}\, t)\} \tag{1}$$

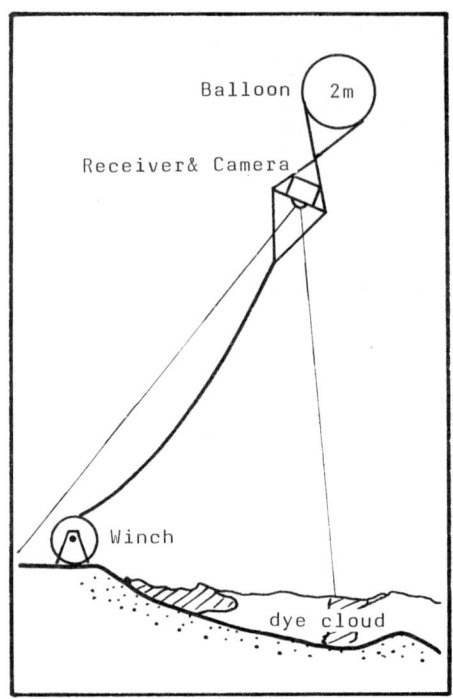

FIG.5. Schematic diagram of photographic system

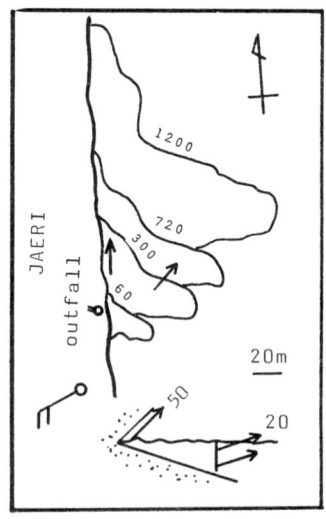

FIG. 6. Horizontal dye distribution traced out from color photograph, and current direction and velocity (9 Jun. 1978)

where $\overline{K}_x$ and $\overline{K}_y$ are the mean values of diffusion coefficient with time in the x and y directions respectively, t is time, Q is a discharging rate and C is the dye concentration of at edge of the cloud. Assuming $\overline{K} = \overline{K}_x = \overline{K}_y$, from Fig. 13 and equation (1), the mean diffusion coefficient is estimated. The relationship between the mean diffusion coefficient and the area of dye cloud is shown in Fig. 14.

In the case of a sea condition with the smallest vertical shear of current, such as in Fig. 8, the smallest horizontal mean diffusion coefficient is obtained. On the other hand, highest value in the case of the highest vertical shear of current, the largest horizontal diffusion coefficient is obtained, such as in Fig. 10, that is, the apparent horizontal mean diffusion coefficient depends strongly on the vertical shear of current [4].

If one wants to assess the shape of dye cloud and its movement with the horizontal diffusion coefficient, he must consider the three-dimensional model for estimation, and the current at each depth and the volume of dye discharged at each depth must be known. After the horizontal diffusion and the movement at each depth are estimated by substituting current velocity and the diffusion coefficient into the Fickian equation, all the distribution of dye are superimposed and the distributions of dye are vertically summed up. In assessing some diffusion problems by the method, good results were obtained in offshore dye experiments [5]. The problem in Fig. 11 may be analyzed by this method.

In the case of continuous discharge, the diffusion coefficient is estimated as follows.

Assuming the concentration is distributed by the error function in a orthogonal direction to the current, the concentration is

$$C = Q/(\overline{u}\sqrt{2\pi} \cdot \sigma) \cdot \exp(-y^2/2\sigma^2) \tag{2}$$

where $\overline{u}$ is the mean current and σ is the standard deviation.

The diffusion coefficient is defined as

$$K = \frac{1}{2}\frac{d\sigma^2}{dt} \tag{3}$$

Therefore,

$$\sigma^2 = 2\int_0^T K dt = 2\overline{K}T \tag{4}$$

where $\overline{K}$ is the mean diffusion coefficient with time from 0 to T seconds. From equation (2) and (4)

$$\overline{K} = \frac{1}{4\pi\overline{u}T}(\frac{Q}{C\overline{u}})^2 = \frac{1}{4\pi\overline{u}x}(\frac{Q}{C})^2 \tag{5}$$

With equation (5) and Fig. 4, the mean horizontal diffusion coefficient can be estimated. Consequently, the diffusion coefficient fluctuates largely. It is considered that the high concentration water approaching to the shore by horizontal shear of current is the cause for the fluctuation.

5. Conclusion

In the experiments, it has been revealed that the longshore current caused by waves and the surface offshore current caused mainly by winds were dominant currents in coastal areas, which made vertical and horizontal shears of current. To estimate the diffusion coefficient which is used to assess the distribution of liquid waste concentration in coastal areas, the current structure in the areas must be known, because the vertical and horizontal shears cause fluctuation of the horizontal diffusion coefficient and cause fluctuation of concentration.

The authors wish to thank Dr. H. Amano and Mr. K. Imai for their advices and Mr. S. Yamada for his assistance in the experiments.

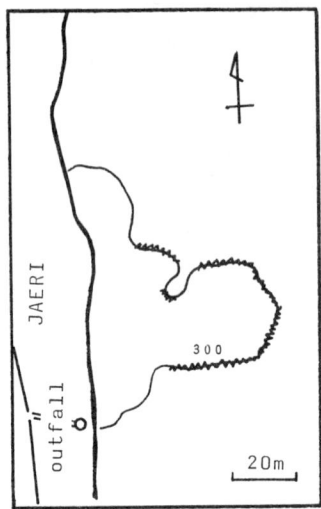

FIG. 7. Horizontal dye distribution traced out from color photograph (5 Jun. 1979) ~~~ : White bubble

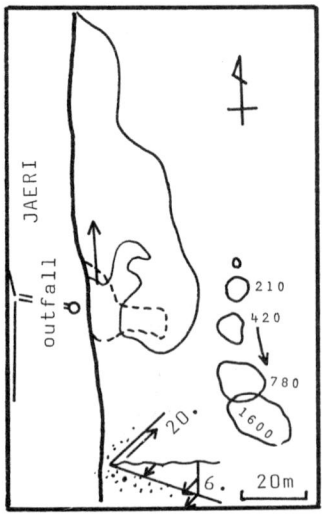

FIG. 8. Horizontal dye distribution traced out from color photograph, and current direction and velocity (22 May 1979)

REFERENCES

[1] OKUBO, A., KARWEIT, M., Diffusion from a continuous source in a uniform shear flow, Limnology and Oceanography, $\underline{14}$(1969), 514-520.
[2] SHEPARD, F. P. and INMAN, D.L., Nearshore circulation, Proc. First Conf. Coastal Eng., Council on Wave Research, Univ. Calif. (1951), 50-59.
[3] FUKUDA, M., ITOH, N. and SAKAGISHI, S., Diffusion phenomena in coastal areas, Proc. 2nd Int. Conf. on Water Pollution Res., (1964), 193-204.
[4] PRITCHARD, D. W., OKUBO, A. and CARTER, H. H., Observation and theory of eddy movement and diffusion of an introduced tracer material in the surface layer of the sea, Proc. Int. Symposium, Disposal of Radioactive Waste into Sea, Oceans and Surface Waters. IAEA, Vienna, (1966), 397-424.
[5] FUKUDA, M., Diffusion pattern in coastal areas, (unpublished).

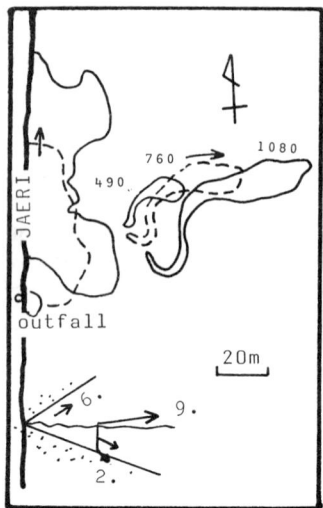

FIG. 9. Horizontal dye distribution traced out from color photograph, and current direction and velocity (23 Jun. 1979)

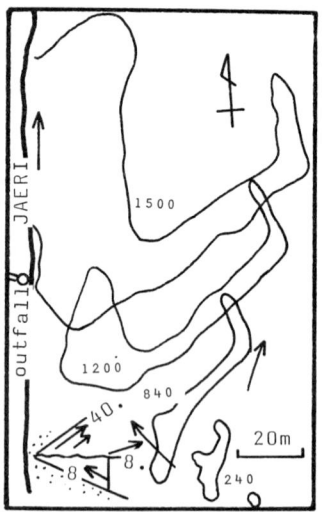

FIG. 10. Horizontal dye distribution traced out from color photograph, and current direction and velocity (9 Jul. 1979)

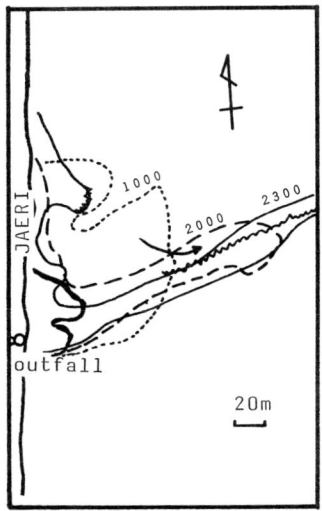

FIG. 11. Horizontal dye distribution traced out from color photograph (28 May 1978) ～～～ : White bubble

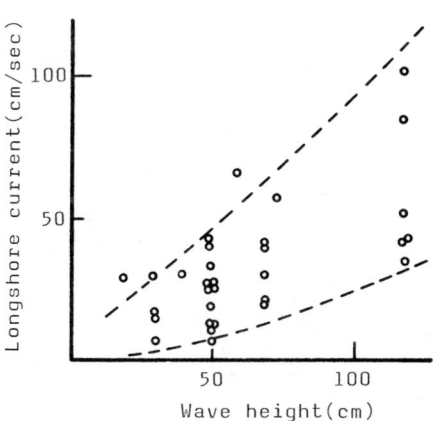

FIG.12. Relationship between longshore current and wave height

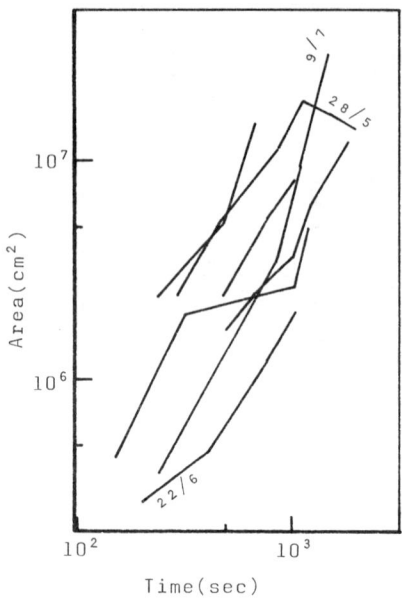

FIG.13. Area of dye cloud as a function of time

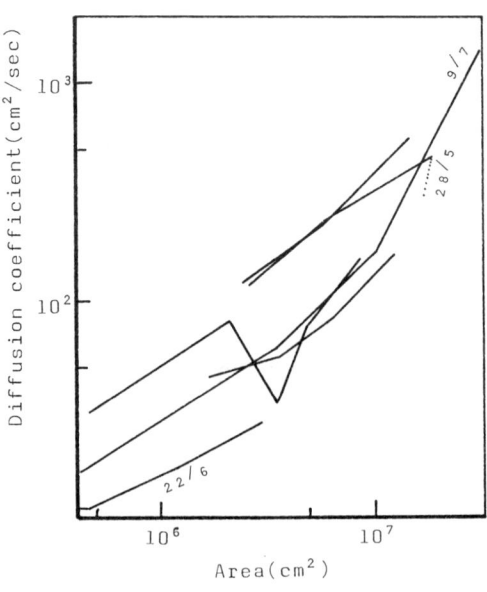

FIG.14. Apparent diffusion coefficient as a function of area

RADIOECOLOGY OF ^{60}Co IN URAZOKO BAY

T.Ueda, Y.Suzuki and R.Nakamura
Division of Marine Radioecology, National Institute of
Radiological Sciences, Nakaminato, Ibaraki, Japan

ABSTRACT

In order to know factors which influence Co-60 contamination of sargassoes in Urazoko Bay, where a nuclear power plant is located, correlations between the contaminations of the algae and of the marine environment were investigated, based on field data observed since January 1971.

It was found that correlation between the Co-60 content in sargassoes (Y_{algae} pCi/kg raw weight) and distance from discharge outlet of the nuclear power plant to the sampling points (d km) was expressed by a function, $Y_{algae} = C \cdot \exp(-kd)$. Where C and k are constants, and k fluctuated from 0.388 in November 1971 to 1.200 in November 1974, whereas C decreased from 120 to 27 during the same period.

Based on the results of the investigation and the laboratory tracer experiments, the destiny of the radionuclides released into coastal sea was discussed.

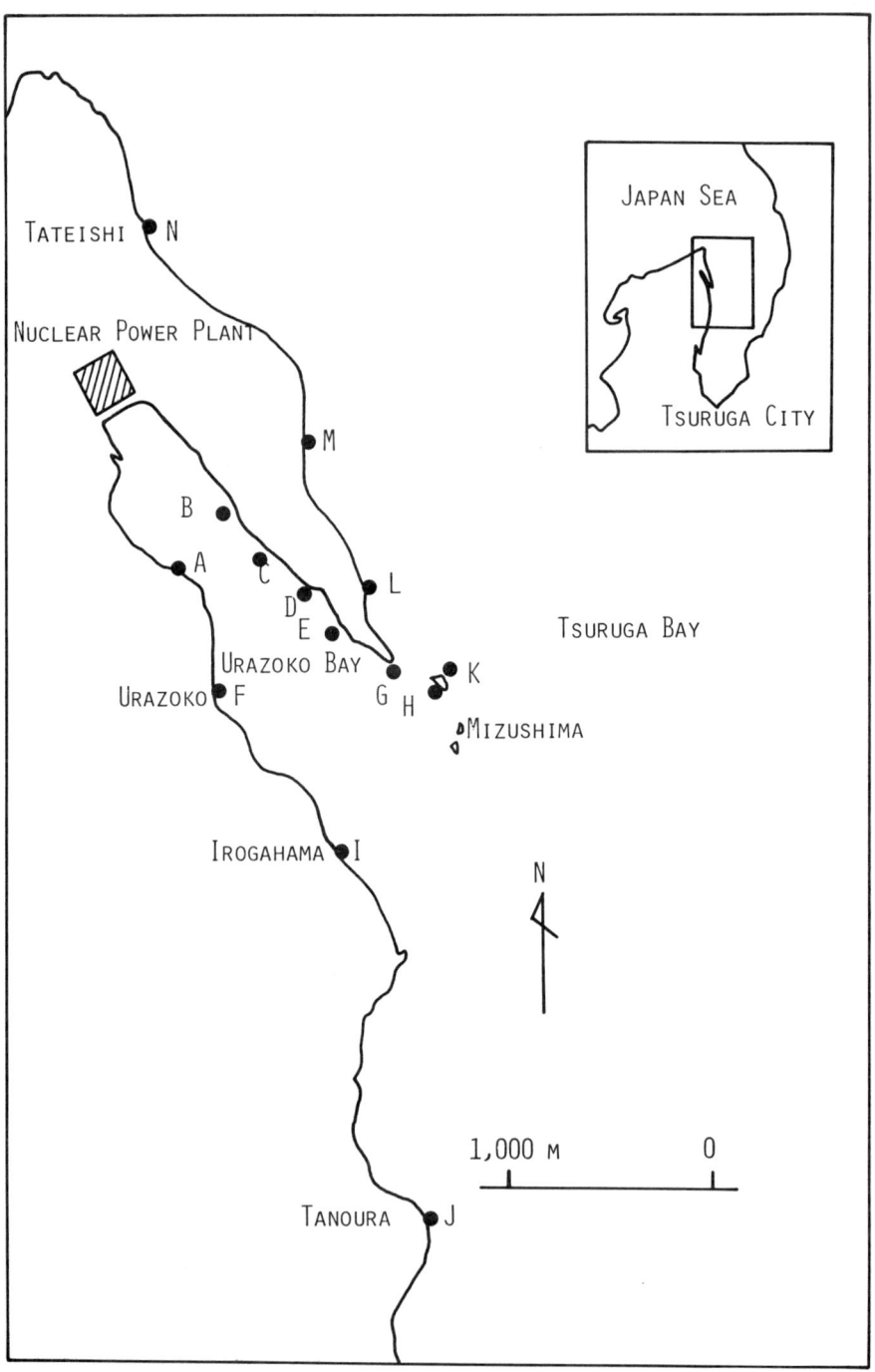

Fig. 1 Location of sampling points for sargassoes

INTRODUCTION

Tsuruga Nuclear Power Plant(BWR-type, 357 MWe) at Fukui Prefecture is a contributor of artificial radionuclides to Urazoko Bay, located along the west coast of Tsuruga Bay. The production of artificial radionuclides by this power plant makes it possible to use Urazoko Bay and also a portion of Tsuruga Bay as a field for the radioecological study of closed coastal marine environment. The tidal range in Urazoko Bay is small as elsewhere on the Japan Sea coast. Fishery production in Urazoko Bay are mainly jack mackerel, flathead and sand smelt as fish, octopus and mussel as molluscus but these catches are extremely small. A sargasso(Sargassum fluvellum) is one of the most abundant seaweeds in Urazoko Bay, grows in 15-20cm in length about the middle of November and withers to flow out in the summer of the next year. In this article, correlation between the concentration of Co-60 in this alga, a non-edible seaweed but a good indicator of radioactive contamination in Urazoko Bay, and the environment, which was examined in the view of distance from discharge outlet of the power plant to sampling points. Furthermore, our concept concerning the destiny of the radionuclide released into coastal sea was discussed including the results of the investigation.

MATERIALS AND METHODS

During January 1971 to May 1975, the samples of sargasso (Sargassum fluvellum) collected for radiochemical analysis were washed with fresh water, weighed after blotting off the water, and then ashed at 450°C for 48 hours in an electric muffle furnace after drying in an electric oven at 110°C for 24 hours. Ashed samples were sifted out by a chemical sieve(0.5mmϕ) to remove shells and sand, and then weighed. The ratio of ashed weight to raw weight of sargassoes used for this investigation was in the range of 3.8 and 7.2%.

The method of determination of Co-60 was essentially the same as that reported by Tsuruga[1], that is, the beta activity of Co-60 was finally measured in the form of cobalt anthranilate using a low-background gas flow counter(Tracer lab., Omni/Guard). The chemical yield of this procedure was in the range of 76-83%, and the counting efficiency was 15-19%.

RESULTS

Location of the sampling points is shown in Fig. 1. The points A-I are in Urazoko Bay and the points J-N in Tsuruga Bay. Concentrations of Co-60 in sargassoes during the period from January 1971 to May 1975 are summarized in Table I expressed in unit of pCi per kilogram of the raw weight of sargassoes. Detection limit for Co-60 is approximately 3.0 pCi/kg raw weight. The maximum concentration observed is 178.6 pCi/kg at point F in January 1971. No seasonal variation of the concentrations could be observed. When samples were obtained at three or more points in Urazoko Bay at the same time, it could be seen from Fig. 2 calculated by the least-squares method that correlation between concentration of Co-60 in sargassoes (Y_{algae} pCi/kg raw weight) and distance from discharge outlet of the plant to sampling point(d km) was expressed by an exponential function, $Y_{algae} = C \cdot \exp(-kd)$, where C and k are constants, and k was 0.388 for November 1971, 0.770 for August 1973, 0.700 for March 1974, 0.869 for June 1974 and 1.200 for November 1974, respectively, and the average of them was 0.787. This value was similar to the value (0.74) in the sediment in Urazoko Bay which was reported by Nakamura and Nagaya[2] of our stuffs. It could be said that the concentration of Co-60 in sargassoes in Urazoko Bay is reduced to one-half at the distance of 1 km from the discharge outlet. The numerical value of C varies mainly with the concentration factor of marine organisms and the amount of discharged Co-60 from the nuclear power plant, on the other hand, the value of k is influenced by the topography of the bay(the current of sea water), the character of marine

Table I The concentration of ^{60}Co for sargassoes in Urazoko Bay

St.	Dist.[a]	'71-Jan.	'71-Jul.	'71-Nov.	'72-May	'72-Aug.	'73-Aug.	'74-Mar.	'74-Jun.	'74-Nov.	'75-May
A	0.5									27.1	
B	0.5			106.9			54.1	35.5	21.5	12.6	8.3
C	0.7							29.6	22.0	15.2	
D	1.0					12.0		19.5	13.3	6.8	
E	1.3				13.1			29.8	8.4	7.7	
F	1.6	178.6	60.1	44.9	12.8	9.9	19.2	25.7	7.7	3.4	7.6
G	1.8									*	
H	1.9			75.9				8.3	6.5	*	
I	2.0	156.3	33.2		*[b]	*	20.0	*	6.9	*	
J	3.3				3.9		17.6		*	3.4	
K	3.4										*
L	3.9				6.4		38.2	9.6	5.3	3.8	
M	4.6								*	4.8	*
N	6.0				*		6.6	*	*	*	

a Distance from the discharge outlet (km).
b Asterisk(*) denotes the value less than detection limit.

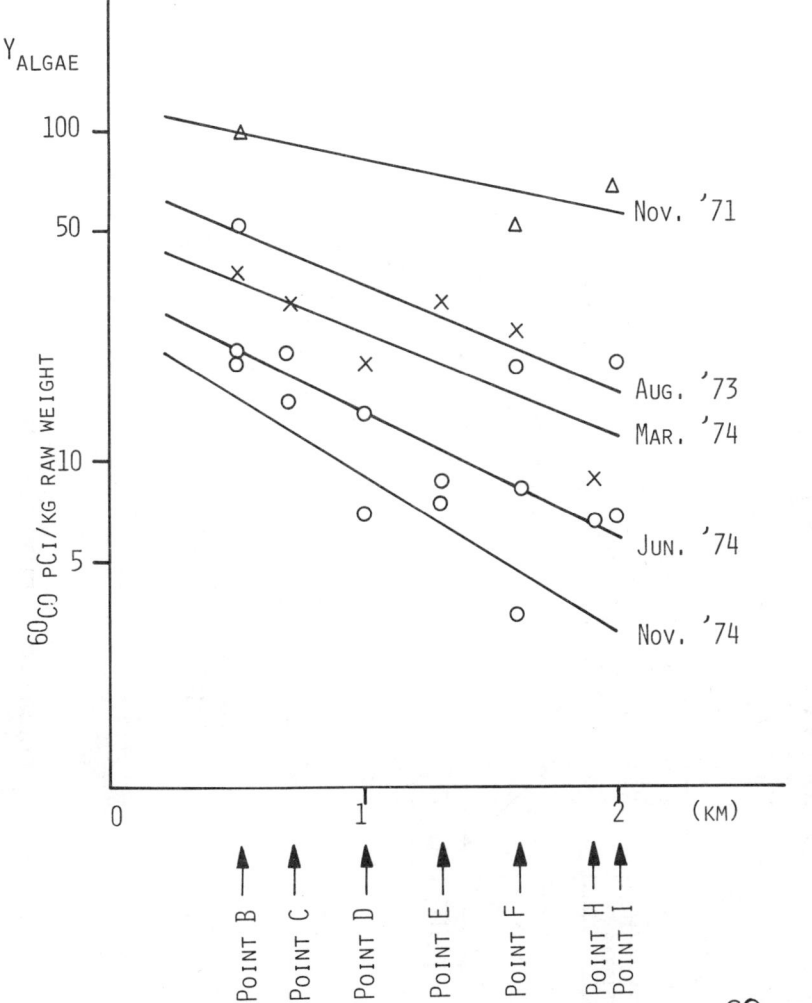

Fig. 2 Correlation between concentration of ^{60}Co in sargassoes and distance from the discharge outlet of the plant to sampling points.

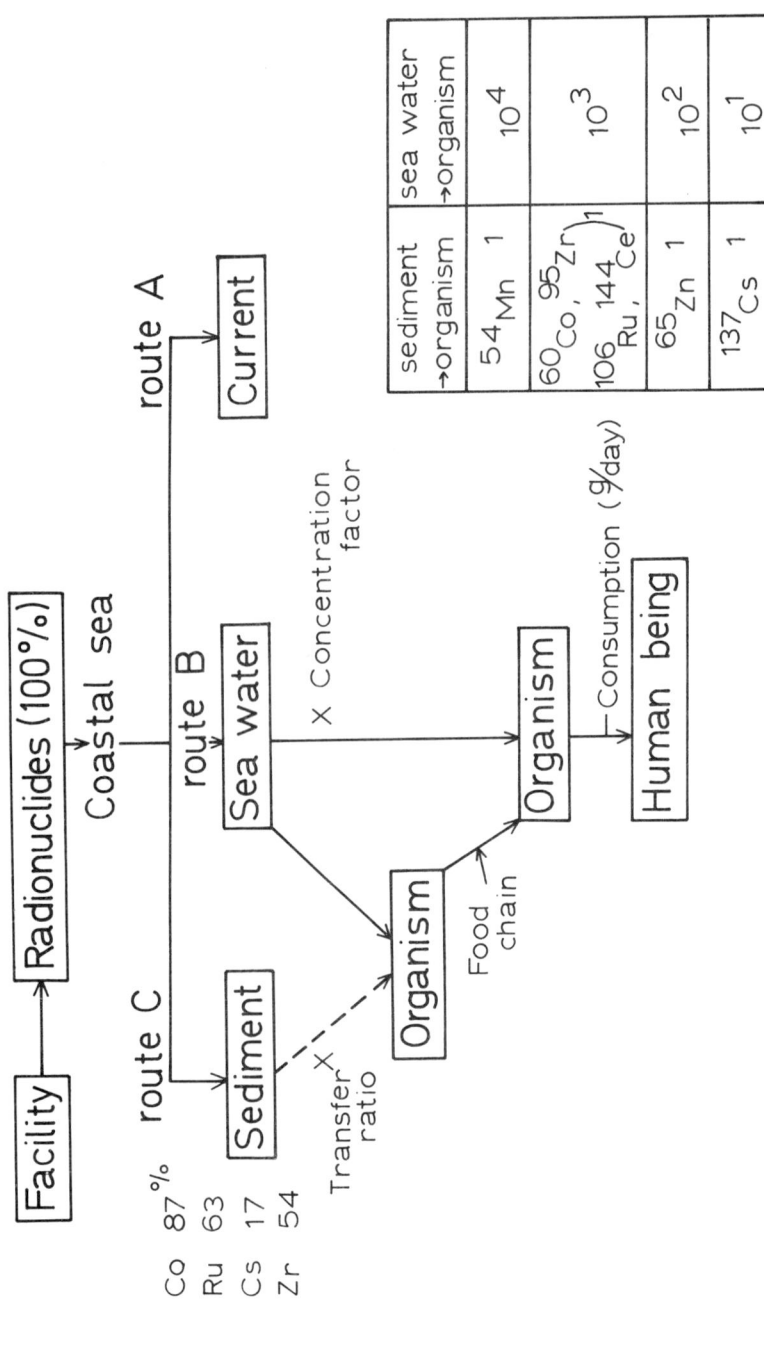

Fig. 3 Destinies of radionuclide released into coastal sea

sediment and so on.

As it can be generally considered that the concentration of Co-60 in sargassoes is dominated by the Co-60 concentration in the environmental sea water, it would be reasonably expected that there should be a correlation between the amount of discharged Co-60 and the concentration of Co-60 in sargassoes. In Urazoko Bay, however no correlation was observed quantitatively. Therefore, it may be a main reason for this anomaly that heavily contaminated marine sediments influence the concentration of Co-60 in sargassoes. Seymour and Nelson[3] pointed out the effect of Zn-65 recycled from sediments and biota upon mussels. Thus, indeed, the radioactive contamination of the sediment could not be neglected on the radioactive contamination of the coastal marine organisms.

DISCUSSION

On the destinies of the radionuclides released into the coastal sea, we have considered as shown in Fig.3 based on the laboratory tracer experiments [4,5,6,7,8,9].

    A) a portion is carried away by current.

        The amount of released radionuclide which takes route A) considerably differs depending on whether the topography of the coast is of the closed type or the open type facing the ocean.

    B) a portion remains in the coastal sea water and

    C) the balance is adsorbed on the sediment.

It is well known that in the B) and C) routes, large amounts of radionuclides are held in the sediment [10,11,12,13,14]. By our study[4,5], the amount of radionuclide adsorbed onto the sediments is 87 for Co-60, 63 for Ru-106, 17 for Cs-137 and 54% for Zr-95, respectively, to the released radionuclide(100%). The radionuclide recycled from the contaminated sediments effects on the contamination of the coastal marine organisms. In order to estimate quantitatively the effect of the contaminated sediments, we examined the transfer of the radionuclide from the contaminated sediments to the organism and expressed the result as the transfer ratio(cpm in g of organism/cpm in g of sediment). Then the transfer ratio was compared with the concentration factor and the ratio(concentration factor/transfer ratio) was described as Biological Factor of the Sediments(BFS). The values for each nuclide are shown in Fig.3. Thus, when we consider the routes by which radionuclides released into sea water reaches the human beings via marine organisms, the main route is B) and a secondary route is C). The sum of the amounts which take the routes B) and C) is varied by the amount of the radionuclide which takes the route A). As mentioned above, the partition of the radionuclide to the route C) considerably differs depending on the topography of the coast. As seen in Fig.4.(B), in an extremely closed type bay which has no outlet to ocean, all of the released radionuclide remains in the bay and would not be decreased from 100% without physical decay of the radionuclide. This can be represented by the line (closed type) parallel to the X-axis meaning time. On the contrary, in the open type coast the released radionuclide goes down to 0 at once owing to dispersion by current. In a closed coastal sea such as type I shown in Fig.4.(B), the released radionuclide decreases along the line (type I) shown in Fig.4.(A). In a more open sea(type II in Fig. 4.(B)), the radionuclide decreases faster than in type I, along the line(type II) in Fig.4.(A). Therefore, the pattern of the contamination of the coastal sea by radionuclide and/or heavy metals can be considered to be restricted to the range(right angle) of combinations possible between the line(closed type) and the Y-axis as seen in Fig. 4.(A), although it may become complicated depending on the various topographies of the coastal sea.

Naturally, the character of the marine sediments also effects on the amount of radionuclide remaining in the coastal sea, and the contamination of the coastal sea could not be classified from the view point of the topography alone. Therefore, further study should be continued on whether the X-axis can represent distance as well as time, or not, and other factors including character of the sediments.

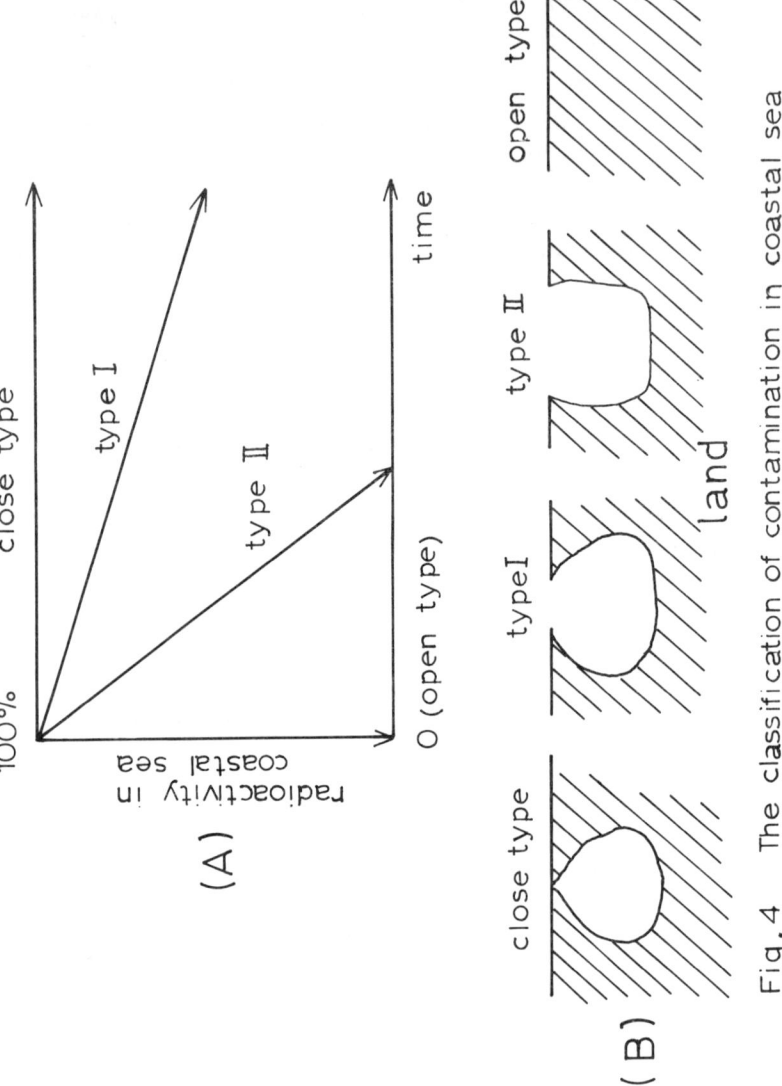

Fig. 4 The classification of contamination in coastal sea

However, as preliminary approach to the classification of the contamination of the coast, we consider as mentioned above.

For details of the field study of Urazoko Bay our previous paper [15] should be referred to.

REFERENCES

[1] Tsuruga,H.:"Sequential Analysis of Radionuclides in Marine Organisms", Bull.Japan.Soc.Sci.Fish., 31, 651-658(1965).
[2] Nakamura,K. and Nagaya,Y.:"Dispersion and Accumulation of Radionuclides in Sediment of Urazoko Bay (1)", J.Oceanogr.Soc.Japan 31, 145-153(1975).
[3] Seymour,A.H. and Nelson,V.A.:"Decline of Zn-65 in Marine Mussels Following the Shutdown of Hanford Reactor", Proc.Symp.on Radioactive Contamination of the Marine Environment, pp.277-286,IAEA, Vienna(1973).
[4] Nakamura,R.,Suzuki,Y. and Ueda,T.:"Influence of Marine Sediment on the Accumulation of Radionucldes by Green Alga(Ulva pertusa)", J.Radiat.Res. 16, 224-236(1975).
[5] Nakamura,R.,Suzuki,Y. and Ueda,T.:"Distribution of Radionuclides among Green Alga, Marine Sediments and Sea Water", ibid, 18, 322-330(1977).
[6] Ueda,T.,Nakamura,R. and Suzuki,Y.:"Comparison of Cd-115m Accumulation from Sediments and Sea Water by Polychaete Worms", Bull. Japan.Soc.Sci.Fish. 42, 299-306(1976).
[7] Ueda,T.,Nakamura,R. and Suzuki,Y.:"Comparison of Influences of Sediments and Sea Water on Accumulation of Radionuclides by Worms", J.Radiat.Res. 18, 84-92(1977).
[8] Ueda,T.,Nakamura,R. and Suzuki,Y.:"Comparison of Influences of Sediments and Sea Water on Accumulation of Radionuclides by Marine Organisms", ibid, 19, 93-99(1978).
[9] Ueda,T.,Nakamura,R. and Suzuki,Y.:"Position of Sediments in Transfer of Radionuclides Released into Coastal Sea to Human Beeings",(Submitted).
[10] Duke,T.W.,Willis,J.N. and Price,T.J.:"Cycling of Trace Elements in the Estuarine Environment", Chesapeake Sci., 7, 1-10(1966).
[11] Murray,C.N. and Murray,L.:"Adsorption-desorption Equilibria of Some Radionuclides in Sediment-Freshwater and Sediment-Seawater Systems", Proc.Symp.on the Interaction of Radioactive Contaminants with the Constituents of the Marine Environment, pp.105-124,IAEA,Vienna(1973).
[12] Phelps,D.K.:"Partitioning of the Stable Elements Fe,Zn,Sc and Sm within a Benthic Community,Anasco Bay, Puerto Rico", Radioecological Contamination Processes, pp.721-734(1966).
[13] Wolfe,D.A.,Cross,F.A. and Jennings,C.D.:"The Flux of Mn,Fe andZn in an Estuarine Ecosystem", Proc.Symp.on the Interaction of Radioactive Contaminants with the Constituents of the Marine Environment,pp.159-175, IAEA,Vienna(1973).
[14] Nelson,J.L.,Perkins,R.W.,Nielsen,J.N. and Haushild,W.L.:"Reactions of Radionuclides from the Hanford Reactors with Columbia River Sediments", Proc.Symp.on Disposal of Radioactive Wastes into Seas, Oceans and Surface Waters,pp.139-161,IAEA,Vienna (1966).
[15] Suzuki,Y.,Nakamura,R. and Ueda,T.:"Radioecology of Co-60 in Urazoko Bay;Correlation between Levels of Co-60 in Sargassoes and Marine Sediments", J.Radiat.Res., 17,115-126(1976).

SOME ASPECTS OF THE POSSIBLE FORMATION OF THE METAL ORGANIC
COMPLEXES IN A SEAWATER BASED ON THE LABORATORY WORKS

Y. Honda and Y. Kimura
Faculty of Science and Technology, Kinki University
Higashi-Osaka, Osaka 577, Japan
T. Ishiyama and T.Matsumura
Radiation Center of Osaka Prefecture
Shinke-cho, Sakai, Osaka 593, Japan

ABSTRACT

The possible formation of cabalt complexes with some amino acids were demonstrated in artificial seawater and their physico-chemical characteristics were determined by means of adsorption on chelating resin, solvent extraction with dithizone, gel filtration or paper electrophresis. Additionally a new mono-glycinato complex of nitrosylruthenium which was formed by the reaction of trichloro-nitrosylruthenium with glycine was also studied by isotachophoresis. The experimental results suggest that the complex formation of Co ions or the nitrosylruthenium complex with amino acids proceeds slowly, unless the reactants are highly concentrated.

Table I Interfering effects of amino acids on the adsorbtion and extraction of radiocobalt in seawater
(After 36days ageing)

Amino acid	Adsorption on Chelex 100 (%)		Extraction with dithizone (%)	
	Retained	Not retained	Extracted	Not extracted
0 (control)	97.2	2.8	97.5	2.5
Glycine $4.48 \times 10^4 M$	85.5	14.5	51.4	48.6
Alanine $5.61 \times 10^4 M$	54.1	45.9	55.7	44.3
Aspartic acid $3.76 \times 10^4 M$	72.0	28.0	74.9	25.1

Table II Fractionation of radiocobalt in seawater by Sephadex G-10 gel chromatography

Column size : 15Φ×830mm
Flow rate : 20ml/hr

Control			Glycine $4.48 \times 10^4 M$			Alanine $5.61 \times 10^4 M$			Aspartic acid $3.76 \times 10^4 M$		
Ageing time (days)	Kd	Recovery (%)	Ageing time (days)	Kd	Recovery (%)	Ageing time (days)	Kd	Recovery (%)	Ageing time (days)	Kd	Recovery (%)
2	1.0	93.8	1	0.3 / 1.1	0.9 / 84.5	1	1.0	86.9	2	0.4 / 1.0	1.9 / 88.6
4	1.0	98.0	3	0.3 / 1.0	4.2 / 84.2	3	1.0	89.6	4	0.3 / 1.0	5.2 / 86.2
8	1.0	96.7	7	0.2 / 1.0	8.9 / 73.7	7	0.3 / 1.0	10.5 / 72.8	8	0.3 / 1.0	6.8 / 78.3
11	1.1	99.3	10	0.2 / 1.0	15.3 / 55.7	10	0.3 / 1.0	14.0 / 63.0	11	0.3 / 1.1	8.6 / 60.0
18	1.1	93.2	17	0.3 / 1.2	19.9 / 50.5	17	0.3 / 1.0	13.1 / 75.5	18	0.3 / 1.2	10.4 / 65.9
25	1.0	96.5	24	0.2 / 1.2	23.6 / 42.2	24	0.3 / 1.0	22.1 / 58.8	25	0.3 / 1.2	11.0 / 61.8
32	1.0	94.7	31	0.2 / 1.2	27.4 / 40.9	38	0.3 / 1.0	29.0 / 48.0	32	0.3 / 1.1	12.6 / 62.9
39	1.2	92.1	38	0.2 / 1.2	29.2 / 50.6	43	0.3 / 1.0	24.2 / 60.4	39	0.3 / 1.2	15.1 / 57.3

Standard deviation of Kd value : ±0.1

Standard deviation of Recovery : <10%

1. INTRODUCTION

It is very important to elucidate the physico-chemical states of the radioisotopes released from nuclear fuel reprocessing or reactor operation into coastal waters for the assessment of the environmental impacts. Among the radioisotopes, strontium and cesium are of generally simple physico-chemical state[1], while the behaviour of transition metals such as cobalt and ruthenium are extremely complicated, since these elements would be able to form the wide varieties of complexed species, depending on the organic matters dissolved in natural waters[2],[3],[4]. However, the occurrences of cobalt or ruthenium complexes with amino acids in natural seawater have not yet been clearly identified due to the lack of suitable preconcentration methods for these complexes. So, the authors demonstrated the possible formation and some physico-chemical characteristics of cobalt complexes with some amino acids such as glycine, alanine or aspartic acid dissolved in an artificial seawater by means of adsoption on chelating resin, solvent extration with dithizone, gel filtration chromatography or paper electrophoresis[5],[6]. Also, the behaviour of a new mono-glycinato complex of nitrosylruthenium which was formed by the reaction of trichloronitrosylruthenium with glycine was studied by isotacophoresis [7].

2. MATERIALS AND METHODS

The experiments on the formation of cobalt complexes with the amino acids such as glycine, alanine or aspartic acid dissolved in an artificial seawater and their physico-chemical characteristics were conducted according to the descriptions by Kimura, Y. et al. [5] and Honda, Y. et al. [6]. The nitrosylruthenium complexes, trichloronitrosylruthenium complex, $RuCl_3(H_2O)_2NO$ and mono-glycinato complex of nitrosylruthenium, $K[Ru(gly)(OH)_3NO]$ were prepared by the methods of Fletcher, J.M. et al.[8] and Ishiyama, T. et al.[7], respectively. The composition of the prepared nitrosylruthenium complexes were ascertained from elemental analyses. The isotachophoresis of the nitrosylruthenium complexes were carried out by the methods of Ishiyama, T. et al.[7] using Shimadzu Model IP-1B capillary type isotachophretic analyzer.

3. RESULTS AND DISCUSSION

The distribution coefficient of cobalt on the resin of Chelex-100 gradually decreased with time even in the absence of organic matter dissolved in the seawater. However, more decreases were observed in the presence of amino acids such as glycine, alanine or aspartic acid in the seawater. Among these three amino acids, glycine showed less interfering effect on the adsorption. The interfering effects of amino acids on the adsorption and extrction of radiocobalt in a seawater are shown in Table I. The Chelex-100 has an unusual selectivity for copper, cobalt and other heavy metals ever in the presence of sodium, potassium, magnesium and calcium which are the major cations in seawater[9]. Cobalt ions could be converted to chelate compounds in contact with the Chelex-100 resin. Lowman, et al.[10] reported that the chelating resin, such as Chelex-100 retained only 0.6% of the organically bound cobalt such as cyanocobalamin, while retained over 90% of the ionic forms in seawater. Callahan, et al.[11] also suggested that the cobalt(III) is tied up as a strong complex and is not available for exchange by the chelating resin. However, Loewenschuss, et al.[12] demonstrated that the chelating resin Dowex A-1 which has an iminodiacetic acid similar to Chelex-100 as a chelating group could form the mixed metal-ligand-resin complexes as well as metal-resin complexes. Thus less interfering effect of glycine on the adsorption of cobalt on the resin might be attributed to the possibility of the formation of cobalt-glycine-resin complexes. Furthermore, taking account of the stability constants of cobalt (II) complexes with amino acids and dithi-

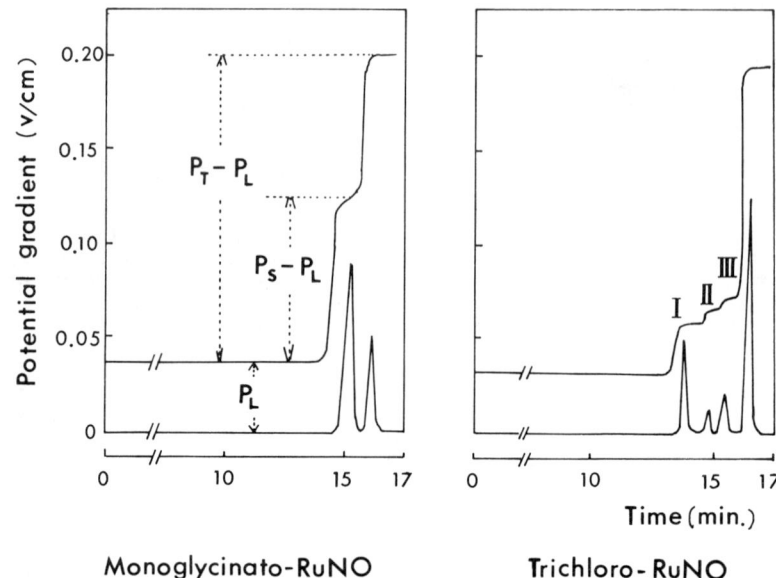

FIGURE 1

ISOTACHOPHORESIS OF MONOGLYCINATO- AND TRICHLORO-NITROSYLRUTENIUM COMPLEXES

zone(log K_1,K_2:aspartic acid $>$ glycine $\geq$ alanine $>$ dithizone)[13], the cobalt complexes with the amino acids are more stable than cobalt dithizone complex. Therefore, the cobalt complexes with the amino acids would not be extracted with dithizone in carbon tetrachloride. From the results of gel filtration chromatography of cobalt in the seawater, the organic cobalt supposed to be associated with the amino acids was separated from the inorganic cobalt according to their molecular size under similar conditions to those of the complex formation in the seawater. As is shown in Table II, the higher molecular cobalt-binding species increased together with decreasing in the lower molecular species with ageing time of cobalt in the presence of the amino acids in the seawater. The results of paper electrophoresis of cobalt in the seawater containing glycine also showed that the dominant cationic species with the apparent mobility of 4.82±0.95 decreased together with increasing the electroniutral and anionic species which had the apparent electrophoretical mobilitris of 0±0.5 and 1.73±0.01 respectively. It was found from the electrophreograms of nitrosylruthenium complexes shown in Figure 1 that the mono-glycinato complex behaved like an anionic species in electrophretic migration, while the trichloronitrosylruthenium complex behaved like three anionic species (I,II and III) which were assumed to be hydrolysis products in a weak alkaline solution at pH 8.0. A potential gradient inversely proportional to the mobility was obtained, the potential unit value(PU value) of the sample being calculated from the relative ratio of the potential gradient of leading electrolyte and the sample zone[14]. The PU value of 0.531 for the mono-glycinato nitrosylruthenium complex was obtained, while the PU values of 0.148, 0.180 and 0.211 for the hydrolysis products I, II and III of trichloronitrosylruthenium complex, respectively. The mobility of the mono-glycinato complex anion was smaller than those of the hydrolysis products of trichloronitrosylruthenium complex, indicating that the negative charge decreases by the interaction of trichloronitrosylruthenium complex with glycine. However, no obvious evidence was not obtained for the formation of the mono-glycinato complex of nitrosylruthenium under similar conditions to those for cobalt-glycinato complexes. All these experimental results might suggest that the complex formation of ionic cobalt or nitrosylruthenium with the amino acids dissolved in seawater would proceed slowly, unless heterogenous surfaces such as inorganic particles or detritus where the reactants are highly concentrated, are involved.

REFERENCES

[1] Robertson, D.E.: "Influence of the Physico-Chemical Forms of Radionuclides and Stable Trace Elements in Seawater in Relation to Uptake by the Marine Biosphere", MARINE RADIOECOLOGY, Proc. the Second ENEA Seminar, Hamburg, pp.21-76, OECD, Paris 1971.
[2] Siegel, A.: "Metal-Organic Interactions in the Marine Environment", in Organic Compounds in Aquatic Environments, Faust, S.D. and Hunter, J.V. eds., pp.265-295, Marcel Dekker, Inc., New York 1971.
[3] Stumm, W. et al.: "Aquatic Chemistry, An Introduction Emphasizing Chemical Equilibria in Natural Waters", Chapter 6, pp.238-299, John Wiley & Sons, Inc., New York 1970.
[4] Belot, Y. et al.: "Comportement du Nitrosyl-Ruthenium dans Une Eau de Surface Simulee en Presence de Cheletants Naturels", Health Physics 15(5), 443-450 (1968).
[5] Kimura, Y. et al.: "Interactions of Radionuclides with Organic Matter Dissolved in Seawater, ---Interaction of Radiocobalt with Some Amino Acids---", RADIOISOTOPES 25(9), 527-533 (1976).
[6] Honda, Y. et al.: "Interactions of Radionuclides with Organic Matter Dissolved in Seawater, ---Studies of the Interaction of Cobalt with Glycine by Paper Electrophoresis---", in Abstracts of 21st Meeting on Radiochemistry (in Japanese), 140-141(1977).

[7] Ishiyama, T. et al.: "Preparation of Potassium Glycinatotri-hydroxonitrosylruthenate(II)", Bulletin of the Chemical Society of Japan, $\underline{52}$(2), 619-620(1979).

[8] Fletcher, J.M. et al.: "Nitrato and Nitro Complexes of Nitrosylruthenium", Journal of Inorganic Nuclear Chemistry $\underline{1}$, 378-401 (1955).

[9] Samuelson, O.: Cited from Lai, M.G. et al. "A Rapid Ion Exchange Method for the Concentration of Cobalt from Seawater", USNRDL-TR-67-11, U.S. Naval Radiological Defense Laboratory San Francisco 1966.

[10] Lowman, F.G. et al.: "The State of Cobalt in Seawater and its Uptake by Marine Organisms and Sediments", IAEA-SM-158/23, Radioactive Contamination of the Marine Environment. Proc. Symp., Seattle, 1972, pp.369-384, IAEA Vienna 1973.

[11] Callahan, C.M. et al.: "The Concentration of Trace Elements from Sea Water by Ion Exchange", USNRDL-TR-67-10, U.S. Naval Radiological Defence Laboratory, San Francisco 1966.

[12] Loewenschuss, H. et al.: "Chelating Properties of the Chelating Ion Exchanger Dowex A-1", Talanta $\underline{11}$, 1399-1408(1964).

[13] Martell, A.E.: "Stability Constants of Metal-Ion Complexes", Part II, Special Publication No.25, The Chemical Society, Burlington House, London 1971.

[14] Akiyama, J. et al.: "A Guide to the Capillary Type Isotachophoresis", (in Japanese) Shimadzu Review $\underline{34}$(1), 111-120(1977).

Session 6

Chairman - Président
M. J. ANCELLIN
(France)

Séance 6

THE NORTH SEA REGION TAKEN AS AN EXAMPLE FOR THE BEHAVIOUR OF ARTIFICIAL RADIOISOTOPES IN NEARSHORE SEA AREAS

H. Kautsky
Deutsches Hydrographisches Institut
Hamburg (Federal Republic of Germany)

Abstract

Different radioisotopes entering the sea with the waste waters from nuclear fuel reprocessing plants are used as tracers to observe transport processes in the sea.
It could be stated that primarily the transport of water masses occurs within narrow regions along the coast. Virtually the whole water mass leaving the Irish Sea to the North flows along the Scottish coast into the North Sea and crosses her central region to the East. The water entering the southern North Sea through the English Channel analogously moves along the coast to the North. The two water masses meet in the region of the Skagerrak. One can follow them in a narrow strip along the Norwegian coast up to the Barents Sea.

Within the framework of the Deutsches Hydrographisches Institut's (DHI) legal duties for the protection of the environment, the monitoring of the seas for radioactivity is also carried out. Thereby, naturally, in the first place, artificial radio isotopes - which are set free during the utilization of nuclear energy - are of interest. Until the end of the 'sixties, practically only isotopes could be observed which originated from the atomic bomb experiments and which entered the sea via the global fallout. After the cessation of those experiments in 1962, the activity concentrations of the artificial radioactive isotopes in seawater gradually decreased.

From 1969 onwards, we then ascertained, once again, a slow increase of the concentrations of different radio nuclides in the water of the North Sea. These originated from the waste waters of the nuclear fuel reprocessing plants, situated on the coasts, at the Centre de la Hague near Cherbourg, Windscale near the Irish Sea, and Dounreay on the northernmost point of Scotland. In general, such slight amounts of activity are released into the environment from normal nuclear fuel power plants that, so far, we have not been able to observe their influence in the sea.

A useful monitoring of the sea regions for released harmful substances requires knowledge about the spreading and transport processes. The simplest method of doing that is to track them with the aid of clearly labelled water masses. Nowadays, a suitable labelling in the North Sea region is provided automatically by the radioactive waste waters which flow into the sea from the aforementioned nuclear fuel reprocessing plants. These waste waters contain isotopes created by the operation of nuclear reactors and which are not present in nature. The principle amount of these isotopes consists of Cs 134, Cs 137, and Sr 90. In addition, considerable amounts of Tritium, Ru 103/106, Zr 95, Ce 144, Pu 239/240, are released and - to a lesser degree - a diversity of other radio isotopes.

Cs 137 is primarily used for the investigations in question. As the result of its good solubility, its half-life of 30 years, its good detectability, and the relatively large amounts given off (some 10,000 Ci/year), it is an excellent tracer. It can be tracked in water over long periods of time, and thereby over long distances. 45 l of seawater per analysis are worked-off for the investigations. Thereby, it was still possible for us in routine operation to detect, in the mean, 0.1 pCi Cs 137/l with a 1σ propagated error of less than ± 5 %.

The North Sea represents the typical case of a flat shelf sea. Even if it is similar to a closed basin, there still exists a constant water exchange with the Atlantic from the South and from the North. The mean water depth in the southern North Sea lies at 20 to 40 m, in the northern part at 60 to 100 m. Only in the Norwegian Deep, along the Norwegian coast, do depth reach down to 700 m.

On the basis of these shallower depths, we can assume that during the cold part of the year a widespread homogeneous vertical distribution of Cs 137 exists in all parts of the North Sea, with the exception of the Norwegian Deep.

In the warmer part of the year, that is - in summer, the formation of the thermocline - that is, of layers of different temperature and density in the water column between surface and bottom - can be observed, to a certain extent, in different regions of the North Sea - mainly those with greater water depths. During that time, in many cases, a reduction of the Cs 137 concentrations - from the surface down to the near-bottom layers - can be measured.

The first discharge of a larger amount of Cs 137 (circa 5.000 Ci) in February/March 1971, with the waste waters of the Centre de la Hague, made it possible for us to track a well-labelled water mass, over a little less than two years, on its way through the North Sea. Those measurements gave the following transport times:-

Cherbourg to Dover about 2 to 3 months;
Cherbourg to the German Bight about 15 months;
Cherbourg to Skagerrak about 21 months.

The mean transport velocity over the whole of the stretch was about 1 to 1.1 nautical miles per day, with here and there deviations of between 0.7 to 1.7 nautical miles per day /_1_/.

For the first time, in March 1971, we measured in the northwestern North Sea also increased Cs 137 activity concentration values. Investigations by the Fisheries Radiobiological Laboratory (FRL) in Lowestoft revealed that those activities originate from the waste waters of the nuclear fuel reprocessing plant at Windscale situated near the Irish Sea /_2_/.

The Cs 137 labelled water reached the North Sea in the vicinity of the Orkney Islands and flowed along the English East coast towards the South. According to our measurements taken during the years 1971 to 1973, the water required about a year to flow from circa 57° N, 1°30' W (Aberdeen) to 53°30' N, 02° E with a mean transport velocity of 0.6 to 0.7 nautical miles per day.

Further investigations during the past years have provided us with a good impression concerning the general transport processes in the North Sea. In principle, two main transport routes could be observed (Fig. 1). One leads from the English Channel along the Belgian, Netherlands, Federal German, and Danish coasts into the Skagerrak. Thereby, the water coming from the South moves in a relatively narrow band along the coasts towards the North. The other route leads from the Orkney Islands along the East coast of England towards the South and then straight across the central and southern North Sea in the direction of the Skagerrak. Thereby, one can observe, in different years, certain deviations of current direction - possibly as the result of different meteorological conditions - between the eastern and northeastern directions.

The rest of the labelled water coming from the North veers round at about 53° N towards eastern directions and flows, parallel to the waterbody coming from the Channel, in the direction of the West coast of Jutland. Thereby, one can observe a steep gradient in the Cs 137 activity concentration values between the two water masses. Mixing of the two water bodies, according to our measurements, first takes place to the North of the German Bight. That signifies that the impurities contained in the water, which enters the North Sea through the English Channel, on its way northwards are now introduced into the region of the German Bight. The water masses which come from the North resp. West normally flow by to the North of the German Bight.

Both water masses unite in the region of the Skagerrak. The bulk of the water then flows through the Skagerrak in a narrow band (in the mean, not broader than 60 to 90 nautical miles) along the Norwegian coast into the northern North Atlantic. During 1972 we measured, in the Barents Sea to about 72° N and 45° E, Cs 137 activity concentration values which - at 0.35 pCi/l - were about three times the value of the measurable value of about 0.1 pCi/l that was taken at the same time in the surface water of the open North Atlantic.

A smaller part of the water carried to the Skagerrak flows into the Baltic Sea. That signifies that, in the final analysis, water from the Irish Sea reaches the Baltic Sea. As, in accordance with measurements on the transport time from Windscale to the Pentland Firth (in the mean, 2 years) taken by our British colleagues from Lowestoft /_2_/, the transport time over the whole of the stretch from the Irish Sea to entry into the Baltic Sea should lie at about 4 to 5 years. These result clearly show to what extent even sea regions which appear to be far distant from another can be connected. Our establishment of the fact that apparently the whole of the water which flows out of the Irish Sea towards the North reaches the North Sea in the vicinity

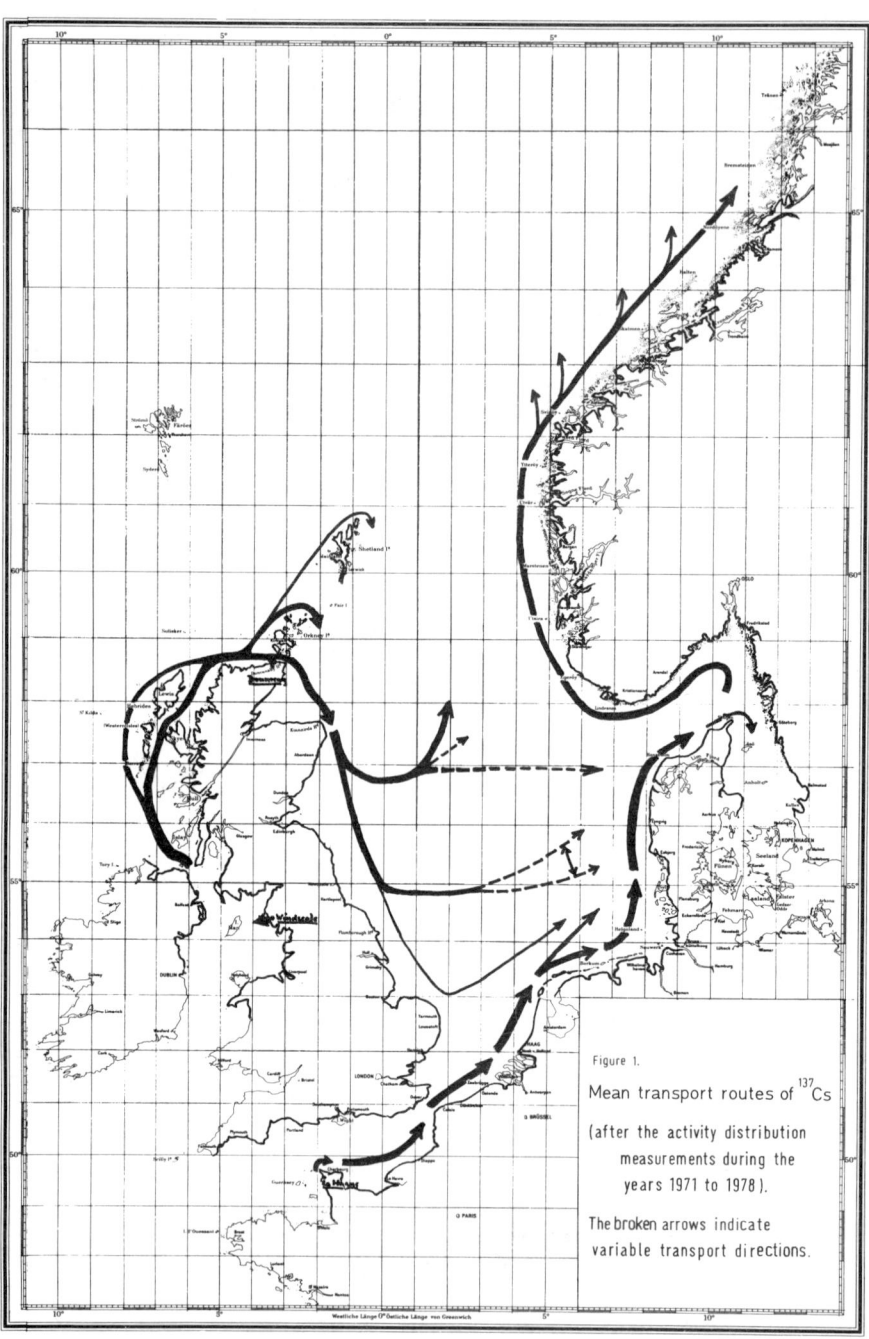

Figure 1.

Mean transport routes of ^{137}Cs (after the activity distribution measurements during the years 1971 to 1978).

The broken arrows indicate variable transport directions.

of the Orkney Islands was also astounding. A direct transport into the Norwegian Sea, with the Atlantic Current (an extension of the Gulf Stream) which flows West of the Hebrides towards the North, has not so far been observed.

The transport pattern of the Sr 90, Ru 106 and the transuranic nuclides in the North Sea is comparable to that of the Cs 137. The main amount of Ru 106 can be observed in the water of the southern North Sea, whereas the Sr 90 is transported in the main with the water entering the North Sea from the northwest. On the contrary the activity concentration values of Pu 239/240 are similar to one another as well in the southern and northern and western parts of the North Sea. A few activity concentration values of the different radioisotopes during spring of 1977 in some typical regions of the North Sea are listed in Table I.

The distinctly different distribution of the Sr 90 and Ru 106 activity concentration enables us to distinguish the water masses coming from the south in comparison to those flowing in from the North on their way through the North Sea over long distances.

Table I

Distribution of different radioisotopes in the North Sea in March/April 1977

Isotope	Street of Dover	German Bight	Pentland Firth	Central Western North Sea
Cs 137	0.58 pCi/l	0.75 pCi/l	11.7 pCi/l	2.8 pCi/l
Sr 90	0.35 pCi/l	0.48 pCi/l	1.9 pCi/l	0.77 pCi/l
Ru 106	197 fCi/l	42 fCi/l	12 fCi/l	5,9 fCi/l
Pu 239/240*	1.7 fCi/l	0.21 fCi/l	2.8 fCi/l	1.5 fCi/l

* Pu values after H.-F. Eicke, Deutsches Hydrographisches Inst.

In recapitulation, one can state that the principle water mass transport in shelf seas fundamentally takes place predominantly in relatively narrow bands alongside the coast. This can be clearly recognized from the measuremants taken by the DHI and the FRL during the years from 1971 to 1978 in the vicinity of the Hebrides, the whole of the North Sea as well as along the Norwegian coast, and in both the regions of the Norwegian Sea and the Barents Sea.

Therefore, in a case of discharge of liquid wastes into coastal waters, one must always reckon that such water can be transported over long distances alongside the coasts.

A dilution of the substances dissolved in the water thereby takes place only very slowly, as - in many cases - the speed of the transport processes appears to clearly exceed those of the mixing processes. That can signify a transport of relatively severely polluted waterbodies over long periods of time and over very large distances, whereby the highest concentrations of harmful substances introduced into the sea, and therewith their greatest effectiveness upon the biosphere, is to be expected predominantly in the region near the coasts.

References

1. Kautsky, H.: "The distribution of the radionuclide caesium 137 as an indicator for North Sea water mass transport", Deutsche Hydrographische Zeitschrift 26, 241-246 (1973).

2. Jefferies, D.F., Preston, A., Steele, A.K.: "Distribution of caesium 137 in British coastal waters", Mar. Pollution Bull. 4, 118-122 (1973).

Discussion

J. WEBER, Netherlands

Your slide shows a current going into the Baltic Sea. How does the water leave this sea ?

H. KAUTSKY, Federal Republic of Germany

Through the Kattegat the surface water of the Baltic Sea flows out to the North Sea, whereas the water in the deeper layers originates from the North Sea, and to some extent even from the Atlantic flowing through the Norwegian Deep, and enters the Baltic Sea.

J. WEBER, Netherlands

How stable are the currents in the North Sea as a function of time ?

H. KAUTSKY, Federal Republic of Germany

As far as I know the currents in the North Sea are stable with minor variations due to seasonal influences.

J. WEBER, Netherlands

Is it possible to relate the concentrations of radionuclides in the different water volumes of the North Sea to measured concentrations in fish ? This seems important in order to obtain the realistic collective dose calculations proposed by ICRP and discussed in the first paper by Dr. Webb.

H. KAUTSKY, Federal Republic of Germany

As we do not perform any biological work in our institute I cannot give a correct answer to this question. But if the discharges of Windscale and La Hague do not change in the near future, the amounts of different isotopes - originating from these sources - now existing in the North Sea will not change. On this basis, I feel some calculation on the concentration of the named radionuclides in fish should be possible.

J. WEBER, Netherlands

Is the water influx from the Rhine into the North Sea contained within the coastal current shown on your slide ?

H. KAUTSKY, Federal Republic of Germany

Fresh water influx into the coastal areas of the North Sea fallows the transport pattern I have shown to you. I see no reason as to why any water entering the sea from the coast should go in the opposite directions to the existing currents.

R.J. PENTREATH, United Kingdom

The collective effective dose equivalent for both the population of the UK, and of other West European countries, resulting from the Windscale discharges is calculated by FRL and published in its annual reports. The major contribution does result from radio-caesium in fish. For 1977 the value for the UK population - from both fish and shell fish - was 85 man-Sv, and 77 man-Sv for other West European countries.

G.A.M. WEBB, United Kingdom

What dilution occurs in the movement of the current along the coast, especially from Cherbourg ? Does this agree with theoretical predictions ?

H. KAUTSKY, Federal Republic of Germany

I am not familiar with the theoretical predictions of dilution processes in the area in question. But I know from discussions with my physical oceanographic colleagues that our findings of transportation mechanisms are not fully in agreement with their calculations. It is astonishing how slow any dilution of contaminants in the water in the region under investigations occurs. The current velocities - in normal cases - seem to go faster than the diffusion processes. The latter may be diminished too by the fact that almost all of the regions under investigation are bordered, at least on one side, by coasts.

J. ANCELLIN, France

Estimez-vous, d'après vos observations mentionnées sur le ruthénium 106 et le plutonium, que les voies et vitesses de transfert de ces deux radionucléides en Manche et Mer du Nord sont les mêmes que celles constatées pour le césium 137 ?

H. KAUTSKY, Federal Republic of Germany

As long as the isotopes of Ru-106 and plutonium are in solution (they may be in colloidal form) they will be transported in the same way as the Cs-137, or other contaminants, in the water.

A PRELIMINARY ASSESSMENT OF SOME NATURALLY-OCCURRING
RADIONUCLIDES IN MARINE ORGANISMS (INCLUDING DEEP
SEA FISH) AND THE ABSORBED DOSE RESULTING FROM THEM

R. J. Pentreath, D. S. Woodhead, B. R. Harvey and
R. D. Ibbett
Ministry of Agriculture, Fisheries and Food
Directorate of Fisheries Research
Fisheries Radiobiological Laboratory
Hamilton Dock
Lowestoft
Suffolk NR32 1DA
England

ABSTRACT

An understanding of the radiation regimes experienced by aquatic organisms as a result of the presence of naturally-occurring radionuclides is essential if the potential impact of introduced radionuclides is to be viewed in perspective. As part of an overall assessment, the contribution of internally accumulated radionuclides, and particularly of ^{210}Po, has been calculated for a number of coastal water and deep sea fish.

INTRODUCTION

An assessment of the possible effects on the marine fauna of increased radiation exposure following the disposal of radioactive wastes into the sea cannot realistically be made without a knowledge of both the magnitude of such exposure and the exposure regime already known to exist as a result of the naturally-occurring radiation field. Studies of certain freshwater and marine coastal environments have established the ranges of dose rates which might be experienced by a variety of aquatic organisms from both internal and external sources of radiation (1-6). *In situ* measurements of the environmental dose rate, using LiF dosimeters, have been made at two aquatic sites: in the north-east Irish Sea the results obtained from dosimeters attached to the plaice (*Pleuronectes platessa*) have confirmed the calculated estimates of the external exposure from contaminant radionuclides discharged from the Windscale reprocessing plant (7); in the Columbia River similar confirmatory data were obtained for the exposure of periphyton communities (8). In some cases the incremental exposure has been shown to be small relative to the natural radiation background and therefore of trivial consequence. Where a substantial increase in exposure has been determined the data have provided a firm basis for a prediction of the magnitude and significance of resultant effects in individual organisms and populations.

The placement of solidified, high-level radioactive wastes arising from fuel reprocessing, either on or beneath the bed of the deep ocean, is being considered as one of a number of disposal options. Although the adequate data base required to support a realistic assessment of the impact of such disposal operations on the deep ocean fauna does not yet exist, there has been some conjecture as to the sensitivity, or otherwise, of the fauna to an increase in radiation exposure (9,10). As a first step in developing the necessary basis for a credible assessment of the radiation aspects of the ecological impact of this disposal option, some preliminary data on the natural radiation regime experienced by the deep ocean fauna are presented here, and compared with analogous situations in shallow, coastal waters.

METHODS

Coastal water fish were obtained from commercial vessels. Only freshly caught specimens were used, and these were dissected immediately. The deep sea species were caught by RRS CHALLENGER in depths of 500-1250 m off the west coast of Ireland; these were also dissected immediately.

Uranium and thorium analyses were made by spark source mass spectrometry, on muffle-ashed samples, using a peak-switching technique to determine the isotopic ratios. Polonium analyses were made on wet samples oxidized in nitric and perchloric acids, diluted with water, and neutralized with sodium hydroxide. The solutions were made 0.5 molar with respect to hydrochloric acid and the polonium removed by spontaneous deposition on to silver discs at 90°C in the presence of ascorbic acid. The coastal water samples were plated on to large discs and total α-counted; the deep sea samples were analysed by α-spectrometry, using ^{208}Po as a yield tracer. Propagated counting errors were better than $\pm$ 10%.

COASTAL WATER ORGANISMS: EXTERNAL SOURCES

The natural, external, radiation background of coastal water organisms arises from three different sources: cosmic radiation, the natural radioactivity in sea water, and the natural radioactivity in marine sediments. The dose rate from cosmic radiation is only of significance for fauna inhabiting the surface layers of the sea. The most recent report of the United Nations Scientific Committee on the Effects of Atomic Radiation (11) estimates the dose rate at sea level from the cosmic radiation flux to be 3.2 μrad.h^{-1} (0.032 μGy.h^{-1}). The attenuation data quoted by Folsom and Harley (in 1) indicate that this would be reduced by over two-thirds at 10 m depth in water.

Potassium-40 is the predominant radionuclide in surface sea water, both in terms of concentration - which at 320 pCi.l^{-1} (11.8 Bq.l^{-1}) represents almost 97% of the total natural radioactivity - and also because it constitutes the major source of external radiation exposure for marine organisms living clear (> 1 m) of the sea bed $(2-4,12)$. The γ-ray dose rate to fish is of the order of 0.1 μrad.h^{-1} (0.001 μGy.h^{-1}) and β-radiation contributes an additional 0.15 μrad.h^{-1} (0.0015 μGy.h^{-1}) to the surface tissue, decreasing to zero at 0.6 cm beneath the body surface.

In shallow coastal waters, and on the continental shelf, the sediments are of terrigenous origin and therefore have uranium, thorium and potassium contents typical of terrestrial rocks. The dose rate at the sediment surface from γ-rays has been estimated to be in the range 1.8-17.0 μrad.h^{-1} (0.018-0.17 μGy.h^{-1}) and that from β-radiation in the range 2.0-23.4 μrad.h^{-1} (0.02-0.234 μGy.h^{-1}) $(2,3)$. These dose rates only apply to epibenthic organisms; the contribution to the exposure from β-radiation being eliminated by a seawater layer of 1 cm thickness, and that from γ-rays attenuated by a factor of 10^{-3} by a seawater layer of 1.5 m. Thus the external radiation regime experienced by free-swimming demersal organisms can be quite variable.

COASTAL WATER ORGANISMS: INTERNAL SOURCES

The available data on the typical concentrations of naturally-occurring radionuclides in a variety of marine organisms have been summarized previously and used as a basis for dose rate calculations using simplified models $(2-4)$. At that time it was noted that α-emitting nuclides, and in particular ^{210}Po, were the major internal source of exposure, with ^{40}K contributing most of the remainder. For ^{40}K, the whole-body dose rate to the larger organisms, mainly due to the β-radiation, shows relatively little variation in the range 1.4-2.8 μrad.h^{-1} (0.014-0.028 μGy.h^{-1}). The relative uniformity of exposure derives mainly from the fairly constant tissue concentrations which result from the physiological requirement for potassium in multicellular organisms, and which are therefore under metabolic control.

In contrast, the measured concentrations of certain of the radionuclides in the natural uranium and thorium decay series show large variations between and within marine species. Such variation probably arises, in part, from the absence of any biological function for these elements and therefore the lack of specific homeostatic mechanisms to control tissue concentrations. As a consequence, the calculated dose rates from these radionuclides show wide variations. It has also been noted $(2,3)$ that the assumption of a uniform distribution of α-emitting nuclides within the organisms could lead to underestimates of the dose rates to specific tissues because the measured whole-body concentrations usually used do not reflect the various degrees of nuclide accumulation by these tissues. This is particularly true for ^{210}Po, which shows a marked tendency to accumulate in the hepatopancreas of certain molluscs and crustaceans, and in the viscera of some fish. Indeed, due to the short range of the α-particles emitted during radioactive decay - typically less than 70 μm in tissue - variations in the distribution of the radionuclides on this scale can significantly influence the spatial distribution of dose.

A further factor which emphasizes the significance of the exposure from α-radiation relates to the much greater biological effect per unit absorbed dose resulting from the dense ionization along the track of an α-particle in tissue compared with the sparse ionization produced by β-particles and the recoil electrons generated by γ-rays. The most recent recommendations of the International Commission on Radiological Protection (13) suggest that a quality factor of 20 is required to convert the absorbed dose from α-particles to the biologically effective dose equivalent unit. It is the latter which should be used as a basis for assessing the relative significance of the various sources of radiation.

There are few published data on the occurrence of alpha-emitting nuclides in marine coastal water fish (14). In order to extend the amount of data available, a number of species have been analysed in this laboratory for uranium and thorium by mass spectrometry. Uranium is readily detectable in fish bone; the means and range of concentrations - expressed in units of radioactivity - obtained from a variety of species are given in Table I. None of the bone samples contained thorium (^{232}Th) above the limits of detection (5 μg.kg^{-1} wet). Muscle samples from a variety of species were also analysed for uranium and thorium, but all were below

TABLE I CONCENTRATIONS (WET WEIGHT) OF URANIUM IN BONE SAMPLES OF EUROPEAN COASTAL AND SHALLOW WATER FISH, ASSUMING THE SAME ISOTOPIC RATIO AS IN SEA WATER

Sample	^{238}U fCi.g^{-1}	(mBq.g^{-1})	^{235}U fCi.g^{-1}	(mBq.g^{-1})	^{234}U fCi.g^{-1}	(mBq.g^{-1})
Pelagic teleosts[a]						
Mean	45.2	(1.67)	2.2	(0.08)	49.2	(1.82)
Range	23.1-79.9	(0.85-2.96)	1.1-3.8	(0.04-0.14)	25.2-87.1	(0.93-3.22)
Demersal teleosts[b]						
Mean	6.3	(0.23)	0.3	(0.01)	6.8	(0.25)
Range	1.6-22.6	(0.06-0.84)	0.1-1.1	(0.004-0.04)	1.8-24.7	(0.07-0.91)
Elasmobranchs[c]						
Mean	3.8	(0.14)	0.2	(0.007)	4.2	(0.16)

a) *Scomber scombrus.*
b) *Pleuronectes platessa, Solea solea, Lophius piscatorius, Anarhichas lupus, Merlangius merlangus, Gadus morhua, Melanogrammus aeglefinus, Pollachius virens, Pollachius pollachius, Molva molva, Dicentrarchus labrax.*
c) *Squalus acanthias.*

TABLE II   CONCENTRATIONS (WET WEIGHT) OF ^{210}Po IN EUROPEAN COASTAL AND SHALLOW WATER FISH

Sample	^{210}Po			
	Mean		Range	
	fCi.g^{-1}	(mBq.g^{-1})	fCi.g^{-1}	(mBq.g^{-1})
Pelagic teleosts [a]				
Muscle	61.9	(2.29)	5.1- 128	(0.19- 4.74)
Liver	2490	(92.1)	1070 -4770	(39.6 -176)
Bone	87.3	(3.23)	86.9- 87.7	(3.22- 3.24)
Demersal teleosts [b]				
Muscle	28.0	(1.04)	1.7- 87.3	(0.06- 3.23)
Liver	1670	(61.8)	280 -6520	(10.4 -241)
Bone	44.6	(1.65)	3.2- 102	(0.12- 3.77)
Elasmobranchs [c]				
Muscle	22.6	(0.84)	5.1- 62.0	(0.19- 2.29)
Liver	353	(13.1)	111 - 484	(4.11- 17.9)
Cartilage	33.6	(1.24)	14.8- 46.1	(0.55- 1.71)

a) *Scomber scombrus, Thunnus thynnus* (liver and muscle only from (15)).
b) *Pleuronectes platessa, Solea solea, Scophthalmus maximus, Scophthalmus rhombus, Lophius piscatorius, Merlangius merlangus, Gadus morhua, Eutrigla gurnardus, Aspitrigla cuculus, Trigla lucerna.*
c) *Squalus acanthias, Raja clavata.*

TABLE III RANGE OF DOSE EQUIVALENT RATES TO COASTAL AND SHALLOW WATER FISH IMPLIED BY THE RADIONUCLIDE CONCENTRATIONS GIVEN IN TABLES I AND II

Organ	Radionuclide	Dose rate $\mu rem.h^{-1}$ ($\mu Sv.h^{-1}$)
Pelagic teleosts		
Bone	^{238}U	4.2 - 14.5 (0.042 - 0.145)
Bone	^{235}U	0.2 - 0.7 (0.002 - 0.007)
Bone	^{234}U	5.2 - 18.0 (0.052 - 0.18)
Muscle	^{210}Po	1.2 - 29.5 (0.012 - 0.295)
Liver	^{210}Po	247 -1100 (2.47 -11.0)
Bone	^{210}Po	20.0 - 20.2 (0.20 - 0.202)
Demersal teleosts		
Bone	^{238}U	0.3 - 4.1 (0.003 - 0.041)
Bone	^{235}U	0.02- 0.2 (0.0002- 0.002)
Bone	^{234}U	0.4 - 5.1 (0.004 - 0.051)
Muscle	^{210}Po	0.4 - 20.1 (0.004 - 0.201)
Liver	^{210}Po	64.5 -1500 (0.645 -15.0)
Bone	^{210}Po	0.7 - 23.5 (0.007 - 0.235)
Elasmobranchs		
Cartilage	^{238}U	0.7 (0.007)
Cartilage	^{235}U	0.04 (0.0004)
Cartilage	^{234}U	0.9 (0.009)
Muscle	^{210}Po	1.2 - 14.3 (0.012 - 0.143)
Liver	^{210}Po	25.6 - 112 (0.256 - 1.12)
Cartilage	^{210}Po	3.4 - 10.6 (0.034 - 0.106)

the limits of detection (1 $\mu g.kg^{-1}$ wet for ^{238}U and 0.5 $\mu g.kg^{-1}$ wet for ^{232}Th).

A variety of coastal water fish species has also been analysed for ^{210}Po, by α-counting techniques, and the results are given in Table II which also includes data from Heyraud and Cherry (15) for liver and muscle of the tuna (*Thunnus thynnus*). The range of values obtained in muscle is generally similar from one group of fish to another, but the liver concentrations in the elasmobranch fish are clearly lower than those in the teleosts.

Only one coastal water invertebrate (*Sepia officinalis*) has so far been analysed in this laboratory for ^{210}Po - the concentration in mantle was about 20 $fCi.g^{-1}$ wet (0.74 $mBq.g^{-1}$ wet) - but Heyraud and Cherry (15) have recently analysed a large variety of invertebrate species from the Mediterranean. These results indicate that ^{210}Po concentrations in the muscle of benthic crustaceans, which vary from about 30 — 140 $fCi.g^{-1}$ wet (1.11 — 5.18 $mBq.g^{-1}$ wet), are not very much different from those of fish but concentrations in the hepatopancreas, which vary from about 6 — 37 $pCi.g^{-1}$ wet (0.22 — 1.37 $Bq.g^{-1}$ wet), are very much higher than those in fish livers. Concentrations of ^{210}Po in benthic molluscs were generally of the same order as those obtained for crustaceans.

The dose equivalent rates implied by the data in Tables I and II are given in Table III. The relative importance of ^{210}Po is immediately apparent, and it should be noted that the dose rates to the liver from this internally accumulated radionuclide are orders of magnitude greater than the dose rates arising from the combined external sources. In comparison, the absorbed dose rates inferred by the ^{210}Po concentrations in Mediterranean invertebrates (15) - again assuming a quality factor of 20 - vary from about 7 — 32 $\mu rem.h^{-1}$ (0.07 — 0.32 $\mu Sv.h^{-1}$) in the muscle of benthic crustaceans, and from about 1.38 — 8.51 $mrem.h^{-1}$ (13.8 — 85.1 $\mu Sv.h^{-1}$) in the hepatopancreas. Even higher dose equivalent rates are inferred by the ^{210}Po concentrations observed by Heyraud and Cherry in pelagic crustaceans caught in coastal waters.

DEEP WATER ORGANISMS: EXTERNAL SOURCES
―――――――――――――――――――――――――――――――――――

The dose rate arising from cosmic radiation is negligible at depths in excess of 4000 m, and although the potassium content of sea water varies in proportion to the salinity, the dose rate from this source in the deep ocean is not greatly different from that in surface or coastal waters. The dose rates to be expected from the sediments, however, are notably different.

Most of the data on the content of natural radionuclides in deep sea sediments have been acquired during the course of investigations into sedimentation and geochemical processes. The concentrations of both uranium and thorium are found to be dependent on the calcium carbonate content of the sediment and there are varying degrees of disequilibria between the longer-lived members of the uranium series. Specifically, the $^{230}Th/^{234}U$ activity quotient is considerably in excess of unity due to the precipitation of ^{230}Th arising from the decay of ^{234}U in solution in the overlying water column; in addition, although the $^{226}Ra/^{238}U$ quotient is greater than unity, the $^{226}Ra/^{230}Th$ quotient is less than unity due to the loss of radium by leaching. Table IV gives data sets for two deep ocean cores taken in the North Atlantic (16). Although it is known that there is an excess of ^{222}Rn with respect to the parent ^{226}Ra in the water of the benthic boundary layer (17) the efflux of radon from the sediment necessary to maintain this situation is not sufficient to disturb significantly the equilibrium between ^{226}Ra and the daughter radionuclides in the sediment. Thus the calculation of the γ-ray dose rate from the uranium series in the sediment can safely assume equilibrium between ^{226}Ra and the γ-emitting daughter nuclides - principally ^{214}Bi. For the thorium series the assumption of equilibrium is less certain because there may be significant leaching of ^{228}Ra from the sediment. However, in the absence of data as to the concentrations of the γ-emitting daughters, equilibrium must be assumed and the probable overestimation of the dose rate from this source recognized.

The α-, β- and γ-ray dose rates at the sediment surface from these concentrations of natural radionuclides calculated using the simple models described previously (3) are summarized in Table V. Insofar as these cores are typical of deep sea sediments it can be seen that, although the dose rates from the β- and

TABLE IV   CONCENTRATIONS OF RADIONUCLIDES OF URANIUM AND THORIUM IN TWO DEEP OCEAN CORES FROM THE ATLANTIC (CALCULATED FROM [16])

Core details	Depth in core (cm)	^{238}U pCi.g^{-1} (mBq.g^{-1})		^{234}U pCi.g^{-1} (mBq.g^{-1})		^{230}Th pCi.g^{-1} (mBq.g^{-1})		^{226}Ra pCi.g^{-1} (mBq.g^{-1})	
Red clay 5190 m 26°31'N 51°47'W	0-6	0.79	(29.2)	0.75	(27.8)	28.29	(1047)	11.31	(418)
	30-35	0.83	(30.7)	0.73	(27.0)	12.42	(460)	11.05	(409)
	75-80	0.86	(31.8)	0.77	(28.5)	5.97	(221)	6.33	(234)
	92-97	0.84	(31.1)	0.74	(27.4)	3.22	(119)	3.06	(113)
Globigerina ooze 4675 m 3°47'N 34°37'W	0-5	0.18	(6.7)	0.18	(6.7)	4.06	(150)	2.31	(85)
	50-57	0.34	(12.6)	0.33	(12.2)	2.17	(80)	2.88	(107)

TABLE V   DOSE EQUIVALENT RATES AT THE SURFACE OF DEEP SEA SEDIMENTS AS INFERRED FROM THE DATA IN TABLE IV

Core	α μrem.h^{-1} (μSv.h^{-1})		β[a] μrem.h^{-1} (μSv.h^{-1})		γ[a] μrem.h^{-1} (μSv.h^{-1})	
Red clay						
U-series	10000	(100)	16	(0.16)	20	(0.20)
Th-series	2200	(22)	3.1	(0.031)	8.4	(0.084)
Globigerina ooze						
U-series	1900	(19)	3.3	(0.033)	4.0	(0.04)
Th-series	310	(3.1)	0.4	(0.004)	1.2	(0.012)

a) For the β- and γ-radiation a quality factor of unity has been assumed to convert absorbed dose rate to dose equivalent rate.

γ-radiation overlap those given for coastal sediments, there is a tendency to higher values, particularly for the red clay. The same is true for α-radiation which, for coastal sediments, is estimated to deliver dose rates in the range 240-3300 μrem.h^{-1} (2.4-33 μSv.h^{-1}). The radiological consequences of these α-radiation dose rates at the sediment surface for marine organisms is difficult to assess; but it is to be expected that they are of significance for the benthic micro- and meio-fauna, and also for the gut epithelium of detrital feeders.

DEEP WATER ORGANISMS: INTERNAL SOURCES

A number of deep water fish have recently been obtained and are being analysed for naturally-occurring radionuclides. The results obtained to date for ^{210}Po are given in Table VI. The range of values observed for different organs are generally similar to those observed for coastal water fish. The concentrations in the livers of the sharks are again lower than those of the teleosts; they are even lower than those observed in the coastal water elasmobranchs. The range of dose equivalent rates implied by these data are given in Table VII.

Only one species of deep sea invertebrate has been analysed to date, the amphipod *Eurythenes gryllus*. These specimens were obtained from depths of 4700 m and deep frozen before being analysed. The ^{210}Po concentrations obtained, and the calculated dose equivalent rates, are given in Table VIII. These values are well within the range of those obtained for the Mediterranean benthic crustaceans discussed above.

DISCUSSION

It would be most unwise to infer too much from the preliminary data presented here, but there does appear to be sufficient evidence to make at least one general statement: the radiation regime obtaining for organisms living in the deep ocean, from both internal and external sources, is unlikely to be any lower than that obtaining for coastal water organisms. Of equal importance would be the observation that the radiation regime is unlikely to be constant; and this is likely to be the case, at least for internal sources of exposure such as that arising from ^{210}Po. The maximum range of ^{210}Po observed to date in a single species is that of 172 — 3664 fCi.g^{-1} wet (6.36 — 135.6 mBq.g^{-1} wet) in the liver of *Coryphaenoides rupestris*. Further data are clearly required to substantiate these tentative statements.

REFERENCES

(1) Anon.: "The Effects of Atomic Radiation on Oceanography and Fisheries", 137 pp., US National Academy of Sciences Publication No.551, Washington 1957.

(2) Woodhead, D. S.: "Levels of radioactivity in the marine environment and the dose commitment to marine organisms", Radioactive Contamination of the Marine Environment, pp.499-525, IAEA, Vienna 1973.

(3) Anon.: "Effects of Ionizing Radiation on Aquatic Organisms and Ecosystems", IAEA Technical Reports Series No.172, 131 pp., IAEA, Vienna 1976.

(4) Woodhead, D. S.: "The estimation of radiation dose rates to fish in contaminated environments, and the assessment of the possible consequences". Population Dose Evaluation and Standards for Man and His Environment, pp.555-575, IAEA, Vienna 1974.

(5) Pentreath, R. J., Woodhead, D. S. and Jefferies, D. F.: "Radioecology of the plaice (*Pleuronectes platessa* L.) in the northeast Irish Sea", Radionuclides in Ecosystems, edited by D. J. Nelson, pp.731-737, USAEC, Oak Ridge National Laboratory, CONF-710501-P2, 1973.

TABLE VI CONCENTRATIONS (WET WEIGHT) OF ^{210}Po IN FISH CAUGHT AT DEPTHS OF 500 TO 1250 METRES

Sample	^{210}Po			
	Mean		Range	
	fCi.g^{-1}	(mBq.g^{-1})	fCi.g^{-1}	(mBq.g^{-1})
Teleosts				
Muscle[a]	61	(2.26)	7.2- 233	(0.27- 8.62)
Liver[b]	1842	(68.2)	172 -5273	(6.36-195)
Bone[c]	90	(3.33)	41 - 131	(1.52- 4.85)
Gonad[b]	377	(13.9)	39 -1107	(1.44- 41)
Elasmobranchs				
Muscle[d]	5.5	(0.20)	3.0- 12.9	(0.11- 0.48)
Liver[d]	9.1	(0.34)	2.7- 22.8	(0.10- 0.84)
Cartilage[e]	60	(2.22)	41 - 99	(1.52- 3.66)
Gonad[e]	4.5	(0.17)	2.4- 7.2	(0.09- 0.27)

a) *Coryphaenoides rupestris, Trachyrhynchus trachyrincus, Alepocephalus bairdi, Phycis blennoides.*
b) *Coryphaenoides rupestris, Trachyrhynchus trachyrincus, Alepocephalus bairdi.*
c) *Coryphaenoides rupestris, Trachyrhynchus trachyrincus.*
d) *Centroscymnus coelolepis, Deania calcea.*
e) *Centroscymnus coelolepis.*

TABLE VII RANGE OF DOSE EQUIVALENT RATES TO DEEP SEA FISH IMPLIED BY THE ^{210}Po CONCENTRATIONS GIVEN IN TABLE VI

Organ	Dose rate	
	μrem.h^{-1}	(μSv.h^{-1})
Teleosts		
Muscle	1.7- 53.7	(0.017- 0.537)
Liver	39.6-1220	(0.396-12.2)
Bone	9.5- 30.2	(0.095- 0.302)
Gonad	9.0- 255	(0.09 - 2.55)
Elasmobranchs		
Muscle	0.7- 3.0	(0.007- 0.03)
Liver	0.6- 5.3	(0.006- 0.053)
Cartilage	9.5- 22.8	(0.095- 0.228)
Gonad	0.6- 1.7	(0.006- 0.017)

(6) Woodhead, D. S.: "The assessment of the radiation dose to developing fish embryos due to the accumulation of radioactivity by the egg", Radiation Research 43, 582-597 (1970).

(7) Woodhead, D. S.: "The radiation dose received by plaice (*Pleuronectes platessa*) from the waste discharged into the northeast Irish Sea from the fuel reprocessing plant at Windscale", Health Physics 25, 115-121 (1973).

(8) Lappenbusch, W. L., Watson, D. G. and Templeton, W. L.: "*In situ* measurement of radiation dose in the Columbia river", Health Physics 21, 247-251 (1971).

(9) Rice, A. L.: "Radioactive waste disposal and deep-sea biology", Oceanologica Acta 1, 483-491 (1978).

(10) Anon.: "The Oceanographic Basis of the IAEA Revised Definition and Recommendations Concerning High-Level Radioactive Waste Unsuitable for Dumping at Sea", Technical Document IAEA-210, 59 pp., IAEA, Vienna 1978.

(11) Anon.: "Sources and Effects of Ionizing Radiation", United Nations Scientific Committee on the Effects of Atomic Radiation, 1977 Report to the General Assembly, 725 pp., United Nations E.77.IX.1, 1977.

(12) Joseph, A. B., Gustafson, P. F., Russell, I. R., Schuert, E. A., Volchok, H. L. and Tamplin, A.: "Sources of radioactivity and their characteristics", Radioactivity in the Marine Environment, pp.6-41, US National Academy of Sciences, Washington 1971.

(13) Anon.: "Recommendations of the International Commission on Radiological Protection", Annals of the ICRP, Publication 26, pp.53, Pergamon Press, Oxford 1977.

(14) Pentreath, R. J.: "Radionuclides in marine fish", Oceanography and Marine Biology Annual Review 15, 365-460 (1977).

(15) Heyraud, M. and Cherry, R. D.: "Polonium-210 and lead-210 in marine food chains", Marine Biology 52, 227-236 (1979).

(16) Ku, T. L.: "An evaluation of the $^{234}U/^{238}U$ method as a tool for dating pelagic sediments", Journal of Geophysical Research 70, 3457-3474 (1965).

(17) Broecker, W. S., Li, Y. H. and Cromwell, J.: "Radium-226 and radon-222: concentrations in Atlantic and Pacific Oceans", Science 158, 1307-1310 (1967).

TABLE VIII AVERAGE CONCENTRATIONS (WET WEIGHT) OF ^{210}Po IN THE DEEP SEA AMPHIPOD *Eurythenes gryllus*, AND THE INFERRED DOSE EQUIVALENT RATES RESULTING FROM THEM

Sample	^{210}Po $fCi.g^{-1}$	$(mBq.g^{-1})$	Dose rate $\mu rem.h^{-1}$	$(\mu Sv.h^{-1})$
Gill	1016	(37.6)	234	(2.34)
Viscera	530	(19.6)	122	(1.22)
Exoskeleton	205	(7.6)	47	(0.47)

Discussion

J. WEBER, Netherlands

What is the quality factor for alpha radiation you used to convert dose to dose equivalent ? Would a RBE factor not be more appropriate ?

R.J. PENTREATH, United Kingdom

We used a factor of 20. You are perfectly correct in pointing out that we are being presumptious in using the term quality factor, which of course applies only to man, in calculating dose equivalent rates to aquatic organisms.

ACCUMULATION OF TRACE METALS IN COASTAL MARINE ORGANISMS.

A.W.van Weers
J.G.van Raaphorst
Netherlands Energy Research Foundation (ECN)
Petten, The Netherlands.

ABSTRACT.

ECN at Petten carries out a survey on the occurrence of trace metals in coastal marine organisms. The survey is aimed to provide an estimate of concentration factors in local marine organisms for neutron activation products released as low-level liquid radioactive waste into the North Sea. The organisms studied are red and brown seaweed, edible mussels and shrimp. A summary of the results of analyses of iron, cobalt, zinc, silver and antimony in these organisms is presented. Concentration factors derived from mean stable-element concentrations range from about 50 for Sb in red seaweed and shrimp to about 10^4 for Fe in red seaweed and mussels. The largest variation is shown for zinc in seaweed, which variation is seasonal and most pronounced in brown seaweed. A discussion of the data is presented in relation to data from other West-European coastal areas and to data used for the radiological assessment of deep sea disposal of radioactive waste.

Tabel I : Summary of material and methods.

	Fe	Co	Zn	Ag	Sb
Porphyra umbilicalis red seaweed	DA a) INAA b)	DA INAA	DA AAS	DA INAA	DA INAA
Fucus spiralis brown seaweed	DA INAA	DA INAA	DA AAS	DA INAA	DA INAA
Mytilus edulis mussel	DA INAA	DA INAA	DA AAS	DA INAA	DA INAA
Crangon crangon shrimp	WA c) AAS d)	DA INAA	WA AAS	DA INAA	DA INAA

a) DA   : Dry ashing, 500 $^{\circ}$C, overnight.
b) INAA : Instrumental neutron activation analysis.
c) WA   : Wet ashing of freeze-dried material with HNO_3 and $HClO_4$ at 150 $^{\circ}$C.
d) AAS : Atomic absorption spectrometry.

INTRODUCTION.

In models developed to estimate the radiological impact of radioactivity released into the sea concentration factors are used to derive radionuclide concentrations in marine organisms from estimated equilibrium concentrations in sea water. The compilations of concentration factors used in these models are largely based on the occurrence of stable elements in organisms and in sea water. However for many elements only limited data are available which often pertain to widely differing species from various localities. Therefore part of the environmental survey carried out by ECN at Petten is aimed to provide an estimate of concentration factors in local marine species for those trace elements which occur as neutron-activation products released by the Centre as low-level liquid radioactive waste into the North Sea. A summary and a discussion of the results of this survey is presented in this paper.

MATERIAL AND METHODS.

Red seaweed (Porphyra umbilicalis), brown seaweed (Fucus spiralis), mussels (Mytilus edulis) and shrimps (Crangon crangon) were all collected at the North Sea coast in the vicinity of the Centre at Petten. Red and brown weed were oven dried at 100 °C. Mussels were killed with boiling sea water and the soft parts were removed from the shells and oven-dried at 100 °C. Shrimps were freeze-dried. A summary of the analytical methods used is given in table 1. In those cases where atomic adsorption spectrometry (AAS) was used on dry-ashed material, the ash was dissolved with boiling conc.HCl (Zn) or with conc.HNO_3 in a teflon bomb at 120-130 °C (Fe). Dry ashing at 500 °C was shown to cause no significant losses of Co, Zn, Ag and Sb |1|, |2|. Loss of Fe during dry ashing was not tested but is not likely to occur.

RESULTS.

The results of the trace-element survey are given in table II to VI as the range, arithmetric mean and standard deviation of the concentrations in dry material. Relatively large variations are found for iron, silver and antimony in Phorphyra and for zinc in Porphyra and Fucus. Concentrations of cobalt, silver and antimony are generally between two and three orders of magnitude lower than zinc and iron concentrations. To convert these data to concentration factors the concentrations of the elements have been expressed on a wet weight basis and divided by the trace-element concentrations for sea water given in table VII. The values given in this table are representative for the concentrations in near shore sea water of the Northern part of the Netherlands. The mean concentration factors derived in this way are presented in table VIII. Average dry to wet weight ratio's are 0.20 for Porphyra, 0.13 for Fucus, 0.28 for Mytilus and 0.25 for Crangon.

DISCUSSION.

Relatively large variation occur in the concentrations of Fe, Zn, Ag and Sb in the red seaweed samples and in the concentration of Zn in the brown seaweed. The zinc concentration range in the latter seaweed reflects a seasonal variation with maximum zinc levels in late winter and early spring |5|. The large range of the zinc concentration in red seaweed is mainly due to high concentrations in samples from 1966 to 1968. Maximum zinc levels in this seaweed generally occur in winter. No clear seasonal dependence has been observed for the other elements and species. Zinc concentrations in Fucus from the North Sea coast at Petten are in about the same range as has been published for the same genus from various locations on the coast of the U.K. |6|, |7|, |8|, on the Norwegian coast |9| and the Atlantic coast of Spain and Portugal |10|. A considerably lower and smaller range (39-113 μg/g) has been reported for Fucus from Caernarvon Bay, U.K. |3|. Data published on iron in Fucus |6|, |8|, |11|, |12| are in good agreement with the results of the present study. Data on cobalt and silver are very limited. The ranges and mean concentrations of Co in Fucus species from Cardigan Bay and Bristol Channel are about a factor of 3 higher than those found in the present survey |8|, |11|. Ranges and mean concentrations of Ag in Fucus from other coastal areas of the British Isles |6| are similar to those for Fucus from the location near Petten. Levels found for Sb in the present survey do agree well with the concentrations reported for Sb in Fucus from Portland, Dorset and Severn Estuary, U.K. |13|.

Even without the high zinc levels in samples of Porphyra from 1966-1968 the range of 58-658 μg/g for the samples of 1969-1973 still exceeds considerably the range of 36-174 μg/g reported for zinc in the same species from two areas of the

Table II: Concentration of iron in marine organisms from the North Sea coast near Petten.

organisms	number of samples	sampling period	concentration in dry material (µg/g)		
			range	mean	stand.dev.
Porphyra	30	1966,1972	130-1470	338	280
Fucus	38	1966,1972	64- 280	134	45
Mytilus	30	1966,1972	130- 680	371	156
Crangon	34	1975-1978	45- 240	122	47

Table III: Concentration of cobalt in marine organisms from the North Sea coast near Petten.

organisms	number of samples	sampling period	concentration in dry material (µg/g)		
			range	mean	stand.dev.
Porphyra	29	1966,1972	0.23-0.66	0.40	0.11
Fucus	38	1966,1972	1.1 -3.8	1.9	0.60
Mytilus	30	1966,1972	0.41-3.4	0.95	0.63
Crangon	8	1975	0.21-0.34	0.29	0.04

Table IV: Concentration of zinc in marine organisms from the North Sea coast near Petten

organisms	number of samples	sampling period	concentration in dry material (µg/g)		
			range	mean	stand.dev.
Porphyra	103	1966-1975	50-2480	305	430
Fucus	148	1966-1975	88-1080	399	187
Mytilus	121	1966-1973	115- 560	236	82
Crangon	34	1975-1978	96- 165	105	25

Table V: Concentration of silver in marine organisms from the North Sea coast near Petten.

organisms	number of samples	sampling period	concentration in dry material (µg/g)		
			range	mean	stand.dev.
Porphyra	30	1966,1972	0.05-2.2	0.51	0.54
Fucus	38	1966,1972	0.10-0.45	0.24	0.03
Mytilus	12	1966	0.18-1.1	0.47	0.26
Crangon	7	1975	1.2 -1.5	1.3	0.1

Irish Sea |6|. The ranges and mean for iron in Porphyra of those two areas agree
well with the range and mean concentration reported in the present study. The mean
silver concentration is a factor of about 5 higher than those reported for the two
Irish Sea areas.
 The trace-element concentrations in the mussel samples do not differ widely
from those reported for samples of the same species from various areas in the
West-European coastal waters except those from areas known to be considerably polluted with heavy metals from anthropogenic sources. This applies to Fe, Co, Zn and
Ag for which elements data have been summarized by Phillips |14|. The zinc
levels are the same as reported for mussels from the Gulf of la Spezia in the Me-
diterranean and from Scottish waters |15|, |16|. Zinc concentrations found in the
present survey are within a factor of 2-3 of those reported for samples from Kat-
tegat and Skagerrak |17|. With the exception of the higher zinc levels the concen-
tration ranges in our samples are within the range reported by Karbe et al |18|
for Fe, Co, Zn and Ag in mussels from German coastal waters. Mean concentrations given in table II to VI are for Co within the range of and for Fe, Zn and Ag
less than a factor of three higher than the mean concentrations reported by Karbe
et al for repeatedly sampled different sites along the German North Sea coast. Fe
Co, Zn, Ag and Sb concentrations have also been reported for soft parts of mussels
from the Belgian North Sea coast |19|. Compared with the mean concentrations found
in the present survey the average values reported for the Belgian location are for
Co and Fe a factor of about two higher, for Ag a factor of about two lower and
similar for Sb. Zn concentrations reported in that study are remarkably low with
a mean value of 40 μg/g. Ranges and mean concentration of the present survey agree
very well with data published on Fe, Co, Zn and Ag in the closely related Mytilus
galloprovincialis from the north-western Mediterranean |20|.
 Fe and Zn concentrations in the shrimp samples show similar ranges and mean
values. The mean concentrations do not differ by more than a factor of two from
the values reported for samples of the same species from the Belgian North Sea
coast |19|. The same applies for Co with respect to concentrations reported for
Crangon from the Mediterranean |21| and from the Belgian coast |19|. Carapaces of
shrimps from the latter location contained as much Sb as samples of whole shrimps
from the present survey. However, Sb levels in shrimp meat were a factor of about
3 to 4 higher.
 From the results of this survey estimates are obtained of the average concentration factors that may be reached for radionuclides of the trace elements in the
organisms studied. Although much higher trace-element concentrations in sea water
may occur particularly in estuaries and lower concentrations in off-shore waters,
evidence from ref.|6| and |22| indicates that the concentrations given in table
VII can be used as estimates for shoreline sea water in West-European coastal
waters. Comparison of the trace-element concentrations with the levels reported
for the same or closely related species from other locations in the West-European
coastal waters shows that the concentration factors presented in table VIII are
reasonably general estimates for these organisms in coastal waters bordering the
North East Atlantic.
 Taking into account a possible overestimate of the concentration factor of zinc
in seaweed, the concentration factors derived from this survey are with a few exceptions within a factor of 5 of the concentration factors used for seaweed, molluscs and crustacea in the radiological basis of the IAEA revised definitions and
recommendations concerning dumping of radioactive waste at sea |23|. Significantly
different values are given in the latter IAEA-report for Fe, Zn and Ag in molluscs,
for which elements concentration factors are used which are respectively a factor
of 20 lower (Fe), a factor of 15 higher (Zn) and a factor of 100 higher (Ag). With
respect to Fe the concentration factor of 10^3 from ref.|23| is probably a too low
general estimate of the ratio between the concentration of Fe in molluscs and in
filtered sea water. The higher Zn and Ag concentration factors are based on the
high stable-zinc concentrations observed in oysters from various locations |24|,
|25|, |26| and the high concentration factors of ^{65}Zn and ^{110m}Ag in oysters from
the vicinity of a nuclear power station in the Blackwater Estuary U.K. |27|. The
combination of the high Zn and Ag concentration factors, typical for oysters, to-
gether with a molluscs consumption rate of 100 g/day in ref.|23| shows that irre-
spective of the assumptions on dispersion, the models used to establish derived
limits for release by dumping are for radionuclides of Zn and Ag indeed of the
hypothetical and maximising type. Concentration factors as compiled in the Radio-
logical Basis |23| should therefore not without further consideration be used for
estimates of collective dose commitments from deep sea dumping. Although radio-

Table VI: Concentration of antimony in marine organisms from the North Sea coast near Petten.

organisms	number of samples	sampling period	concentration in dry material (µg/g)		
			range	mean	stand.dev.
Porphyra	30	1966,1972	0.01-0.20	0.05	0.04
Fucus	37	1966,1972	0.04-0.11	0.07	0.02
Mytilus	12	1966	0.05-0.14	0.08	0.03
Crangon	7	1975	0.02-0.06	0.04	0.02

Table VII: Concentrations of trace-elements in Dutch coastal sea water used to derive the concentration factors in marine organisms.

element	concentration in µg/l	reference
iron	6	|3| a)
cobalt	0.1	|4|
zinc	10	|3|
silver	0.1 b)	
antimony	0.2	|4|

a) Derived on the basis of a total-Fe to particulate leached-Fe ratio of 3:1 and a ratio of total Fe to solved Fe of 50:1.
b) Assumed to be similar to levels reported in ref.|6| and |22|.

Table VIII: Concentration factors for Fe, Co, Zn, Ag and Sb in marine organisms from the Dutch North Sea coast near Petten.

	Red seaweed Porphyra umbil.	Brown seaweed Fucus spiralis	Mussels Mytilus edulis	Shrimp Crangon crangon
Fe	1×10^4	6×10^3	2×10^4	3×10^3
Co	8×10^2	4×10^3	3×10^3	7×10^2
Zn	6×10^3	7×10^3	7×10^3	3×10^3
Ag	1×10^3	4×10^2	1×10^3	3×10^3
Sb	5×10^1	7×10^1	1×10^2	5×10^1

nuclides of Zn and Ag are rather short-lived and their contribution to collective dose commitments therefore small, a similar overestimation may result for radionuclides with longer half-lifes.

Whether or not the concentration factors observed for elements and radionuclides in organisms from shallow waters may in fact be reached for radionuclides in deep sea fauna is unknown. There is virtually no information on the occurrence of elements in deep sea fauna relative to their occurrence in deep sea water. Moreover it may be doubted that direct correlations exist between the concentrations of elements in marine organisms and those in sea water. For deep dea organisms organic material from higher layers may not only be the source of organic carbon but of other elements as well.

REFERENCES.

|1| Van Raaphorst,J.G., Van Weers,A.W., Haremaker,H.M., Loss of zinc and cobalt during dry ashing of biological material, Analyst, 99, 523 (1974)

|2| Van Raaphorst,J.G., Van Weers,A.W., Haremaker,H.M., On the loss of cadmium, antimony and silver during dry ashing of biological material, Fres.Z.Anal. Chem. 293, 401 (1978).

|3| Duinker,J.C., Nolting,R.E., Dissolved and particulate trace metals in the Rhine Estuary and the Southern Bight, Mar.Poll.Bull. 8, 65 (1977).

|4| Van der Sloot,H.A. Personal communication.

|5| Van Weers,A.W., Zinc and cobalt uptake by the brown seaweed Fucus spiralis(L), Proc.Symp.Radioecology Applied to the Protection of Man and his Environment, Rome 7-10 Sept'71. EUR-4800, p 1357.

|6| Preston,A. et al, British Isles coastal waters: the concentrations of selected heavy metals in sea water, suspended matter and biological indicators - a pilot survey, Environ.Poll. 3, 69 (1972).

|7| Ireland,M.P. Result of fluvial zinc pollution on the zinc content of littoral and sub-littoral organisms in Cardigan Bay, Wales, Environ.Poll. 4, 27(1973).

|8| Fuge,R., James,K.H., Trace metal concentrations in brown seaweeds, Cardigan Bay, Wales, Marine Chem. 1, 281 (1973).

|9| Stenner,R.D., Nickless,G. Distribution of some heavy metals in organisms in Hardangerfjord and Skjerstadfjord, Norway, Water, Air, Soil Poll. 3, 279 (1974).

|10| Stenner,R.D., Nickless,G.,Heavy metals in organisms of the Atlantic coast of South-West Spain and Portugal, Mar.Poll.Bull. 6(6), 89 (1975).

|11| Fuge,R., James,K.H., Trace metal concentrations in Fucus from the Bristol Channel, Mar.Poll.Bull. 5(1), 9 (1974).

|12| Foster,P. Concentrations and concentration factors of heavy metals in brown algae. Environ.Poll. 10, 45 (1976).

|13| Leatherland,T.M., Burton,J.D., The occurrence of some trace metals in coastal organisms with particular reference to the Solent region, J.Mar.Biol.Ass.UK, 54, 457 (1974).

|14| Phillips,D.J.H, The use of biological indicator organisms to monitor trace metal pollution in marine and estuarine environments - A review, Environ. Pollut. 13, 281 (1977).

|15| Bernard,M., Studies on the Radioactive Contamination of the Sea, Annual Report 1965, EUR-3274.e 1966.

|16| Topping,G., Heavy metals in shellfish from Scottish waters, Aquaculture 1, 379 (1973).

|17| Phillips,D.J.H.,The common mussel Mytilus edulis as an indicator of trace metals in Scandinavian waters. 1. Zinc and cadmium, Marine Biol. 43, 283 (1977).

|18| Karbe,L., Schnier,C., Niedergesäss,R., Trace Elements in Mussels (Mytilus edulis) from German Coastal Waters, GKSS 78/E/52, 1978.

|19| Bertine,K.K., Goldberg,E.D., Trace elements in clams, mussels and shrimp, Limnol.and Oceanogr. 17(6), 877 (1972).

|20| Fowler,S.W., Oregioni, Trace metals in mussels from the N.W.Mediterranean, Mar. Poll.Bull. $\underline{7}$(2), 26 (1976).

|21| Fukai,R., Distribution of cobalt in marine organisms, IAEA Radioactivity in the Sea, Publication nr 23, 1968.

|22| Dutton,J.W.R., Jefferies,D.F., Folkard,A.R., Jones,P.G.W., Trace metals in the North Sea, Mar.Poll.Bull. $\underline{4}$(9), 135 (1973).

|23| The Radiological Basis of the IAEA Revised Definitions and Recommendations Concerning High-Level Radioactive Waste Unsuitable for Dumping at Sea, IAEA, Technical Document IAEA-211, Vienna 1978.

|24| Brooks,R.R., Rumsby,M.G., The biogeochemistry of trace element uptake by some New Zealand bivalves, Limnol.and Oceanogr. $\underline{10}$, 521 (1965).

|25| Shuster,C.N., Pringle,B.H., Trace metal accumulation by the American Eastern Oyster., Crassostreavirginica, Proc.Nat.Shellfish Ass. $\underline{59}$, 91 (1969).

|26| Coombs,T.L., The distribution of zinc in the oyster Ostrea edulis and its relation to enzymic activity and to other metals, Marine Biol. $\underline{12}$, 170 (1972).

|27| Preston,A., Mitchell,N.T., Evaluation of public radiation exposure from the controlled marine disposal of radioactive waste (with special reference to the United Kingdom). Proc.Symp.Radioactive Contamination of the Marine Environment, Seattle 10-14 July 1972, IAEA, Vienna 1973.p 575.

Discussion

<u>G.A.M. WEBB</u>, United Kingdom

A comment rather than a question. I am pleased to see the distinction being made between pessimistic values suitable for calculation of maximum individual doses. This is a welcome trend in several of the papers at the meeting.

<u>M. ISHIKAWA</u>, Japan

You showed data on sea water given by a third party, and suggested it applicable to the coastal sea of your country. However, for such research, original data done by you is indispensable. My question arises from this : do you have data carried out by your laboratory ?

<u>A.W. VAN WEERS</u>, Netherlands

The data on the concentrations of trace elements in sea water used to derive concentration factors are, with the exception of Ag, based on analyses of sea water from the same area as that from which the organisms were taken. The values given with reference (4) are preliminary results of a study carried out in close cooperation between ECN and the Netherlands Institute of Sea Research. These results will be published in the near future. With respect to silver I have indeed used data for sea water from other areas [UK-data (ref. 6 and 22)] because of the lack of a sufficiently firm data base for the area involved. For the time being the UK-data seem a better basis than the often quoted value of 0.3 µg/l appearing, for instance, in the RIME-report.

RADIONUCLIDE ACCUMULATION BY MARINE DEMERSAL FISHES

T. Koyanagi, M. Nakahara and M. Matsuba

Division of Marine Radioecology
National Institute of Radiological Sciences
Nakaminato-shi, Ibaraki-ken, 311-12 Japan

ABSTRACT

Whole-body retention and excretion of ^{57}Co-cyanocobalamin and ^{60}Co-chloride by marine demersal fish, right-eye flounder, Kareius bicoloratus were observed by rearing the fishes in labelled seawater or administering radio-isotopes orally to the fishes as solution or labelled marine sediment. Sediment-bound radiocobalt was excreted rapidly during the short period after the administration but 60 % of ^{57}Co and 20 % of ^{60}Co were retained and eliminated at the slower rates comparable with those nuclides taken up from seawater by the fishes. Distribution of residual radioactivities in the fish bodies was not identical between two nuclides and significant accumulation of ^{57}Co-cyanocobalamin in liver and gall-bladder of the fishes was observed.

INTRODUCTION

Radionuclides introduced into the ocean as radioactive fallout or radioactive effluent from nuclear facilities are dispersed by ocean currents, surface waves, or turbulent mixing of seawater. However, it has been noticed that many kinds of the radionuclides are adsorbed by suspended matter in seawater or sediment at seabottom with high distribution coefficients and transported or redistributed in marine environment by the various geological and ecological processes.[1]-[7] The radioactivities associated with marine sediment may become the long term contamination sources as reconcentrated forms in the marine environment through desorption or resuspension even after the concentration of radionuclides in seawater is reduced by dilution or dispersion. Transfer of radioactive or stable elements associated with sediment to organisms has been investigated and reported on several kinds of marine organisms[8]-[25] but observations on marine fishes have been scarcely reported. Among the marine organisms, radioactive contamination from sediment can take place by direct contact with the radioactive sediment as well as by ingestion of the sediment with preys in the case of bottom-dwelling or deposit-feeding marine benthic organisms. Marine demersal fishes, for example, are known to ingest their preys and the surrounding mud or sand simultaneously. Therefore, sediment-bound radionuclides may become the source of radioactivity pathway to marine demersal fishes in addition to the other pathways of the radionuclides such as seawater or food. In the present study, retention of sediment-bound radioactive cobalt by marine demersal fishes was observed by the oral administration experiments with the sediment-bound radioisotopes supposing the ingestion of radioactive sediment by the fishes. Accumulation of radiocobalt from seawater or by the direct administration of radioisotope were also observed on the same species of the fishes and compared each other.

MATERIALS AND METHODS

Adsorption and desorption of radioisotopes by the sediment:

Seashore sediment was collected at the beach of Ibaraki Prefecture, the Pacific coast of Mid-Japan, and the fraction of grain size from 0.1 to 0.5 mm which represents 98.6% of the sediment sample was sifted out in seawater before use. Twenty grams of the wet sediment were shaken with 200 ml of seawater labelled with radioisotope (^{57}Co-cyanocobalamin or ^{60}Co-chloride) in flasks at room temperature for two weeks. Radioactivity in the sediment was calculated by subtracting the radioactivity in the seawater after two weeks from the initial radioactivity of the seawater and distribution coefficient (Kd) was calculated as follows.
$$\text{Distribution coefficient (Kd)} = \frac{\text{radioactivity of sediment (cpm/g)}}{\text{radioactivity of seawater (cpm/ml)}}$$
Desorption of the radioisotopes from the sediment was examined by shaking the labelled sediment with ten-fold volume of seawater which was acidified to the appropriate pH values by adding HCl solution for simulating the acidic conditions of stomach of the fishes by the secretion of gastric juice.

Administration of radioisotopes to the fishes:

Right-eye flounder, Kareius bicoloratus (200-250 mm length and 165-280 g weight) obtained from the same area with the sediment collection was selected as a deposit-feeding marine demersal fish and aclimated in aquariums with running seawater before the experiments. Gelatine capsules containing 0.4 g of the labelled sediment or 0.25 ml of radioisotope solution were administered into stomach of the fishes orally after anesthetizing the fishes in 0.01% solution of meta-amino benzoic acid ethylester methanesulfonate in seawater for a few minutes. The administration was repeated every 48 or 72 hours for about a week and then the change of radioactivity was followed for a week to four weeks. Four fishes were used for each radioisotope and each fish was reared in 20 l of seawater in the aquarium without feeding. Temperature of the seawater was controlled within $18 \pm 1°C$ and aeration was performed in each aquarium by air stones and air pumps. Radioactivity of the whole-body of the fishes was measured with a scintillation counter (Armac Scintillation Detector, model 446 with Tricarb Scintillation Spectrometer, model 3001, Packard) after each administration to get the retention curves of radioisotopes by the fishes. The fishes were transferred into fresh seawater after each radioactivity measurement. Distribution of residual radioactivities in the fish bodies after the elimination was examined by measuring the radioactivity in organs or tissues of the dissected fishes.

RESULTS AND DISCUSSION

Distribution of radionuclides to sediment:

Distribution curves of ^{57}Co-cyanocobalamin and ^{60}Co-chloride to the sediment used in the experiments were shown in Figure 1. The distribution coefficient calculated after two weeks was about 4,000 for ^{60}Co, whereas less than 3 for ^{57}Co. However, these values are considered to be varied by many parameters concerning with geological media, water qualities, content of organic matter in the sediment, and also other experimental or hydrologic conditions.[26],[27] Sediment-bound ^{60}Co-chloride was not leached in fresh seawater of natural pH around 8.2, but leachability increased with decrease of pH of the seawater and about 70% of ^{60}Co were leached at pH 2 suggesting that noticeable fraction of radioactivity may be leached from the sediment in stomach of the fishes by the secretion of acidic gastric juice. In the case of ^{57}Co-cyanocobalamin, on the other hand, leachability was independent of pH of the seawater and around 30% of radiocobalt were leached in the seawater during the range of pH from 1 to 8.2.

Retention of sediment-bound radiocobalt by the fishes:

Whole-body retention curves of radiocobalt by the fishes administered the labelled sediment continuously for about a week and reared in fresh seawater were shown in Figure 2 and 3. The radioactivity retained by the fishes just after the each administration was placed as 100% and each point shows the average and standard deviation for four fishes. Retention patterns of radiocobalt taken up by the fishes from seawater were shown under each figure as the results of observations on the fishes reared in labelled seawater for a week and transferred into the fresh seawater. Radioactivity of the fishes just before the transfer was also placed as 100%. Quite rapid loss of the whole-body radioactivity occurred in the initial short period during one to three days after each administration and suggested the elimination of unassimilated radiocobalt with excretion of the sediment, but apparent biological half-life of ^{57}Co increased by repeating administration of the sediment. After the last administration, the elimination rates of the both nuclides dropped markedly after two days and whole-body elimination of the nuclides proceeded at a much slower rate, nearly exponential rate.

The fraction of radioactivities eliminated by the fishes at the slower rates reached to about 20% of the administered dose for ^{60}Co and near to 60% for ^{57}Co, respectively, suggesting higher absorption efficiency of cyanocobalamin through digestive tract of the fishes. Elimination of radiocobalt taken up by the fishes from seawater showed different patterns and residual rate of ^{60}Co was higher than ^{57}Co. However, biological half-lives of long components were almost similar between two nuclides or between the different mode of radioisotope intake by the fishes though various biological or environmental conditions should be considered to affect on the accumulation or elimination of radiocobalt by the fishes.[28]

Absorption of radiocobalt by the fishes:

Whole-body radioactivity of the fishes contaminated by the continuous oral administration of ^{57}Co-cyanocobalamin solution was measured for a month and the retention curve was shown in Figure 4 in the same manner with the previous figures. Biological half-life of short component increased from 0.8 days to 6 days

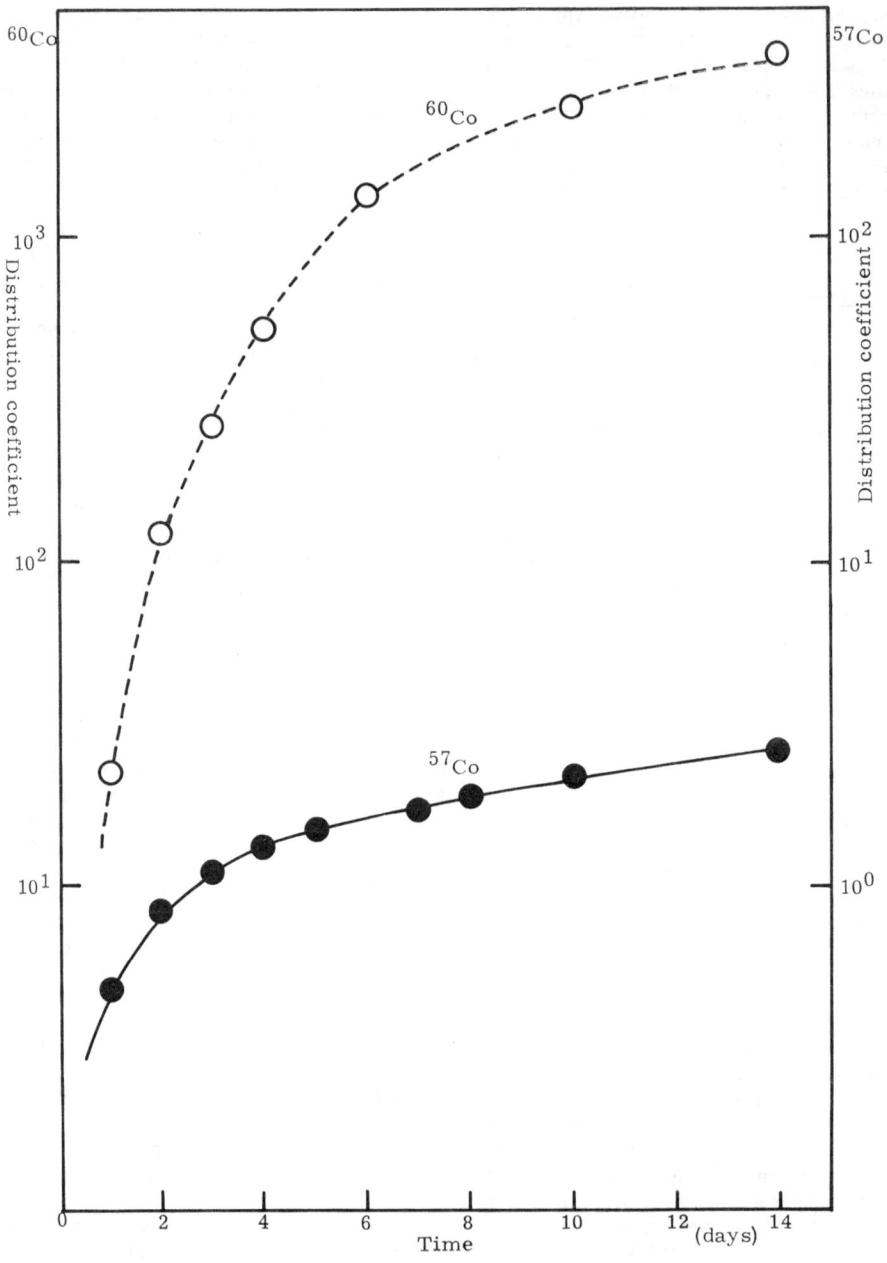

Figure 1. Distribution of ^{57}Co-cyanocobalamin and ^{60}Co-chloride to the sediment.

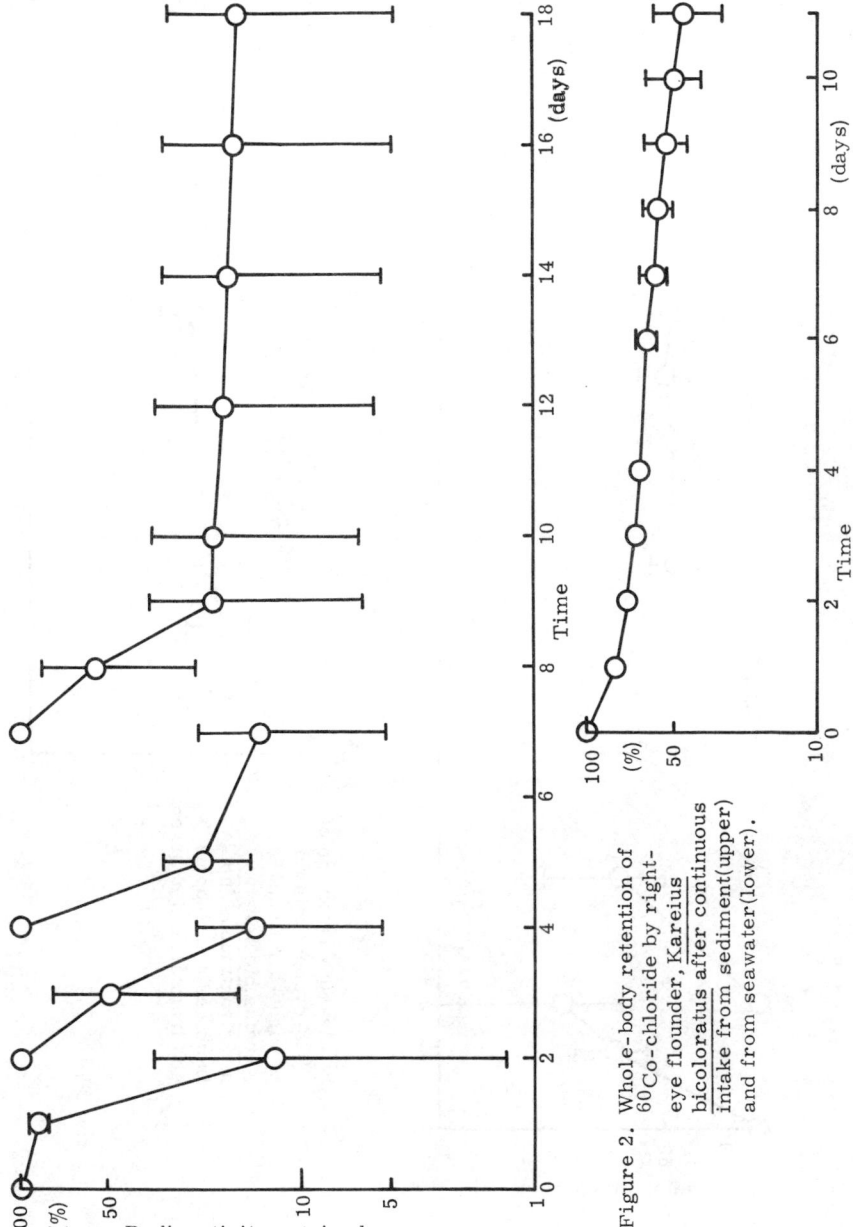

Figure 2. Whole-body retention of 60Co-chloride by right-eye flounder, Kareius bicoloratus after continuous intake from sediment(upper) and from seawater(lower).

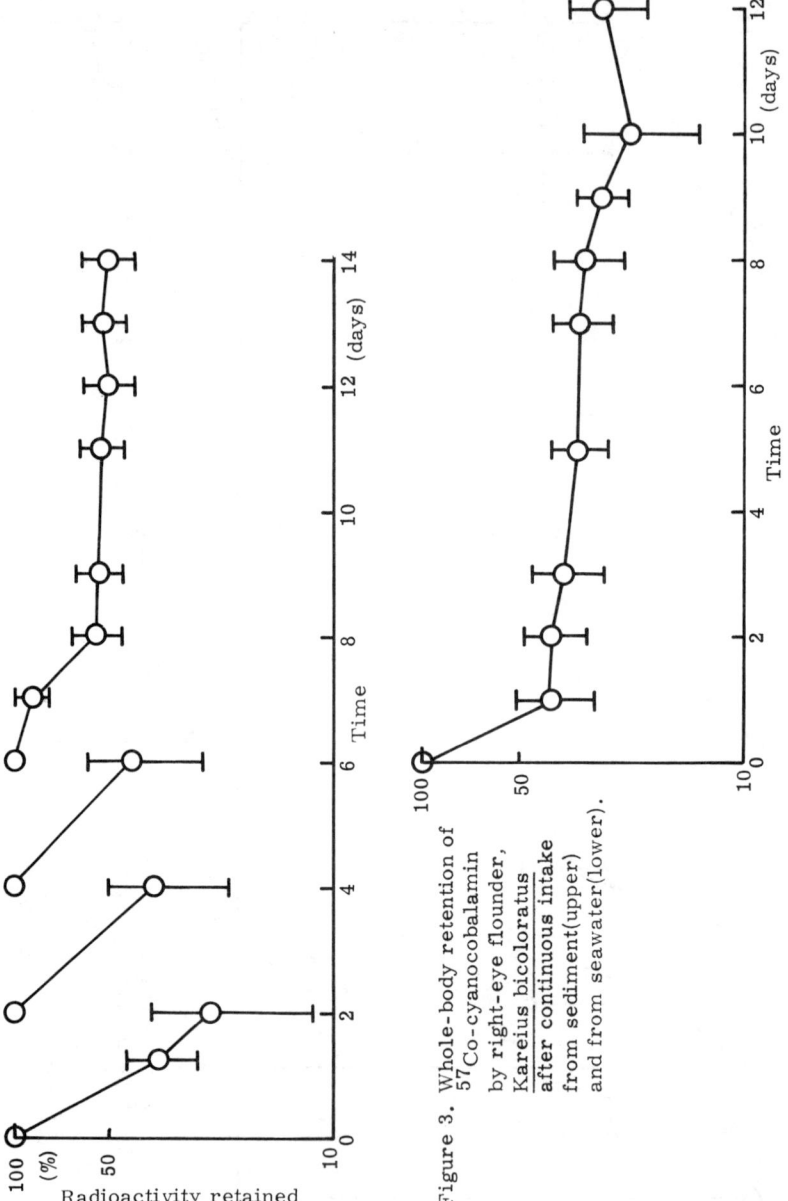

Figure 3. Whole-body retention of ^{57}Co-cyanocobalamin by right-eye flounder, Kareius bicoloratus after continuous intake from sediment(upper) and from seawater(lower).

by the repeating administration and after the last administration ^{57}Co was eliminated by two components elimination pattern as same as those accumulated by the fishes from sediment or seawater. Seventy per cent of the whole-body radioactivity was lost at a slower rate and biological half-life calculated on the long component was 39.7 days which was comparable with the values observed and reported on radiocobalt in marine fishes.[29-33] From the retention patterns obtained it is supposed that absorption of ^{57}Co-cyanocobalamin through digestive tract of the fishes exceeds gill absorption and notable fraction of radioactivity may be retained if radiocobalt associated with sediment in the form of cyanocobalamin is ingested by the fishes and after the unassimilated fraction is excreted within a few days, the absorbed radioactivity is eliminated with the biological half-life almost similar to that absorbed and accumulated from seawater or food by the fishes.

Distribution of radiocobalt in the fish bodies:

Distribution of radiocobalt absorbed through digestive tract and residued in the fish bodies after the elimination experiments was examined and shown in Table I. The fishes were dissected in twelve parts and the distribution rate (%) was calculated by measuring radioactivity and weight per cent of each part of the fishes. Concentration of radiocobalt was shown as the ratio of radioactivity in each organ or tissue to that in intestine which was normalized to 1.00.

Table I. Distribution and concentration of radiocobalt in right-eye flounder, Kareius bicoloratus.

Nuclides		^{57}Co		^{60}Co	
Organ or tissue	Weight per cent (%)	Distribution rate(%)	Concentration	Distribution rate(%)	Concentration
Intestine	1.74	4.49	1.00	8.39	1.00
Blood	2.91	1.15	0.21	16.8	1.23
Liver	2.13	68.9	14.1	12.2	1.17
Kidney	0.40	1.53	2.47	1.99	1.05
Gall-bladder	0.20	6.11	13.7	0.13	0.13
Spleen	0.09	0.26	1.63	0.45	1.05
Gonad	0.18	0.11	0.37	0.47	0.53
Gill	2.23	3.55	0.68	4.01	0.38
Skin	6.39	3.48	0.26	14.9	0.49
Muscle	51.81	5.07	0.05	11.8	0.05
Bone	13.95	1.13	0.05	8.68	0.13
Head	14.56	4.23	0.17	20.2	0.30

Distribution patterns were considerably different between the two nuclides. Most significant accumulation of ^{57}Co-cyanocobalamin was observed in liver and the concentration showed more than an order of magnitude higher value than that in intestine where the isotope was administered. Transfer of ^{57}Co from intestine to gall-bladder was also remarkable and resulted high distribution ratio in spite of very low contribution on weight, whereas quite poor distribution to gall-bladder was shown in the case of ^{60}Co-chloride. The distribution ratio to muscle showed relatively higher values in both nuclides due to the large weight contribution but the concentration was lowest among organs or tissues suggesting very little transfer of the absorbed radiocobalt to muscle of the fishes. Low concentration in edible part of the fishes suggested the insignificant role of the nuclides as the source of radioactivity pathway to men, but peculiar accumulation in the particular organ of the fishes should be examined by further investigations on the physiological properties of marine demersal fishes.

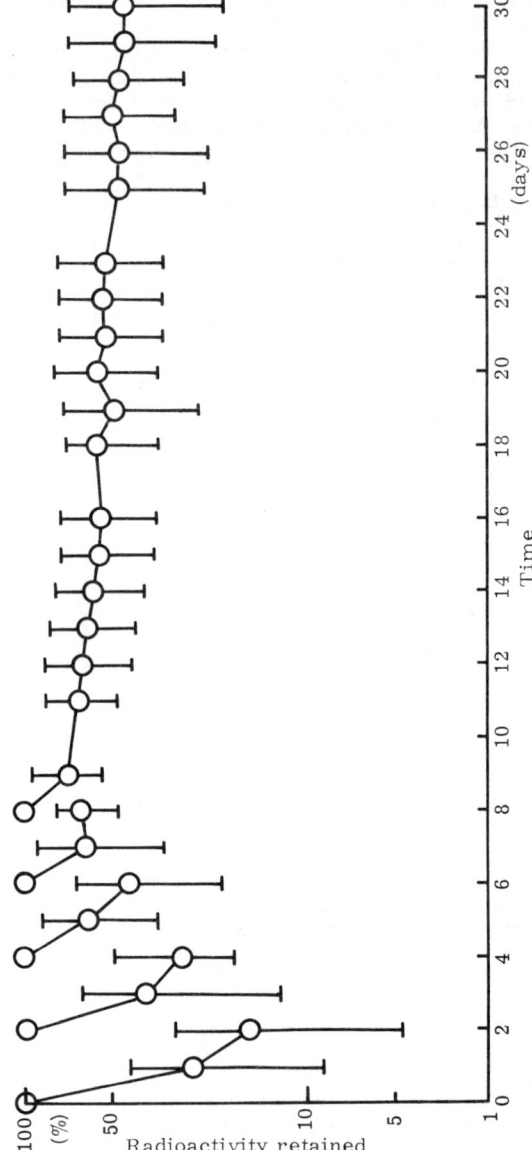

Figure 4. Whole-body retention of ^{57}Co-cyanocobalamin by right-eye flounder, Kareius bicoloratus contaminated by oral administration of radioisotope solution.

REFERENCES

[1] E. K. Duursma, P. Parsi and G. Statham(1974) Fixation of radionuclides with time by marine sediments. In Activities of the International Laboratory of Marine Radioactivity, 1974 Report, IAEA-163, pp. 107-108.

[2] J. A. Hetherington and D. F. Jefferies(1974) Distribution of some fission product radionuclides in sea and estuarine sediments. Neth. J. Sea Res., 8(4): 319-338.

[3] K. C. Pillai, N. N. Dey, E. Mathew and B. U. Kothari(1975) Behaviour of discharged radionuclides from fuel reprocessing operations in the aquatic environment of Bombay Harbour Bay. In Impacts of Nuclear Releases into the Aquatic Environment, IAEA STI/PUB/406, pp. 277-300.

[4] R. Schaeffer(1975) The effect of sediment displacement of the dispersal of radionuclides., ibid, pp. 263-276.

[5] J. A. Hetherington, D. F. Jefferies and M. B. Lovett(1975) Some investigations into the behaviour of plutonium in the marine environment., ibid, pp. 193-212.

[6] K. Nakamura and Y. Nagaya(1975) Accumulation of radionuclides in coastal sediment of Japan(II). Contents of fission products in some coastal sediments collected in 1966-1972. J. Radiat. Res., 16:184-192.

[7] L. D. Labeyrie, H. D. Livingston and V. T. Bowen(1976) Comparison of the distributions in marine sediments of the fall-out derived nuclides ^{55}Fe and ^{239,240}Pu. In Transuranium Nuclides in the Environment, IAEA-SM-199, pp. 121-137.

[8] G. W. Bryan(1974) Adaptation of an estuarine polychaete to sediments containing high concentrations of heavy metals. Pollut. Physiol. Mar. Org. (ed. by F. J. Vernberg and W. B. Vernberg, Academic; New York, N. Y.), pp. 123-135.

[9] C. Amiard-Triquet(1974) Contamination by cerium-144 and iron-59 of a sediment with Arenicola marina. Cah. Biol. Mar. 15;483-494.

[10] C. Amiard-Triquet(1975) Study of radionuclide transfer between invertebrates and their marine sedimentary environment. Report 1975, CEA-R-4705, p. 98.

[11] W. C. Renfro and G. Benayoun(1974) Sediment-worm interactions: transfer of ^{65}Zn, ^{109}Cd, and ^{110m}Ag from sediment by Nereis diversicolor. In Activities of the International Laboratory of Marine Radioactivity, IAEA-163, pp. 33-43.

[12] W. C. Renfro and G. Benayoun(1976) Sediment-worm interaction: transfer of ^{65}Zn from marine silt by the polychaete Nereis diversicolor. In Radioecology and Energy Resources(ed. by C. E. Cushing, Jr.), pp. 250-255.

[13] T. Ueda, R. Nakamura and Y. Suzuki(1976) Comparison of ^{115m}Cd accumulation from sediment and sea water by polychaete worms. Bull. Japan. Soc. Sci. Fish., 42:299-306.

[14] R. J. Huggett, F. A. Cross and M. E. Bender(1975) Distribution of copper and zinc in oysters and sediment from three coastal-plain estuaries. ERDA Symp. Ser., 36:224-238.

[15] C. T. Hess, C. W. Smith and A. H. Price(1975) Model for the accumulation of radionuclides in oysters and sediments. Nature(London), 258, (5532):225-226.

[16] F. L. Harrison, K. M. Wong and R. E. Heft(1976) Role of solubles and particulates in radionuclides accumulation in the oyster Crassostrea gigas in the discharge canal of a nuclear power plant. In Radioecology and Energy Resources (ed. by C. E. Cushing, Jr.), pp. 9-20.

[17] Tin Mo and F. G. Lowman(1976) Laboratory experiments on the transfer of plutonium from marine sediments to seawater and to marine organisms., ibid, pp. 86-95.

[18] S. N. Luoma and E. A. Jenne(1976) Factors affecting the availability of sediment-bound cadmium to the estuarine, deposit-feeding clam, Macama balthica. ibid, pp. 283-290.

[19] R. Nakamura, Y. Suzuki and T. Ueda(1975) Influence of marine sediment on the accumulation of radionuclides by green alga(Ulva pertusa). J. Radiat. Res. 16:224-236.

[20] Y. Suzuki, R. Nakamura and T. Ueda(1976) Radioecology of ^{60}Co in Urazoko Bay. Correlation between levels of ^{60}Co in Sargassoes and marine sediments.

ibid, 17:115-126.
[21] W.R. Schell and R.L. Watters(1975) Plutonium in aqueous systems. Health Phys., 29:589-597.
[22] J.C. Guary and A. Fraizier(1977) Influence of trophic level and calcification on the uptake of plutonium observed, in situ, in marine organisms., ibid, 32:21-28.
[23] V.T. Bowen, H.D. Livingston and J.C. Burke(1976) Distribution of transuranium nuclides in sediment and biota of the North Atlantic Ocean. In Transuranium Nuclides in the Environment, IAEA-SM-199, pp. 107-120.
[24] T. Ueda, R. Nakamura and Y. Suzuki(1978) Comparison of influence of sediments and sea water on accumulation of radionuclides by marine organisms. J. Radiat. Res., 19:93-99.
[25] T. Koyanagi, M. Nakahara and M. Iimura(1978) Absorption of sediment-bound radionuclides through the digestive tract of marine demersal fishes., ibid, 19:295-305.
[26] A. Nissenbaum and D.J. Swaine(1976) Organic matter-metal interactions in recent sediments; the role of humic substances. Geochim. Cosmochim. Acta, 40(7):809-816.
[27] A. Vertacnik, P. Strohal and S. Lulic(1979) Fixation of ^{60}Co on natural sorbent from phenol-polluted water. Health Phys., 36:491-496.
[28] M. Nakahara, S. Hirano, T. Ishii and T. Koyanagi(1979) Accumulation and excretion of cobalt-60 taken up from seawater by marine fishes. Bull. Japan. Soc. Sci. Fish., 45(in press).
[29] J.P. Baptist, D.E. Hoss and C.W. Lewis(1970) Retention of ^{51}Cr, ^{59}Fe, ^{60}Co, ^{65}Zn, ^{85}Sr, ^{95}Nb, 141mIn and ^{131}I by the Atlantic croaker(Micropogon undulatus). Health Phys., 18:141-148.
[30] J.R. Reed(1971) Uptake and excretion of ^{60}Co by Black bullheads, Ictalurus melas(Rafinesque). ibid, 21:835-844.
[31] K. Kimura and R. Ichikawa(1972) Accumulation and retention of ingested cobalt-60 by the common goby. Bull. Japan. Soc. Sci. Fish., 38:1097-1103.
[32] R.J. Pentreath(1973) The accumulation and retention of ^{59}Fe and ^{58}Co by the plaice, Pleuronectes platessa L. J. exp. mar. Biol. Ecol., 12:315-326.
[33] R.J. Pentreath(1973) The accumulation from sea water of ^{65}Zn, 54Mn, 58Co and ^{59}Fe by the thornback ray, Raja clavata L. ibid, 12:327-334.

CONCENTRATION FACTORS OF MESOPELAGIC ORGANISMS

M. Nakahara[*], T. Ueda[*], Y. Suzuki[*], T. Ishii[*] and H. Suzuki[**]

[*] Division of Marine Radioecology
National Institute of Radiological Sciences
Nakaminato, Ibaraki, 311-12 Japan

[**] Health and Safety Division, Tokai Works
Power Reactor and Nuclear Fuel Development Cooperation
Tokai, Ibaraki, 319-11 Japan

ABSTRACT

The amounts of several stable elements in mesopelagic organisms were measured to determine the concentration factors of these elements on the assumption that the contamination by the radioactive materials due to deep sea disposal of radioactive wastes. The concentration factor is one of the important parameters used to evaluate the internal radiation doses to man by the intake of the contaminated organisms.

The levels of Cs-137 and Sr-90 due to fallout in the mesopelagic fishes were also investigated with interest in vertical transfer of the nuclides in the sea.

Table I. Sample organisms

	Japanese name	Scientific name	Average body weight (g)
Mesopelagic fishes	Akoudai	Sebastes matsubarai	2700
	Mutsu	Scombrops boops	2650
	Medai	Hyperoglyphe japonica	2250
	Aodai	Paracaesio caeruleus	1200
	Kinmedai	Beryx splendens	1550
	Himedai	Pristipomoides sieboldi	1250
Mesopelagic cephalopod	Mizudako	Paroctopus dofleini	1560
Coastal fishes	Hirame	Palalichthys olivaceus	200
	Ezoisoainame	Lotella maximowiczi	400
	Suzuki	Lateolabrax japonicus	300
	Katsuo	Katsuwonus pelamis	1700
	Ishimochi	Argyrosomus argentatus	130
Coastal cephalopod	Madako	Octopus vulgaris	1300
	Yariika	Doryteuthis bleekeri	320
	Bakaika	Sthenoteuthis bartrami	380
	Surumeika	Ommastrephes sloanei	390

INTRODUCTION

The deep sea disposal of radioactive wastes is scheduled to begin in the late 1980's in Japan. If the leak of radioactive materials from the wastes to seawater occurs, contamination of deep sea organisms would be unavoidable. In Japan, about 150 species of deep water fishes are known and 9 species of the fishes are caught commercially from the mesopelagic layer: an annual catch is from 500 to 3000 tons per species. [1] Therefore, it would be significant to determine concentration factors for some elements in the edible part of these mesopelagic organisms for evaluation of internal radiation doses to man by intake of the contaminated organisms. The route through the deep water organisms seems to be a major pathway for the entry of contamination from these wastes into the human food chain. Furthermore, it would be of great interest to know the levels of Cs-137 and Sr-90 due to fallout in deep water organisms, with particular reference to the vertical transfer of the nuclides in the sea. With these aims, the amounts of some stable elements Mn, Fe, Cu, Zn, Co and Cs) in the edible part of mesopelagic fishes and cephalopods and the radioactivity of Cs-137 in muscle and of Sr-90 in bone of several species of the mesopelagic fishes were measured and the concentration factors were estimated on these elements. These results were also compared with those of the coastal organisms.

MATERIALS AND METHODS

Samples:
The names and average body weight of the organisms used in the investigation are shown in Table I. and the living depth of the mesopelagic organisms are illustrated in Figure 1. Gathering of these organisms was carried out several times during October 1977 and August 1979. Each sample was constructed from the tissues of from 1 to 66 individual organisms. The samples dissected from the organisms were ashed by an electric furnace at 450 °C for more than 48 hours after being dried in an oven at 110 °C.

Analytical methods:
The concentrations of Mn, Fe, Cu and Zn which were extracted into 8N HCl solution from the ashed samples were measured by using an atomic absorption spectrophotometer (Perkin-Elmer Co., 403 type) after removing interfering elements by solvent extraction (TOA-Xylene). For the determination of Co and Cs, the ashed samples (about 50 mg each) were encapsuled in quarz capillaries of high purity and the prepared samples were irradiated together with comparative sample of the elements in the JRR-3 reactor (Japan Atomic Energy Research Institute) at a neutron flux of 2×10^{13} $n/cm^2/s$ for 260 hours. After cooling for about 2 weeks, the radioactivity of neutron induced isotopes of Co and Cs was counted by a 40 cm^3 Ge(Li) detector equipped with a 1024 channel pulse height analyzer after radio-chemical separation of the irradiated samples.
The radioactivity of Cs-137 in the fish muscle and of Sr-90 in the fish bone originating in radioactive fallout was measured by using a low background gas-flow counter according to the ordinary radio-chemical analysis. [2],[3]

RESULTS AND DISCUSSION

The concentrations of Mn, Fe, Cu, Zn, Co and Cs in muscle of the mesopelagic and coastal fishes are shown in Table II. The figures in brackets indicate the numbers of individual organisms used for each sample. The amounts of each elements excluding Co and Cs was measured twice (in October 1978 and June 1979) for the mesopelagic fishes and the values showed good agreement between the samples of October and June in almost every case. The concentrations of these elements in the mesopelagic fish muscle were as follows:

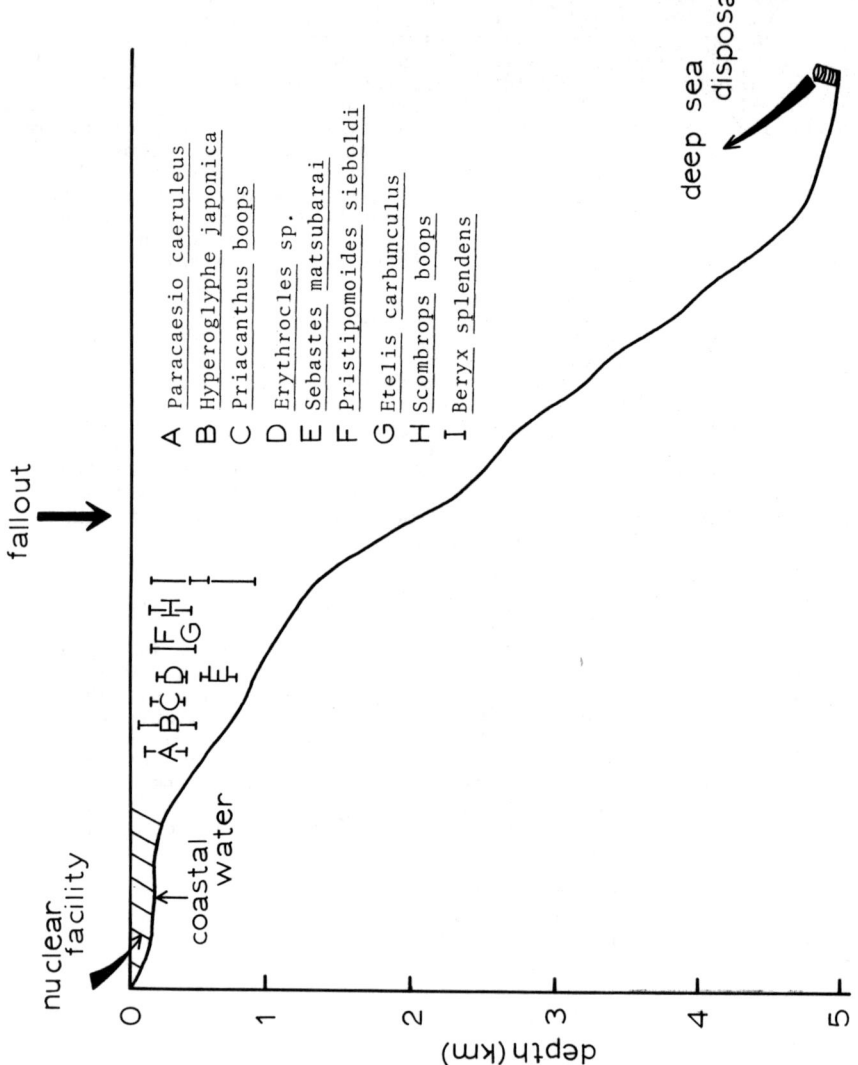

Fig. 1 Inhabiting depth of mesopelagic fishes

Table II. Concentrations of trace elements in fish muscle (mg or μg/Kg wet)

Species	No. in Sample I*	No. in Sample II**	Mn (ppm) I*	Mn (ppm) II**	Fe (ppm) I*	Fe (ppm) II**	Cu (ppm) I*	Cu (ppm) II**	Zn (ppm) I*	Zn (ppm) II**	Co (ppb) I*	Cs (ppb) I*
Mesopelagic												
Sebastes matsubarai	(3)	(1)	0.09	0.06	1.7	2.2	0.19	0.12	4.6	2.6	—	54.9
Scombrops boops	(2)	(2)	0.06	0.06	3.6	3.7	0.19	0.27	3.0	3.0	2.3	29.5
Hyperoglyphe japonica	(3)	(2)	0.05	0.06	2.5	2.5	0.27	0.21	4.1	2.8	3.3	20.6
Paracaesio caeruleus	(6)	(2)	0.05	0.06	3.3	3.9	0.19	0.23	4.3	3.0	4.0	21.0
Beryx splendens	(5)	(2)	0.08	0.06	2.4	2.9	0.24	0.16	8.7	2.6	1.9	35.4
Pristipomoides sieboldi	(2)	—	0.06		2.9		0.26		3.9		5.3	12.1
Average ± SD***			0.06±0.01		2.9±0.7		0.21±0.05		3.9±1.8		3.4±1.4	28.9 ±15.1
Coastal												
Palalichthys olivaceus		(25)		0.13		2.3		0.18		3.6		
Lotella maximowiczi		(14)		0.10		3.2		0.16		3.4		
Lateolabrax japonicus		(10)		0.08		2.9		0.18		4.8		
Katsuwonus pelamis		(3)		0.06		4.0		0.88		4.5		
Argyrosomus argentatus		(41)		0.10		3.3		0.21		3.4		
Average ± SD***			0.09±0.03		3.1±0.6		0.32±0.31		3.9±0.7			

* data of Oct. 1978
** data of Jun. 1979
*** standard deviation

Table III. Levels of ^{137}Cs in fish muscle and of ^{90}Sr in fish bone (pCi/Kg wet)

Species	^{137}Cs in muscle I*	^{137}Cs in muscle II**	^{90}Sr in bone II**
Mesopelagic			
Sebastes matsubarai	4.0	18.4	1.3
Scombrops boops	26.0	37.4	0.8
Hyperoglyphe japonica	—	9.7	2.0
Paracaesio caeruleus	14.4	13.4	0.9
Beryx splendens	—	26.0	2.0
Average ± SD***	18.7 ± 10.7		1.4 ± 0.6
Coastal			
Paralichthys olivaceus		12.0	0.2
Lotella maximowiczi		15.1	0.9
Lateolabrax japonicus		43.8	4.3
Katsuwonus pelamis		12.6	2.1
Argyrosomus argentatus		7.0	1.3
Average ± SD***		18.1 ± 14.7	1.7 ± 1.6

* data of Oct. 1978
** data of Jun. 1979
*** standard deviation

0.05-0.09 ppm of Mn, 1.7-3.9 ppm of Fe, 0.12-0.27 ppm of Cu, 2.6-8.7 ppm of Zn, 1.9-5.3 ppb of Co and 12.1-54.9 ppb of Cs.   These results for Fe, Zn, Co and Cs in the mesopelagic fish muscle were similar to those in the coastal fish muscle we measured and also comparable to the data of coastal and epipelagic fish muscle reported by other authors. [4-10]   However, Mn content in the mesopelagic fish muscle seems slightly lower than those in the coastal and epipelagic fish muscle.   Among the coastal fishes we examined, Katsuwonus showed relatively high Cu concentration in the muscle. Though the reason is not clear, this high value raises the average value from 0.18 mg excluding Katsuwonus to 0.32 mg Cu/Kg wet including the fish.   Cu content in the mesopelagic fish muscle was in accord with the former value of the coastal fish muscle.

The levels of Cs-137 in the fish muscle and of Sr-90 in the fish bone are shown in Table III.   The average values were almost comparable between the mesopelagic fish and the coastal fish as 18.7 and 18.1 pCi/Kg wet, respectively.   According to Nagaya and Nakamura [11], the level of Cs-137 in deeper seawater is lower than in surface water.   Thus, it is curious that the level of Cs-137 in the mesopelagic fish muscle is comparable to the level of the nuclide in the coastal fish muscle.   We estimated by laboratory experiments that the accumulation of the nuclide Cs-137 in marine fishes in a natural ecosystem is contributed to equally by food contamination and by the contamination of seawater. [12]   At this point, the contribution of food contaminated by Cs-137 should be considered as well as the contribution of seawater to explain the phenomenon; the level of Cs-137 in the mesopelagic fish muscle is comparable to that in the coastal fish muscle, though they live in different concentrations of the nuclide.   This requires further investigation.   The concentration factor of Cs-137 in the mesopelagic fish muscle can be estimated at about 100 from the Cs-137 measurement using the value 0.2 pCi Cs-137/l of surface seawater, in lieu of the currently unknown value for the mesopelagic depth. On the other hand, the mesopelagic fishes showed slightly lower Sr-90 level in bone than the coastal fishes.   This seems reasonable because the level of Sr-90 in deeper seawater is lower than that in surface seawater [11] as in the case of Cs-137 and the level of Sr-90 in marine fish is mainly influenced by the level of the nuclide in seawater. [12]

The concentrations of Mn, Fe, Cu, Zn, Co and Cs in several commercially important cephalopods are shown in Table IV.   Of these cephalopods, only Paroctopus dofleini corresponds to mesopelagic organism and it dwells from 100 to 1000 m deep.   O. vulgaris and the other three species of squids live at a depth of less than 100 m. As in the case of the fishes we examined, a difference between mesopelagic and coastal cephalopods could not be observed in the amounts of the elements in muscle tissue (mantle, arms and tentacles). The amounts of the elements in the muscle tissue of the cephalopods were 0.09-0.33 ppm of Mn, 1.1-2.8 ppm of Fe, 1.2-4.0 ppm of Cu, 8-14 ppm of Zn, 2.2-8.8 ppb of Co and 1.9-4.6 ppb of Cs.   These results are similar to the data reported by other authors. [4][5][10][13]

The concentrations of the elements in the liver and the branchial heart of the cephalopods are shown in Table V.   The contents of Fe, Cu, Zn and Co in these organs were much higher than those in the muscle, whereas those of Cs and Mn were almost same.   Since the values of the elements fluctuated widely among the species of the organisms, a difference between the mesopelagic and the coastal cephalopods could not be discriminated.   O. vulgaris showed extremely high Co content in both organs, particularly in the branchial heart.   We have also observed a peculiar accumulation of Co-60 in the branchial heart of O. vulgaris by radiotracer experiment. [14] From these results, it is presumed that the peculiar affinity of the branchial heart of O. vulgaris with Co is a species specificity of the organism.

The concentration factors of the elements in muscle of the mesopelagic fishes and cephalopods were calculated using the rounded

Table IV. Concentrations of trace elements in muscle of cephalopods (mg or μg/Kg wet)

Species	No. in Sample	Mn (ppm)	Fe (ppm)	Cu (ppm)	Zn (ppm)	Co (ppb)	Cs (ppb)
Mesopelagic							
Paroctopus dofleini	(16)	0.17	1.4	4.0	13	2.2	1.9
Coastal							
Octopus vulgaris	(19)	0.33	1.8	2.5	14	8.1	3.5
Doryteuthis bleekeri	(20)	0.12	2.8	1.2	8	4.5	3.2
Sthenoteuthis bartrami	(64)	0.13	1.8	2.6	14	8.8	4.6
Ommastrephes sloanei	(66)	0.09	1.1	1.2	12	3.1	3.6
Average ± SD*		0.14 ±0.03	1.8 ±0.6	2.0 ±1.2	12.2 ±2.5	5.3 ±3.0	3.4 ±1.0

* standard deviation

Table V. Concentrations of trace elements in liver and branchial heart of cephalopods (mg or µg/Kg wet)

Organ and species	Mn (ppm)	Fe (ppm)	Cu (ppm)	Zn (ppm)	Co (ppb)	Cs (ppb)
Liver						
Mesopelagic						
Paroctopus dofleini	1.20	90	190	160	640	2.3
Coastal						
Octopus vulgaris	1.08	110	110	520	3520	2.9
Doryteuthis bleekeri	0.39	33	90	25	48	3.1
Sthenoteuthis bartrami	0.46	81	540	78	—	—
Ommastrephes sloanei	0.73	150	96	43	480	3.0
Branchial heart						
Mesopelagic						
P. dofleini	0.50	5	20	17	420	4.3
Coastal						
O. vulgaris	0.89	69	41	43	9450	5.6
D. bleekeri	—	—	—	—	30	1.8
S. bartrami	0.32	10	66	15	67	2.9
O. sloanei	0.47	16	70	23	—	—

Table VI. Concentration factors of trace elements in muscle of mesopelagic fish and cephalopod

	Mn	Fe	Cu	Zn	Co	Cs
Fish	60	290	70	390	30	60
Cephalopod	200	100	1300	1000	20	10
Seawater* (µg/l)	1	10	3	10	0.1	0.5

* Thompson et al. (1972)

values: 1 μg Mn/l, 10 μg Fe/l, 3 μg Cu/l, 10 μg Zn/l, 0.1 μg Co/l and 0.5 μg Cs/l of seawater [15] and shown in Table VI.

In conclusion, the amounts and the concentration factors of Mn, Fe, Cu, Zn, Co and Cs in muscle tissue of mesopelagic fishes and cephalopods are similar to those for coastal fishes and cephalopods, respectively.

For details of the investigation, our papers [16][17] should be referred to.

REFERENCES

[1] Masuzawa, H. : "Ecology of Beryx splendens and Other Benthic Fishes", Proc. 12th Fisheries Meeting of Soc. Japan. Sci. Fish., pp. 27-38, (1978).
[2] Suzuki, Y., Nakamura, R. and Ueda, T. : "Cesium-137 Contamination of Marine Fishes from the Coasts of Japan", J. Radiat. Res. 14 (4), 382-391 (1973).
[3] Ueda, T., Suzuki, Y. and Nakamura, R. : "Accumulation of Sr in Marine Organisms-I. Strontium and Calcium Contents, CF and OR Values in Marine Organisms", Bull. Japan. Soc. Sci. Fish., 39 (12), 1253-1262 (1973).
[4] Fujii, M., Kudo, T. and Kawashima, M. : in Annual Report of Institute Environmental Pollution and Public Health,Vol. 2, pp. 8-12, Oita Prefecture, Japan 1974.
[5] Ishii, T., Suzuki, H. and Koyanagi, T. : "Determination of Trace Elements in Marine Organisms-I.", Bull. Japan. Soc. Sci. Fish., 44 (2), 155-162 (1978).
[6] Ichikawa, R. and Ohno, S. : "Levels of Cobalt, Cesium and Zinc in Some Marine Organisms in Japan", ibid, 40 (5), 501-508 (1974).
[7] Tsuchiya, Y. and Makuta, T. : "The Occurrence of Cobalt in Marine Products", Tohoku J. Agr. Res. 2, 113-117 (1951).
[8] As, D. V., Fourie, H.O. and Vleggaar, C.M. : "Accumulation of Certain Trace Elements in Marine Organisms from the Sea around the Cape of Good Hope", Proc. Symp. on Radioactive Contamination of the Marine Environment, pp. 615-624, IAEA, Vienna 1973.
[9] Tanaka, Y., Ikebe, K., Tanaka, R. and Kunita, N. : "Contents of Heavy Metals in Foods (1)", Journal of Food Hygienic Society of Japan, 14 (2) 196-201 (1973).
[10] Tanaka, Y., Ikebe, K., Tanaka, R. and Kunita, N. : "Contents of Heavy Metals in Foods (3)", ibid, 15 (5) 390-393 (1974).
[11] Nagaya, Y. and Nakamura, K. : "Distribution of Sr-90 and Cs-137 in Deep Waters around Japan", J. Oceanogr. Soc. Japan (to be published)
[12] Suzuki, Y., Nakahara, M., Nakamura, R. and Ueda, T. : "Roles of Food and Sea Water in the Accumulation of Radionuclides by Marine Fishes", Bull. Japan. Soc. Sci. Fish., (in press)
[13] Tsuchiya, Y. : "Inorganic Elements in Muscle of Marine Organisms", in Fisheries Chemistry, pp. 148-156, Kouseisha-Kouseikaku Co. Ltd., Tokyo 1965.
[14] Nakahara, M., Koyanagi, T., Ueda, T. and Shimizu, C. : "Peculiar Accumulation of Cobalt-60 by the Branchial Heart of Octopus", Bull. Japan. Soc. Sci. Fish., 45 (4), 539 (1979).
[15] Thompson, S.E., Burton, C.A., Quinn, D.J. and Ng, Y.C. : "Concentration Factors of Chemical Elements in Edible Aquatic Organisms", UCRL-50564, Rev. 1 (1972).
[16] Ueda, T., Nakahara, M., Ishii, T., Suzuki, Y. and Suzuki, H. : "Amounts of Trace Elements in Marine Cephalopods", J. Radiat. Res., (submitted).
[17] Ueda, T., Suzuki, Y., Nakahara, M., Ishii, T. and Suzuki, H. : "Concentration Factors of Some Elements for Mesopelagic Fishes", J. Radiat. Res., (submitted).

Discussion

A.W. VAN WEERS, Netherlands

I refer to your last table where you give derived concentration factors based on seawater concentrations taken from Thompson 1972. Have you evidence that, especially with respect to iron and zinc, the values given by Thompson are indeed applicable to sea water from the area where the mesopelagic species were sampled ? I ask this because the concentration of these two elements in open sea water can be considerably below the values used in your last table.

M. NAKAHARA, Japan

Unfortunately we do not have data on the concentration of these elements in mesopelagic sea water. We therefore used the data by Thompson to calculate the concentration factors. We also do not know whether the concentrations of the elements in sea water close to the slopes of sea mountains, where the mesopelagic species are living, are similar to those of open sea water or not.

CONCENTRATION FACTORS OF MARINE ORGANISMS USED FOR THE ENVIRONMENTAL DOSE ASSESSMENT

M. Kurabayashi, S. Fukuda and Y. Kurokawa

Health and Safety Division, Power Reactor and
Nuclear Fuel Development Corporation
Akasaka, Minatoku, Tokyo, Japan

ABSTRACT

Radionuclides contained in the low level liquid wastes discharged from the Tokai fuel reprocessing plant may possibly be accumulated in marine organisms in the coastal environment.

The bioaccumulation or concentration factors of marine organisms are required to assess doses of the public through the intake of seafoods caught near the site.

Published information on the concentration of radionuclides in seawaters and marine organisms was surveyed.

The data obtained by field measurements at the coastal sea of Windscale and Tokai were compiled, and the concentration factors applicable to the dose assessment were calculated.

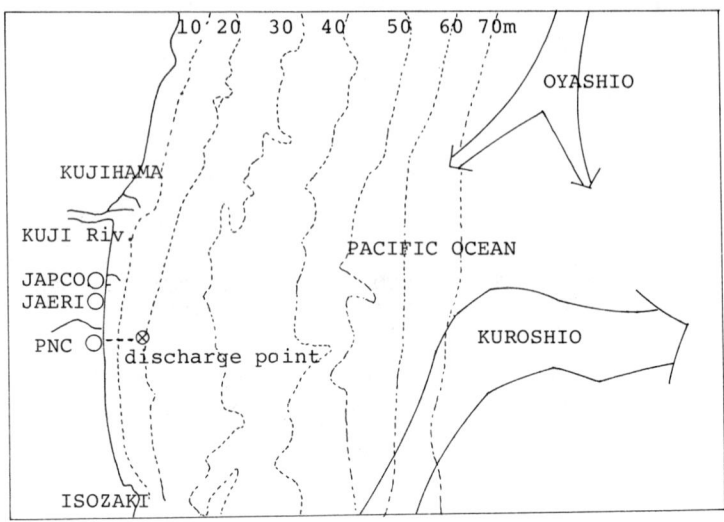

Figure 1 Discharge Area at the Tokai Fuel Reprocessing Plant

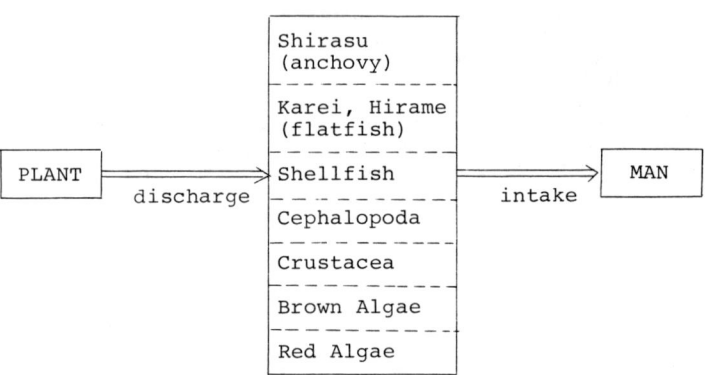

Figure 2 Pathways which PNC adopted to assess doses to the Public through intakes of seafoods caught at Tokai offshore

1. INTRODUCTION

The construction of the Tokai fuel reprocessing plant started in 1971 at the site of Tokai-mura Ibaraki Prefecture and finished in 1974. The plant has now been placed under the hot test operation, after various examinations.

Power Reactor and Nuclear Fuel Development Corporation (PNC) performed the dose estimation for the public through the intake of seafoods caught near the site. PNC investigated the concentration factors of marine organisms which are the important constituents of foods taken by the inhabitants near the site.

The concentration factors of marine organisms have been hitherto reviewed in many references [1, 2, 3, 4, 5].

In general, to derive the concentration factors of marine organisms, there are three methods as follows and each method may have some specialities [6, 7, 8];

(1) from stable elements accumulated in sea water and marine organisms.
(2) by tracer experiments of marine organisms.
(3) from field data of radionuclides in sea water and marine organisms.

We investigated the original references and calculated the concentration factors for dose assessment from the field data at Tokai coastal area and also calculated concentration factors at the Irish Sea.

2. OUTLINE OF THE DISCHARGE AREA

The low level liquid wastes of Tokai plant are released into the coast of the Pacific Ocean at about 1.8 kilometers offshore, about 18 meters in depth, from the vertical nozzle with the initial velocity of about 7 meters per second.

This coastal area is under the influence of two currents, in spring and summer by KUROSHIO current flowing northward and in autumn and winter by OYASHIO current southward [9, 10, 11].

The current direction of the surface waters at this area was observed mainly to be southward and/or northward parallel to the shore line, and the average velocity to be nearly 10 centimeters per second [unpublished].

The temperature of the surface waters observed was in the range of 7°C ∿ 25°C and the chlorinity fluctuated between 14.8‰ ∿ 19.2‰ under the influence of KUJI River at the distance of about 5 kilometers from the discharge point [unpublished].

The sea bottom at Tokai offshore was observed to be covered with sands as a whole and partially by rocks [11, unpublished].

The water depth increases gradually with a gentle slope [11, unpublished].

The abundance of marine benthic organisms observed was very poor in both species and quantity near the discharge point, because the bottom is almost covered with sands and always washed with waves [11, unpublished].

In view of fisheries, Tokai offshore is one of the fishing grounds of Shirasu (young of Anchovy) and other fishes such as Hirame (flounder), Karei (flatfish), shellfish, shrimps, octopus and so on. Brown Algaes such as Hijiki and Wakame are produced in Kujihama and Isozaki at the distance of about 6 kilometers northward and southward from the discharge point. Red algaes such as laver are produced at Isozaki in a small quantity [11].

Table I LEVELS OF RADIOACTIVITY (pCi/ℓ) OF SEA WATER IN THE VICINITY OF TOKAI OFFSHORE

Nuclide	Ru-106	Sr-90	Cs-137	Ce-144	Zr/Nb-95	Pu-239,240
Sumples	73	70	62	58	45	22
Min.	0.01	0.13	0.04	0.01	0.01	0.1×10^{-3}
Max.	0.24	0.93	1.04	0.28	0.40	1.7×10^{-3}
Ave.	0.06	0.27	0.33	0.06	0.12	0.7×10^{-3}

Table II LEVELS OF RADIOACTIVITY (pCi/kg.wet wei.) OF MARINE ORGANISMS IN THE VICINITY OF THE TOKAI OFFSHORE

Marine Organisms		Ru-106	Sr-90	Cs-137	Ce-144	Zr/Nb-95	Pu-239,240
Shirasu (whole)	ave.	1.55	0.93	5.74	1.27	4.44	0.036
	range	0.17 ∿6.40	0.17 ∿2.70	3.88 ∿11.1	0.17 ∿2.80	0.59 ∿7.81	0.015 ∿0.085
Fish(adult) (flesh)	ave.	1.50	0.86	7.45	3.47	6.22	0.099
	range	0.16 ∿5.09	0.26 ∿1.86	1.59 ∿12.4	0.14 ∿9.20	1.30 ∿18.0	0.002 ∿0.371
Shellfish (flesh)	ave.	15.7	1.39	2.36	11.9	6.10	0.14
	range	3.8 ∿34.5	0.57 ∿3.40	1.16 ∿4.76	2.91 ∿27.1	3.47 ∿12.1	0.03 ∿0.23
Cephalopoda (flesh)	ave.	1.22	0.44	3.33	1.28	6.01	0.032
	range	0.49 ∿2.30	0.17 ∿1.00	2.78 ∿3.98	0.20 ∿3.25	0.88 ∿13.9	0.008 ∿0.054
Crustacea. (whole)	ave.	10.6	7.57	3.68	7.35	5.84	0.26
	range	1.3 ∿35.0	2.95 ∿12.8	2.41 ∿5.70	1.03 ∿15.0	2.11 ∿15.1	0.13 ∿0.36
Brown algae	ave.	5.82	2.92	6.54	4.12	3.53	0.20
	range	0.10 ∿14.5	1.43 ∿6.84	2.29 ∿12.7	0.42 ∿10.3	1.09 ∿28.2	0.021 ∿0.41
Red algae	ave.	26.5	4.06	3.95	6.81	2.80	0.54
	range	13.1 ∿40.0	3.26 ∿4.83	2.80 ∿6.04	0.93 ∿12.7		

3. THE PREMISES OF THE PREOPERATIONAL DOSE ASSESSMENT

3-1 Pathways of exposure

Pathways by which man may be exposed to radiation and radioactivity following the release of radioactive materials to the environment are shown in ICRP PUBLICATION 7 [12].
In our case, we adopted the pathway where the public takes mainly marine organisms caught near the discharge area.
The diagram is shown in Fig. 2.

3-2 Radionuclides

The radionuclides used for the dose assessment are Ru, Sr, Cs, Ce, Zr/Nb, Pu and I.
Ru, Sr, Cs, Ce and Zr/Nb are main nuclides included in the low level liquid wastes of Tokai plant.
Pu and I are very little included in the low level liquid wastes, but both are very important nuclides in view of health and safety.

3-3 Marine organisms

As mentioned above, this area is the fishing ground of Shirasu and other fishes. In consideration of this fact, we chose the following marine organisms; Shirasu, Karei, Hirame, Shellfish, Cephalopoda, Crustacea and Seaweeds.
Generally, these marine organisms are eaten by the public with other seafoods caught at other sites far from the discharge area, and so food dilution may be considered in assessing doses through the intake of marine organisms caught at the discharge area [13].

4. CONCENTRATION FACTORS OF MARINE ORGANISMS AT THE COASTAL ENVIRONMENT OF TOKAI-MURA

4-1 Elements used to derive the concentration factors

The preoperational surveillance of the marine environmental radioactivity around Tokai plant has been performed for several years. Among the radionuclides discharged into the coastal sea from the plant, Ru-106, Sr-90, Cs-137, Ce-144, Zr/Nb-95 and Pu-239, 240 are the main isotopes of our concern.

(1) Sea water

The sea water data are very important in deriving concentration factors from the field data [8]. The results of measurements performed at PNC from 1971 to 1975 are compiled in Table I.

(2) Marine organisms

As the concentration factor is one of the parameters for dose assessment to the public, concentrations of edible parts of marine organisms should be used to derive concentration factors from in situ data. The results of measurements performed at PNC from 1971 to 1975 are compiled in Table II. Undetectable values were omitted in these data.

4-2 Concentration factors

Concentration factors (C.F.) of marine organisms are defined as follows [6, 7, 8];

$$C.F. = \frac{\text{radionuclide concentration of marine organisms}(\rho Ci/g.wet.weight)}{\text{radionuclide concentration of sea water }(\rho Ci/ml)}$$

Table III   AVERAGE CONCENTRATION FACTORS OF MARINE ORGANISMS DERIVED FROM THE FIELD DATA IN THE VICINITY OF THE TOKAI OFFSHORE

Marine Organisms	Ru-106	Sr-90	Cs-137	Ce-144	Zr/Nb-95	Pu-239,240
Shirasu	26	3.4	17	21	37	51
Fish (adult)	25	3.2	23	57	52	92
Shellfish	260	5.1	7.2	200	51	200
Cephalopoda (octopus.squid)	20	1.6	10	21	50	46
Crustacea (shrimp.crub)	180	28	11	120	49	250
Brown Algae	97	11	20	69	29	290
Red Algae	440	15	12	110	23	770

Table IV.   CONCENTRATION FACTORS DERIVED FROM THE FIELD DATA IN THE VICINITY OF THE TOKAI OFFSHORE MEASURED BY PNC AND ENVIRONMENTAL POLUTION OF RESEARCH CENTER OF IBARAKI-KEN

Marine Organisms	Ru-106	Sr-90	Cs-137	Ce-144	Zr/Nb-95
Shirasu	32	3.7	18	46	37
Fish(adult)	46	3.4	30	47	52
Shellfish	330	5.0	9.2	220	44
Cephalopoda	78	1.8	13	32	46
Crustacea	170	28	20	85	49
Brown Algae	180	15	26	90	59
Red Algae	440	15	12	110	23

Therefore, concentration factors of marine organisms can be derived from both values of radionuclide concentrations of sea water and marine organisms.

The concentration factors of marine organisms in the vicinity of Tokai offshore were calculated from the field data shown in Table I and Table II. The results are shown in Table III.

The marine environmental surveillance at this area has been performed also by Environmental Pollution Research Center of IBARAKI-KEN, and lots of data except Pu have been stored [14, 15]. The concentration factors were calculated from the field data of PNC and IBARAKI-KEN and these are shown in Table IV.

5. THE PRINCIPLE THOUGHTS TO DETERMINE THE CONCENTRATION FACTORS USED TO ASSESS DOSES OF THE PUBLIC

The concentration factors used for the dose assessment were determined based on the following thoughts;

(1) Though the measured radionuclides originated from nuclear bomb tests, the field data reflects the accumulation of radionulides by the food chains of the marine organisms at this area.

Therefore, in principle, the concentration factors calculated from the field data of Tokai offshore were used for the dose assessment.

(2) As we have not the field data of iodine in sea water and marine organisms at this area, the concentration factors of iodine were calculated from the field data of the stable element reviewed in the references [16, 17]. These values concurred with reviewed data [2, 3, 4].

(3) The concentration factors derived from the field data at discharge area of nuclear facilities are different from those at fallout areas in some case.

Radionuclides in seawater transport to seaweeds mainly by the absorption or direct adsorption to these surface [18]. This tendency may be remakable in Ru, Ce, Zr/Nb and Pu.

We observed that the values calculated from the field data in the vincinity of Windscale were larger than those at Tokai offshore as shown in Table V [19, 20, 21].

So, it will be more practical and safer to use the concentration factors at the discharge area instead of those at the fallout area. Therefore, the values obtained at Windscale are used for the dose assessment.

(4) In Plutonium, we have a small number of field data of Cephalopoda and the concentration factors of shellfish are, in general, larger than those of Cephalopoda, so that the concentration factors of shellfish were used from the standpoint of safety.

(5) In order to estimate the annual doses as practical as possible, the average concentration factors to one significant digit were used.

6. THE CONCENTRATION FACTORS USED FOR THE DOSE ASSESSMENT

The concentration factors of marine organisms applied to dose assessment of Tokai plant were determined based on the above thoughts and are compiled in Table VI.

Table V THE CONCENTRATION FACTORS OF SEAWEEDS AT TOKAI AND WINDSCALE [19, 20, 21]

Area	Seaweeds		Ru-106	Ce-144	Zr/Nb-95	PU-239,240
Tokai	Brown Algae (Undaria, Hijiki)	ave.	97	69	29	290
		range	1.6-240	7.0-170	9.1-240	30-590
	Red Algae (Chondrus)	ave.	440	110	23	770
		range	220-670	16-210		
Windscale	Brown Algae (Fucus)	ave.	460	560	1100	—
		range	150-1200	190-1100	570-1500	
	Red Algae (Porphyra)	ave.	1770	630	310	3000
		range	340-4600	70-2700	10-2000	

Table VI CONCENTRATION FACTORS APPLIED TO DOSE ASSESSMENT OF TOKAI PLANT

Species	Ru	Sr	Cs	Ce	Zr/Nb	Pu	I
Shirasu	30	4	20	50	40	100	30
Fish(adult)	50	3	30	50	50	100	30
Shellfish	300	5	9	200	40	200	60
Cepharopoda	80	2	10	30	50	200	3
Crustacea	200	30	20	90	50	400	30
Brown Algae	500	20	30	600	1000	3000	2000
Red Algae	2000	20	10	600	300	3000	1000

REFERENCES

[1] Polikarpov, G.G. Radioecology of aquatic organisms, North Holland Pub. Co., 1966.

[2] A.M. Freke, A Model for the Approximate Calculation of Safe Rates of Discharge of Radioactive Wastes into Marine Environment, Health Physics, Vol-13, 743-758, 1967.

[3] F.G. Lowman, T.R. Rice, F.A. Richards, Accumulation and Redistribution of Radionuclides, Radioactivity in the Marine Environment, Chapter 7, NAS, 1971.

[4] S.E. Thompson et al. Concentration Factors of Chemical Elements in Aquatic Organisms, UCRL-50564 Rev. 1

[5] B. Patel, Field and Laboratory Comparability of Radioecological Studies, IAEA TECHNICAL REPORTS SERIES No. 167, Design of Radiotracer Experiments in Marine Biological Systems, 1975 211-239

[6] Y. Hiyama, M. Shimizu, The Contamination of Aquatic Organisms, The Research of Radioactive Influences, Chapter 5, Press of Tokyo University, 1971 (in Japanese)

[7] R. Ichikawa, Accumulation of Radionuclides to the Fishes, Radioactivity and Fish, Chapter 7, KOSEISHA KOSEI KAKU, 1973 (in Japanese)

[8] M. Shimizu, About the Bioaccumulation of Radioactive Materials in the Environment, RADIOISOTOPES, Vol. 22, No.11 662-671, November 1973, (in Japanese)

[9] M. Fukuda et. al. Report for the Results of Research during 5 years by the Special Committee for the Research on the Discharge of Radioactive liquid wastes into the Sea, Nuclear Safety Research Association of Japan, March 1972 119-187 (in Japanese)

[10] Y. Tanigawa, Y. Kuniya, Report of Marine Research in 1967, JAERI-memo-3384. January 1969 (in Japanese)

[11] Results of Marine Research near Tokai Mura Ibaraki Prefecture (Middle Report), Tokai Regional Fisheries Research Laboratory & Japan Fisheries Resource Conservation Association, May 1970 (in Japanese)

[12] ICRP PUBLICATION 7, Principles of Environmental Monitoring related to the handling of Radioactive Materials 1965

[13] Y. Ohmomo, Survey on Marine Food Consumption, Fundamental Aspects of Marine Radioecology(1), National Institute of Radiological Sciences, October 1975, 60-64 (in Japanese)

[14] J. Koike et. al. Marine radioactivity in the environment around nuclear facilities, Environmental Pollution Research Center of Ibaraki-ken, March 1975 (in Japanese)

[15] S. Morita et. al. The estimation of Internal Doses through the intake of seafoods at the coastal area of Tokai-Mura, Proceeding of 17th meeting on the Research of Environmental Radioactivity, The Science and Technology Agency, 1974, 142-147 (in Japanese)

[16] E. Katsura, R. Nakamichi, The Quantities of Iodine in Foods at Japan, EIYO TO SHOKURYO, 12NO.5.342 1959 (in Japanese)

[17] T. Ohkubo, The Kind and Quantity of Inorganic Matter in Aquatic Products, KOKUMIN EISEI, 14, 1937, 408-419 (in Japanese)

[18] S. Saiki et. al. Accumulation of Radioactive Nuclides by Marine Organisms, Genshiryoku Kogyo, 14(10), 1968, 10-32 (in Japanese)

[19] N.T. Mitchell, Radioactivity in Surface and Coastal Waters of the British Isles, Technical Report FRL 1-2, 5, 7-9 Ministry of Agriculture, Fisheries and Food, Fisheries Radiobiological Laboratory, 1967-1969, 1971-1973

[20] D.S. Woodhead, Levels of Radioactivity in the Marine Environment and the Dose Commitment to Marine Organisms, IAEA-SM-158/31, 1972, 499-525

[21] J.A. Hetherington, D.F. Jefferies, M.B. Lovett, Some Investigations into the Behaviour of Plutonium in the Marine Environment, IAEA-SM-198/29, 1975, 193-212

Discussion

<u>Y. IZUMO</u>, Japan

In your report you use mean concentration factors to evaluate doses. But these values are always very different because of the effects of biological factors, experimental conditions, environment, methods of measurement, physical-chemical state, etc. as you know. Thus I think that the statistical error should always be given with the mean value.

Although the contamination level expressed by the concentration factor has been determined, I think that it cannot be directly related to an MPC value, because this last value was obtained mainly on an experimental basis by administring the radioelements in inorganic form to experimental mammals, instead of feeding them with actual contaminated marine food. In order to approach this problem we have examined, for example, concentration and excretion of Ru-106 contained in the muscle of shrimp fed to the mouse and compared it with the inorganic form. Our results (in preparation) have shown a significant difference (at the 95 % level) between the retention of the two forms of Ru-106 in the food. Thus I think that the MPC values will have to be revised.

<u>R.J. PENTREATH</u>, United Kingdom

In the Windscale area we have noted considerable variation in the concentration of different radionuclides from one species of algae to another. These data were published several years ago. I would, however, encourage you to consider relating the concentrations in algae, at fixed points along the coast, to the rates of discharge. If the data are represented in this way - for example as pCi/kg^{-1} per Ci/day^{-1} discharged - then you may well find such data to be more useful. Such an approach also regates the need for a more detailed estimation of concentration factor data.

Session 7

Chairman - Président
Dr. M. SAIKI
(Japan)

Séance 7

FURTHER REPORT ON
DIETARY SURVEY AROUND NUCLEAR SITE IN THE TOKAI AREA OF JAPAN AND
THEIR RADIOLOGICAL SIGNIFICANCE TO THE RELEVANT POPULATION

M. Sumiya and Y. Ohmomo
Division of Radioecology, National Institute of
Radiological Sciences. Isozaki, Nakaminato-shi,
Ibaraki, Japan

ABSTRACT

Investigations on marine food consumption and on the relevant fishing ground have been conducted in coastal area of Ibaraki prefecture, in order to estimate radionuclide intake and the consequent internal radiation dose due to the ingestion of sea food. Through the survey, the fishermen and their families in Nakaminato were selected as a temporary critical group. Effective consumption was adopted for the estimation of radionuclide intake due to the ingestion of sea foods. Internal radiation doses to the whole body, the bone and the gastrointestinal tracts of the group were estimated.

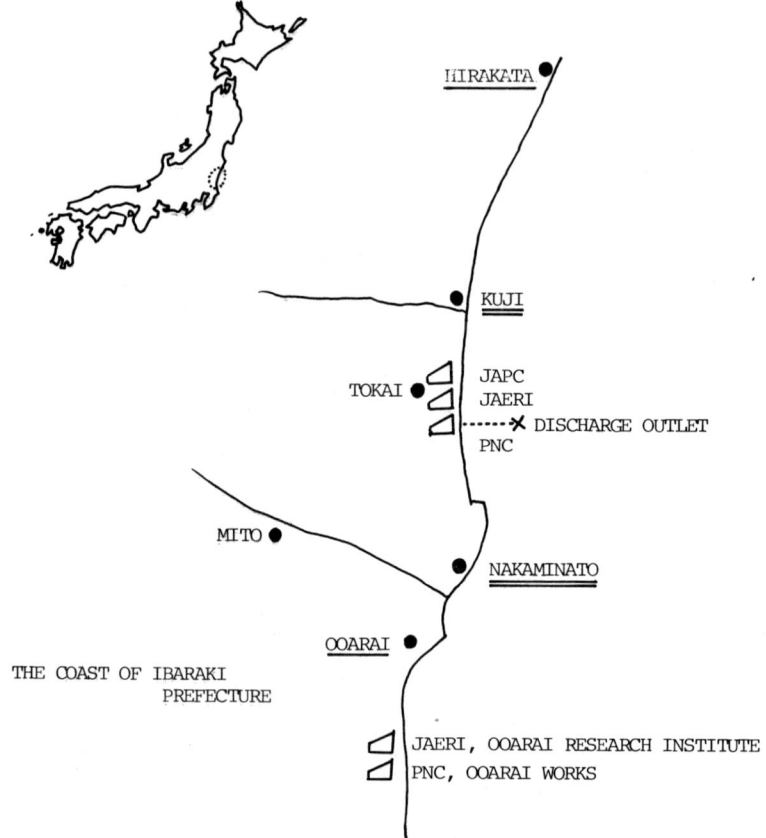

Fig. 1 Map of coastal area of Ibaraki prefecture.

1.   INTRODUCTION

Survey on marine food consumption has been carried out for inhabitants in coastal area of Ibaraki Prefecture for these ten years. Along the coast there are nuclear installations, such as The Japan Atomic Power Company, Japan Atomic Energy Research Institute (JAERI), Power Reactor and Nuclear Fuel Dvelopment Corporation (PNC) and others. A nuclear fuel reprocessing plant is located in the site of PNC in Tokai-mura. The marine fish and other sea foods are the most important protein resources for the Japanese, so release of radioactive nuclides into the coastal sea is of great concern for the Japanese people. This survey was intended to know radionuclide intakes through marine organisms and the consequent estimation of internal radiation dose of the inhabitants. Besides, through the survey, it was expected to select a critical group with respect to internal exposure.

2.   METHOD OF SURVEY

2.1 Locations

Survey has been made in Ooarai, Nakaminato, Mito, Tokai, Kuji and Hirakata as illustrated in Fig. 1. Nakaminato and Kuji, double-underlined, are the nearest fishing towns, about 5 Km far from the discharge outlet of the nuclear fuel reprocessing plant. The outlet is opened in the sea 1.8 Km off the coast of Tokai-mura. Ooarai and Hirakata, single underlined, are also fishing towns located more than 10 Km far from the outlet, but their fishing boats usually cover the sea area including off the coast of Tokai.

2.2 Method

A woman investigator made house-to-house visits to hear of one day's consumption and, as far as possible, to ask for making a note on sea food consumption for successive one or two weeks. Consequently, two kinds of data for sea food consumption can be obtained, that is, from one day's consumption and from one or two weeks' consumption.

2.3 Calculation of sea food consumption

Personal intake is obtained as an arithmetical mean of a whole family's consumption divided by the number of the constituent members above school age (above 6 years old).

3.   RESULTS AND DISCUSSION

3.1 Food consumption by categories

Food consumption by categories of the inhabitants in coastal area of Ibaraki Prefecture are shown in Table I, in comparison with those of national average which is provided by The Welfare Ministry of Japan /I/. From this table, it is clearly recognized that the sea food consumption in these area is higher than that of national average and, as expected, fishermen and their families take more marine products every day. This may be a general trend observed in fishing towns. Therefore, a critical group may be found in fishermen's group.

3.2 Fishermen's organization

Each fishing town has three kinds of fishermen's cooperative associations, that is, of pelagic fishery, of inshore fishery and of catching abalone and algae. Individual fisherman belongs to either of them without exception. It is also observed that

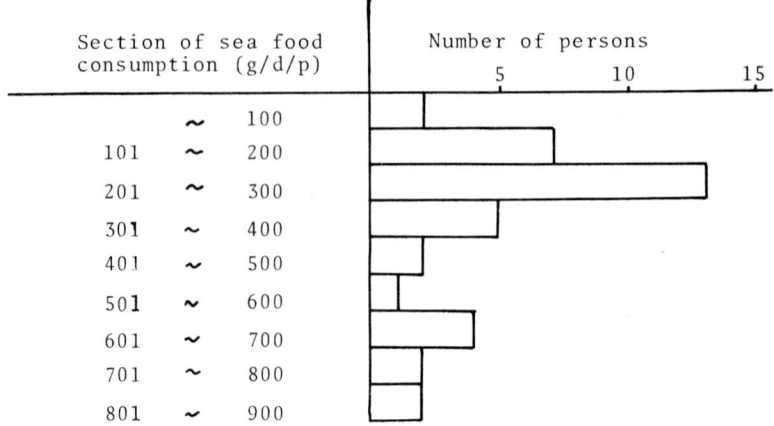

Fig.2 Histogram of sea consumption by inshore fishermen and their families in Nakaminato.

Table I Food consumption by categories

Category of food	Consumption (fresh weight: g/d/p)			
	Fishermen	Farmers	Others	National average
Cereals	412	346	373	356
Bean products	73	60	65	72
Vegetables	260	416	359	336
Fruits	125	51	74	178
Marine products	246	159	166	105
Livestock products	124	133	142	182
Others	52	34	30	40
TOTAL	1292	1199	1209	1269

* Edible part only. Excluding seasoning and table luxuries.

fishermen and their families are uesd to take their catch. Among
them, the fishermen's families of pelagic fishery take much pelagic
fish but less inshore catch. So those people are less concerned with
the nuclear discharge into the coastal sea. Therefore, attention was
directed to the inshore fishermen and their families including those
catching abalone and algae.

3.3 Sea food consumption by inshore fishermen and their families in Nakaminato

Fig. 2 shows the histogram of sea food consumption by
inshore fishermen and their families in Nakaminato, one of the
nearest fishing twons from the discharge outlet. This is made on the
basis of one day's consumption a season and therefore a temporarily
high or low consumption can be directly reflected in the histogram.
The distribution pattrn seems to be composed of two different normal
distributions whose medians are about 250 g/d/p and 650 g/d/p,
respectively, rather than to be lognormal distribution. It is not
unusual to take more than 600g or even more than 1Kg of sea foods a
day as a spot datum. However, it is not understandable to take so
much every day through a year. Then, successive one or two weeks'
seasonal survey was carried out for the inshore fishermen and their
families, specifically for those of high consumption. In general, it
is quite a hard work for housewives to make a note on sea food
consumption every day, because they work together with their husbands
in addition to housekeeping. Besides, the Kinds of sea foods they
usually take are over several tens, so that it must be expecting too
much to make a note without any ommission. So we asked for keeping a
consumption diary on limited kinds of sea foods as a time. We believe
this needs much time to cover the whole but must be a sure method.
Table II shows the sea food consumptions thus obtained in Nakaminato.
Half of the total are fishes, followed by algae and then cephalopods.

3.4 Effective consumption

Diffusion of radionuclides released into the sea was
estimated from dyes diffusion experiment conducted by PNC and from
inflow of river water (Geochemical Lab., Meteorological Research
Institute). It was reported that dyes were quickly diluted upto 10^{-3}
near the outlet and gradually transported by waves and ocean
currents to be further diluted by one or two orders of magnitude.
This suggests that significant sea area of radioactive contamination
is limited within comparatively small area, within 5 Km or at most 10
Km radius from the outlet. Estimation of radioactive nuclide intake
due to the ingestion of sea foods was carried out under the
assumption that all of the products were caught in the sea area
within that range. The results of the survey, however, indicate that
the products they take are not always caught within the sea area.
This fact suggests the radionuclide intake calculated from the sea
food consumption given in Table II should be over estimation. So we
tried to introduce the idea of effective consumption. Of course, the
conversion factors for calculating effective consumptions are
variable according to sea foods, meteorological conditions and even
economical situations of the year. As an example, temporary factors
applicable to the individual sea food consumed by inshore fishermen
and their families in Nakaminato are presented in Table III. Pelagic
fish accounts for about 50% of fish consumption excluding immature
anchovy. Therefore, a factor of 0.5 can be applied to fish
consumption. Immature anchovy, *Engraulis japonica*, was considered
as a imaginary critical marine organism for the initial safety
assessment carried out in advance of the newconstruction of the
nuclear fuel reprocessing plant in Tokai-mura, because its whole body
is edible and it has poor swimming ability to move with the low level
liquid waste. At that time, the internal radiation doses to bone and
whole body of critical people were estimated under the assumption
that they should have 200g of immature anchovy everyday through a
year. Along the coast of Ibaraki Prefecture there are good fishing

Table II Sea food consumption by inshore fishermen and their families in Nakaminato.

Category of sea food	Consumption (g/d/p*)
Fish	127
Immature anchovy	19
Algae	53
Cephalopod	43
Shellfish	19
Crustacea	15
Processed	40
TOTAL	316

* Fresh weight and edible part only.

Table III Conversion factors for calculating effective consumptions applicable to sea food consumption of inshore fishermen and their families in Nakaminato.

Category of sea food	Conversion factor
Fish	0.5
Immature anchovy	0.3
Algae	0.7
Cephalopod	0.5
Shellfish	0.8
Crustacea	0.7

grounds of the fish, which is often eaten raw by fishermen. But only less than one-third is caught near the discharge outlet. Then, a factor of 0.3 may be acceptable. Squid, *Sepiidae*, and Octopus, *Octopus vulgaris*, are mostly caught about 18 Km off the coast. That may be out of the range of significant radioactive contamination. Therefore, a factor of 0.5 may be too much conservative. On the other hand, the main fishing ground of prawn, *macrura*, is near the outlet, but from winter to spring, more than half are supplied from outside the prefecture. So, a factor of 0.7 may be reasonable. Shellfish are distributed widely along the coast of Ibaraki Prefecture, but about 80% are caught in front of the coast of Nakaminato and 20% are from south of Ooarai. Afactor of 0.8 may be acceptable. Algae, such as Hijiki, *Hijikia fusiforme*, and Wakame, *Undaria pinnatifida*, are also widely distributed along the coast. However, harvesting is allowed only a few times a year in Nakaminato because of a preservation of abalone which feeds algae. The harvesting is thus limited and not a little quantity of algae is imported from the outside of Ibaraki Prefecture. So, a factor of 0.7 may be acceptable. Table Ⅳ shows thus obtained effective consumptions.

3.5 Estimation of internal radiation dose

Daily intake of radioactive nuclides through marine products is given as following;
$$A = C_{sw} \cdot F \cdot R \quad \text{-----} \quad (1)$$
where A : daily intake of radioactive nuclide (Ci/d)
  Csw : concentration of radionuclide in sea water (Ci/cm^3)
  F : concentration factor
  R : daily intake of marine product (g/d)

Daily intake of individual radionuclide through sea food (A) was calculated under the assumptions that radioactive lequid waste was continuously released at a rate of 1 Ci/d and that all of the products were caught in the relevant fishing grounds nearest from the discharge outlet. The supposed composition of radioactive liquid waste is presented in Table V as reference. The concentrations of each nuclide in the sea waters of the relevant fishing grounds (Csw) were estimated from following empirical formula [2];
$$C_{sw} = \frac{0.75 Q}{X \cdot Z} \times F \quad \text{-----} \quad (2)$$
where Q : rate of discharge (Ci/sec)
  X : distance from discharge outlet (cm)
  Z : depth" of diffusion (cm); constant value of 460cm is adopted in this paper.
  F : direction frequency of the diffusion of the effluent. For example, 7,5% per year for south-west direction.

The values of X, corresponding to the distance from the outlet to the relevant fishing grounds, are as follows;
  0.5 Km for immature anchovy
  1 Km for crustacea
  2 Km for shellfish
  5 Km for fish and algae
  10 Km for cephalopod

Concerning the concentration factors, many papers have been reported [3]. The factors adopted in this paper are depicted from those compiled by PNC, which are going to be presented by H. Kurabayashi et. al. in this seminar. Table Ⅵ represents internal radiation doses to whole body, bone and gastrointestinal tracts of inshore fishermen and their families in Nakaminato by the ingestion of Ru-106, Sr-90, Cs-137, Zr-95, Nb-95, Ce-144, Pu-239 and Tritium through sea foods, the consumptions of which are given as effective consumptions. Strontium-90 and Pu-239 are related to the doses to bone, Cs-137 and Tritium to whole body and the others to gastrointestinal tracts. As shown in Table Ⅵ, internal radiation doses this obtained are quite limited. It is also observed that every sea food makes similar degree of contribution and therefore it is impossible to select one or two items as critical sea foods.

4. REFERENCE

(1) Nutrifion in Japan : Welfare Ministry (1968).
(2) T. Nomura et. al. Monitoring for the Dispersion of the Discharged Tritium in seawater around the Tokai-mura Fuel Reprocessing Plant, IAEA-SM-232/27 (1978
(3) Saiki et. al. ; Accumulation of Radioactive Nuclides by Marine Organisms, Genshiryoku Kogyo, 14 (10), 10, 1968.

Table IV Effective consumptions of sea foods by inshore fishermen and their families in Nakaminato.

Category of sea food	Effective consumption (g/d/p)
Fish	64
Immature anchovy	6
Algae	37
Cephalopod	22
Shellfish	16
Crustacea	11

Table V Supposed composition of radioactive liquid waste

Nuclide	Composition (Ci/d)
$^{103+106}Ru + {}^{103+106}Rh$	0.5
only ^{106}Ru	0.2
^{90}Sr	0.01
^{137}Cs	0.04
$^{95}Zr + {}^{95}Nb$	0.1
^{144}Ce	0.2
^{239}Pu	2×10^{-4}
^{3}H	200

Table VI Estimated internal radiation doses to critical organ.

	Daily intake of nuclide (uCi/d/p)		Organ	Dose (mrem/y)
^{106}Ru	Fish I.Anchovy Algae Shellfish Cephalopod Crustacea	1.8×10^{-6} 1.0×10^{-6} 1.0×10^{-5} 6.7×10^{-6} 4.9×10^{-7} 6.2×10^{-6}	Gastrointestinal tract	1.8
^{90}Sr	Fish I.Anchovy Algae Shellfish Cephalopod Crustacea	5.4×10^{-9} 6.7×10^{-9} 4.1×10^{-8} 5.6×10^{-9} 6.2×10^{-10} 4.6×10^{-9}	Bone	0.8
^{137}Cs	Fish I.Anchovy Algae Shellfish Cephalopod Crustacea	2.7×10^{-7} 1.3×10^{-7} 4.1×10^{-7} 4.0×10^{-8} 1.2×10^{-8} 1.2×10^{-7}	Whole body	1×10^{-2}
^{95}Zr	Fish I.Anchovy Algae Shellfish Cephalopod Crustacea	4.5×10^{-7} 3.4×10^{-7} 5.2×10^{-6} 2.3×10^{-7} 7.8×10^{-8} 3.9×10^{-7}	Gastrointestinal tract	7.6×10^{-2}
^{95}Nb	Fish I.Anchovy Algae Shellfish Cephalopod Crustacea	4.5×10^{-7} 3.4×10^{-7} 5.2×10^{-6} 2.3×10^{-7} 7.8×10^{-8} 3.9×10^{-7}	Gastrointestinal tract	4.6×10^{-2}
^{144}Ce	Fish I.Anchovy Algae Shellfish Cephalopod Crustacea	1.8×10^{-6} 1.7×10^{-6} 1.2×10^{-5} 4.5×10^{-6} 1.8×10^{-7} 2.8×10^{-6}	Gastrointestinal tract	1.6
^{239}Pu	Fish I.Anchovy Algae Shellfish Cephalopod Crustacea	3.6×10^{-9} 3.4×10^{-9} 6.2×10^{-8} 9.0×10^{-9} 1.2×10^{-9} 6.2×10^{-9}	Bone	2.3×10^{-2}
^{3}H	Fish I.Anchovy Algae Shellfish Cephalopod Crustacea	3.6×10^{-5} 3.4×10^{-5} 2.1×10^{-5} 2.2×10^{-6} 6.2×10^{-6} 3.1×10^{-5}	Whole body	6.8×10^{-3}

FINNISH STUDIES IN THE BALTIC SEA ON THE BEHAVIOUR OF RADIONUCLIDES
DUE TO GLOBAL FALL-OUT AND RELEASES FROM NUCLEAR POWER PLANTS.
A SUMMARY REPORT.

Anneli Salo
Institute of Radiation Protection
P.O. Box 268, Helsinki 10, Finland

ABSTRACT

A summary is given of studies on the behaviour of radionuclides in the Baltic Sea. The balances of ^{90}Sr and ^{137}Cs in water phase are discussed as well as their elimination from water by sedimentation and distribution in sediments. The distribution of ^{239,240}Pu in sediments is also described. In soft bottoms the amount of ^{90}Sr per unit area varies from 6 to 15 % of that deposited on land surface, ^{137}Cs from 17 to 700 % and ^{239,240}Pu from 65 to 1000 %. Accumulation of the above-mentioned radionuclides as well as ^{54}Mn, ^{58}Co, ^{60}Co, ^{110m}Ag and ^{124}Sb by some Baltic mussels, a crustacean different littoral algae, plankton and some fish species is reported.

INTRODUCTION

The present report is a summary of Finnish studies on the behaviour of radionuclides in the Baltic Sea. It is based on work carried out in the Institute of Radiation Protection by E. Ilus, M. Koivulehto, J. Ojala, R. Saxén, K-L Sjöblom, K. Tuomainen and the author. Prof. A. Voipio, Institute of Marine Research,was the expert on the Baltic Sea. Studies on the vertical distribution of ^{239,240}Pu in sediments were also carried out at the Department of Radiochemistry, University of Helsinki, under the leadership of Prof. J.K. Miettinen.

The Baltic Sea has, in many respects, its own special character, differing from both oceans and lakes. This fact raised the question of the possibility of radionuclides also behaving differently in this recipient, thereby questioning the applicability of the data obtained on other recipients.

The brackish water of the Baltic Sea is a mixture of the fresh water received from the rivers and precipitation, and the salt water penetrating from the North Sea.

Typical factors contributing to the special character of the Baltic are:
- the limitation of water exchange caused by the narrowness of the connections with the North Sea and by the sills lying between the Baltic basins resulting in strong density stratification,
- the climate, with its great difference in temperature between summer and winter including partial freezing during the latter season.

The Baltic Sea is in effect an estuary, but not being directly connected with the ocean, it has far more stable salinity conditions than estuaries with a strong tidal flow. Owing to these unusual conditions it also has an exceptional flora and fauna, characterized by both true marine and fresh water as well as brackish-water species, the number of species being, however, very small.

Two of the Swedish reactors (440 and 580 MW since 1971 and 1974 respectively) are in operation on the coast of the Baltic Proper, while four are situated in the region of the Danish Sounds. Two Soviet reactors (1000 MW each, since 1973 and 1975) are in operation in the furthest corner of the Gulf of Finland. One Finnish reactor (440 MW since 1977) is in operation on the coast of the Gulf of Finland and one (660 MW since 1978) is test-operating on the coast of the Bothnian Sea.

The convention on the Protection of the Marine Environment of the Baltic Sea Area - the Helsinki convention - places stricter limitations than the London convention on the disposal and dumping of harmful substances - including radioactive substances - into the Baltic. It expressly prohibits dumping in the Baltic Sea Area excluding dredge spoils.

Thus the sources of artificial radionuclides in the Baltic Sea are global fall-out from nuclear weapons tests either directly deposited on the sea surface, or via run-off or discharges of the above-mentioned nuclear power plants. An additional source is the Atlantic water flowing in through the Danish Sounds, and includes both global fall-out and radionuclides from nuclear industry.

As regards the long-lived radionuclides, their behaviour is especially interesting because of the limited water exchange and the special ecological conditions of the Baltic Sea. The long-lived radionuclides due to global fall-out were used for these studies, which were carried out in different regions of the Baltic Sea.

The short lived radionuclides due to releases from nuclear power plants are mainly of local interest only, and were mostly studied at our first reactor site.

The already published initial results with the methods used and the sampling areas summarized in Fig. 1 are given in more detail in the annual reports of the Institute of Radiation Protection referred to in ref. /1/.

LONG-LIVED RADIONUCLIDES FROM GLOBAL FALL-OUT

Balance in water phase

The water balance of the Baltic Sea is given by the equation:

Run-off	Precipitation	Inflow	Outflow	Evaporation
+	+	=	+	
500 km³/a	200 km³/a	500 km³/a	1000 km³/a	200 km³/a

and the material balance can be described by a general equation of the form

$$c_P V_P + c_R V_R + c_I V_I - c_O V_O - c_E V_E - D + G + B = A$$

where A is the net change in the total amount of a single radionuclide, c is the radionuclide concentration, and V the water volume in the following steps in the hydrological cycle: precipitation (P), run-off (R), inflow through the Danish Sounds (I), outflow through the Sounds (O), and evaporation (E). The decrease due to physical decay is given by D, and the other quantities are the changes in the radionuclide concentrations caused by geochemical (G) and biological (B) processes.

On the one hand the total inventories of ^{90}Sr and ^{137}Cs were calculated by multiplying the concentrations in different water masses by volumes of the latter. On the other hand, inventories were calculated on the basis of the above material balance equation. The calculations and the origin of the data used are described in detail in ref. /2/.

Ten years ago, when the calculations were made for the first time, the only significant source of these radionuclides was atmospheric deposition on land and sea surface due to nuclear weapons tests in autumn 1961, with a deposition maximum in 1963.

As the observation time was rather short the magnitude of the possible bio-geochemical processes eliminating ^{90}Sr and ^{137}Cs from water phase was not expected to be clearly seen at that time. This, along with the possibility of ^{90}Sr and ^{137}Cs originating from nuclear industry being transported to the Baltic Sea via the North Sea, suggested by several authors, e.g. ref. /3/, led to a revision of the balance calculations in 1979 /4/. These results are given in fig. 2. A clear difference between the "observed" (based on sea water analyses) and the "calculated" (based on the material balance calculation) curves for both nuclides, can be seen mainly reflecting the elimination of these nuclides from water by sedimentation. As expected, there is a difference in their sedimentation characteristics. The calculation does not, however, give clear evidence of surplus ^{90}Sr and ^{137}Cs from sources other than global fall-out.

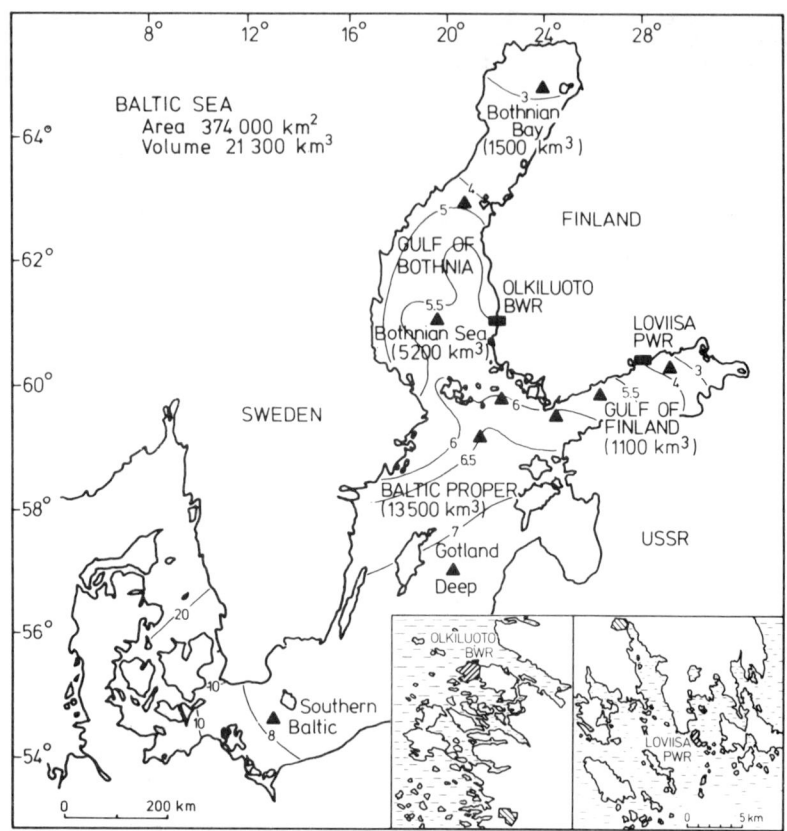

Fig. 1 Sampling areas in the Baltic Sea.

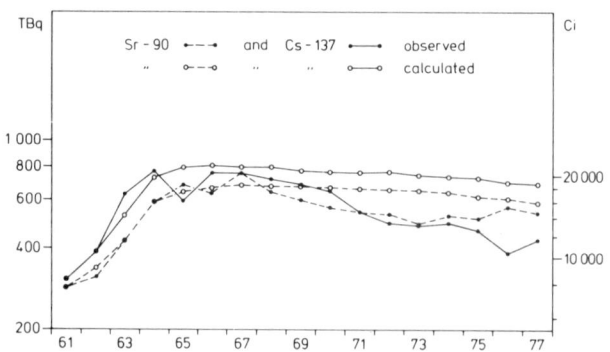

Fig. 2 ^{90}Sr and ^{137}Cs balances in the Baltic sea (Salo and Voipio, 1979)

If the radionuclide amounts corresponding to the differences between the "calculated" and "observed" curves are divided by the surface area of the Baltic Sea the approximate amounts to be expected in the bottom sediments would be $40 - 80$ Bqm^{-2}, ($1 - 2$ mCikm^{-2}) ^{90}Sr and ca. 800 Bqm^{-2} (20 mCikm^{-2}) ^{137}Cs at the end of the '70s.

Horizontal and vertical distribution in sediments

To verify the removal of ^{90}Sr and ^{137}Cs from the water phase caused by sedimentation, both the horizontal and vertical distribution of these nuclides in the bottom sediments are studied. To get a better picture of the sedimentation phenomenon ^{239,240}Pu is also studied. Both nearshore areas, with unstable sedimentation conditions, and areas of undisturbed sedimentation in the Baltic Sea are included in the studies. Examples of the results showing the amount and distribution of the above-mentioned nuclides due to global fall-out near the sites of the Finnish power plants and off shore in the Baltic proper /5,6/ are given in Fig. 3.

The total amounts in sediment samples at the end of the '70s vary from place to place for ^{90}Sr from 74 Bqm^{-2} (2.0 mCikm^{-2}) to 200 Bqm^{-2} (5.3 mCikm^{-2}), for ^{137}Cs from 310 Bqm^{-2} (8.5 mCikm^{-2}) to 13 000 Bqm^{-2} (340 mCikm^{-2}) and for ^{239,240}Pu from 22 Bqm^{-2} (0.6 mCikm^{-2}) to 340 Bqm^{-2} (9.2 mCikm^{-2}). On hard sandy bottoms there are found only traces of fall-out radionuclides /5,6/. Corresponding land deposits are 1 300 Bqm^{-2} (35 mCikm^{-2}), 1 800 Bqm^{-2} (50 mCikm^{-2}) and 34 Bqm^{-2} (0.92 mCikm^{-2}) respectively /1/.

In the undisturbed sedimentation areas these radionuclides are not found below 10 cm. Even in areas where the sedimentation is disturbed by wave action and benthic animals and where sediment material is transported from surrounding areas they are seldom found below 20 cm.

Sedimenting particles in the water column in the coastal areas have the same radionuclide concentrations as the average for the 5 topmost cms of the bottom sediment in recent years.

Besides ^{210}Pb/^{210}Po and other techniques ^{239,240}Pu seems to offer a valuable tool in dating the sediments /6,7/ in the Baltic Sea.

Accumulation by biota

The biota of the Baltic Sea is much poorer in species than that of the oceans. Two euryhaline mussels <u>Mytilus edulis</u> and <u>Macoma baltica</u>, the small crustacean <u>Mesidotea entomon</u>, a postglacial relict, and littoral algae <u>Fucus vesiculosus</u>, <u>Cladophora glomerata</u> and <u>Enteromorpha</u> species were chosen as possible detectors of radioactive contamination.

The species of fish closely studied were the Baltic herring, pike, perch, pikeperch, roach, whitefish and burbot.

Besides the long-lived radionuclides ^{90}Sr, ^{137}Cs and ^{239,240}Pu the stable element analyses were also used in evaluating concentration factors. The concentration factors found for the above species in the Baltic Sea as compared with values found for similar species in fresh water and oceanic environments /10,11/ are given in Tables 1 /8/ and 2 /9/.

Only with plankton was it possible to observe clearly, in the field, the variation in availability of ^{90}Sr and ^{137}Cs with time during the years of heavy fall-out (see Fig. 4 /12/).

Table I. Concentration factors from water (CF = $\frac{\text{act. conc. in organism, fresh wt}}{\text{act. conc. in water}}$)
and sediment (CF$_{sed}$ = $\frac{\text{act. conc. in organism, fresh wt}}{\text{act. conc. in sediment, dry wt(a)}}$) to biota
(Koivulehto, Saxén and Tuomainen, unpublished).

Species	^{239,240}Pu	^{137}Cs	^{90}Sr	Sr	Ca	K	Fe	Mn	Zn
Macoma baltica (whole)	2×10^3	60	5×10^2	7×10^2	2×10^3	10	9×10^3	2×10^3	4×10^3
(CF$_{sed}$)	1×10^{-2}	1.5×10^{-2}	7	30	–	–	7×10^3	1	1
Mesidotea entomon	1.5×10^3	1×10^2	2×10^2	2×10^2	–	50	–	9×10^2	–
(CF$_{sed}$)	5×10^{-3}	1×10^{-2}(b)	2.5 (b)	8	–	–	–	0.15	–
Mytilus edulis	4×10^2	1×10^2	2.5×10^2	3×10^2	1×10^3	10	3×10^3	1×10^3	–
Fucus vesiculosus	3×10^3	2×10^2	1.5×10^2	1×10^2	2.5×10^3	80	4×10^3	2×10^3	1.5×10^3
Cladophora glomerata	4×10^3	60	10	–	–	–	–	–	–
Enteromorpha sp.	1×10^3	50	10	–	–	–	–	–	–
Chara sp.	–	1.5×10^2	70	–	–	–	–	–	–
Reference, CF-values:									
Sea water									
Molluscs [10,11]	340–2000	3–28	0.1–10		0.2–112	3.5–10	1000–13000	170–150000	2100–330000
Crustacea [10]	750–1620	0.5–26	0.1–1.1		0.5–250	8–19	1000–4000	600–7500	1700–15000
Algae [10,11]	85–3000	17–240	0.2–28		1.8–31	4–31	300–6000	2000–20000	80–2500

(a) mean act. conc. to the depth of 25 cm used for Macoma baltica
    mean act. conc. to the depth of 5 cm used for Mesidotea entomon

(b) In 1965 the CF$_{sed}$ values were 6 for ^{90}Sr and 0.2 for ^{137}Cs

Table II. Concentration factors from water to fish flesh (CF, fresh wt). (Saxén and Koivulehto, unpublished).

Fish	137Cs	90Sr	Sr	Ca	K	Fe	Mn	Zn	Cu	Co	Cr
Baltic herring, Clupea harengus membras	1.5×10^2	3	1	5	90	2×10^2	30	5×10^2	10^2	20	8×10^2
Pike, Esox lucius	3×10^2	2	4	8	70	60	10	4.5×10^2	70	2×10^2	1×10^3
Perch, Perca fluviatilis	4×10^2	3	2	11	70	90	30	5×10^2	60	1×10^2	9×10^2
Pike perch, Lucio-perca sandra	4.5×10^2	2	1	10	80	50	80	3×10^2	40	6×10^2	8×10^2
Roach, Rutilus rutilus	1×10^2	10	4	13	70	1.5×10^2	70	9.5×10^2	80	3×10^2	5×10^2
Whitefish, Coregonus lavaretus	2×10^2	10	1	5	70	1×10^2	50	6×10^2	2×10^2	3.5×10^2	5×10^2
Burbot, Lota vulgaris	3.5×10^2	4	2	5	40	80	20	4×10^2	70	70	6×10^2
Average	3×10^2	5	2	8	70	1×10^2	40	5×10^2	90	2.5×10^2	7×10^2

Reference values:

	137Cs	90Sr				Fe	Mn	Zn		Co	
Sea water [10]	5- 224	0.1 -0.5				600-3000	35-1800	280-15500		20-5000	
Fresh water [10]	80-4000	0.85-90				0.1-1225	0.1- 400	10- 7000		60-3450	

- 365 -

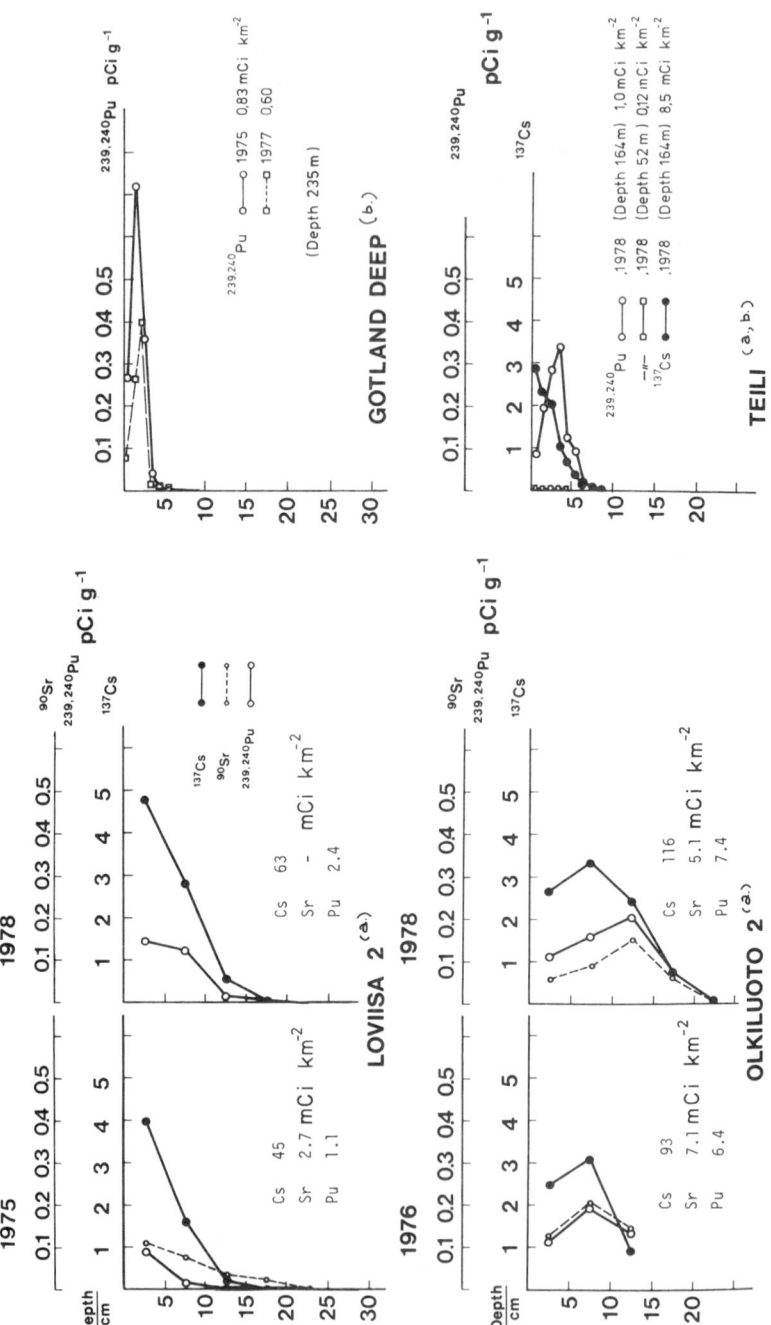

Fig. 3 ^{90}Sr, ^{137}Cs and $^{239,240}Pu$ amounts and their vertical distribution patterns at two nearshore stations, Loviisa and Olkiluoto, two off-shore deeps, Gotland Deep and Teili (/a/ Tuomainen and Koivulehto unpublished, /b/ Simola et at. 1979).

In addition, the concentration factors from sediment to Mesidotea entomon for ^{90}Sr, and especially for ^{137}Cs, show a difference between 1965 and 1975-78, which could be attributed to the difference in availability.

SHORT-LIVED RADIONUCLIDES FROM NUCLEAR POWER PLANTS

Radionuclides released from nuclear power plants are monitored regularly in the recipient. The releases at Loviisa nuclear power plant take place mainly during the revision period in February - March while the recipient is frozen.

Concentrations in environmental samples of gamma emitters with highest releases at Loviisa power plant are given in Fig. 5 /13/ as an example. The sampling station in this figure represents an area where the discharged water is considered to be fairly well mixed with that of the recipient.

Based on the figure it can be estimated that the orders of magnitude of the ^{54}Mn, ^{58}Co, and ^{60}Co concentration ratios of Mesidotea entomon to water are 10^2, that of ^{110m}Ag of both Mesidotea entomon and Fucus vesiculosus to water as well as that of ^{124}Sb of the latter to water 10^3 and of ^{54}Mn, ^{58}Co and ^{60}Co of Fucus to water, 10^4. Also, Enteromorpha and Cladophora species seem to accumulate radionuclides in the same way as Fucus. ^{124}Sb is not readily concentrated by the species above, but it is easily detected in water phase.

^{58}Co, ^{60}Co, ^{124}Sb, and especially ^{110m}Ag, are detected in sedimenting particles. The concentration ratios of Mesidotea entomon to sedimenting particles (dry) for the above radionuclides are 10^{-3} except for ^{110m}Ag, which is one order of magnitude higher.

REFERENCES

1. Studies on environmental radioactivity in Finland 1976-77, Annual Report STL-A26 (1979).

2. Voipio, A. and Salo, A.: "On the balances of ^{90}Sr and ^{137}Cs in the Baltic Sea", Nordic Hydrology II, 57-63 (1971).

3. Kautsky, H.: "The distribution of the radionuclide caesium-137 as an indicator for North Sea watermass transport", Deutsche Hydrographische Zeitschrift, 26, 241 (1973).

4. Salo, A. and Voipio, A.:"The balances of ^{90}Sr and ^{137}Cs in the Baltic Sea. Revised 1978",in Swedish. Paper presented at the second Nordic radioecology meeting, Helsingør, 1979.

5. Tuomainen, K. and Koivulehto, M.,personal communication.

6. Simola, K., Jaakkola, T., Miettinen, J.K., Niemistö, L. and Voipio, A., "Vertical distribution of fall-out plutonium in some sediment cores from the Baltic Sea", Finnish Mar. Res. No. 246 (1979) (in press).

7. Niemistö, L. and Voipio, A.: "Studies on the recent sediments in the Gotland Deep", Finnish Mar. Res., No 238,(1974).

8. Koivulehto, M., Saxén, R. and Tuomainen, K.,personal communication.

9. Saxén, R. and Koivulehto, M.,personal communication.

10. Jinks, S.M. and Eisenbud, M.: "Concentration factors in the aquatic environment", Radiation Data and Reports, May (1972).

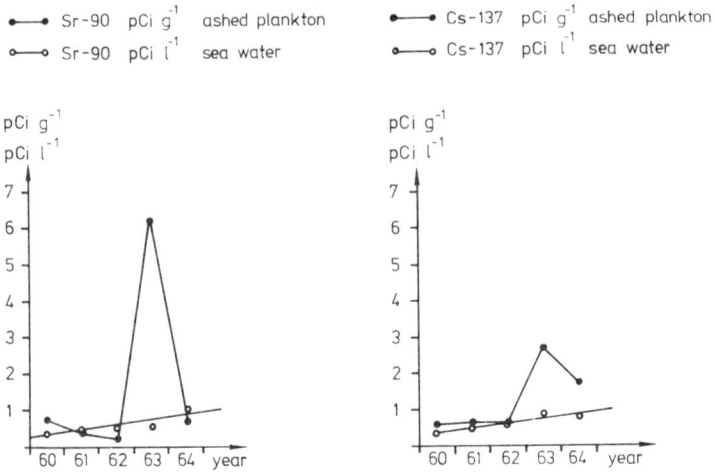

Fig. 4 Variation with time of fall-out ^{90}Sr and ^{137}Cs concentrations in plankton compared with those in sea water. (Bagge and Salo, 1967).

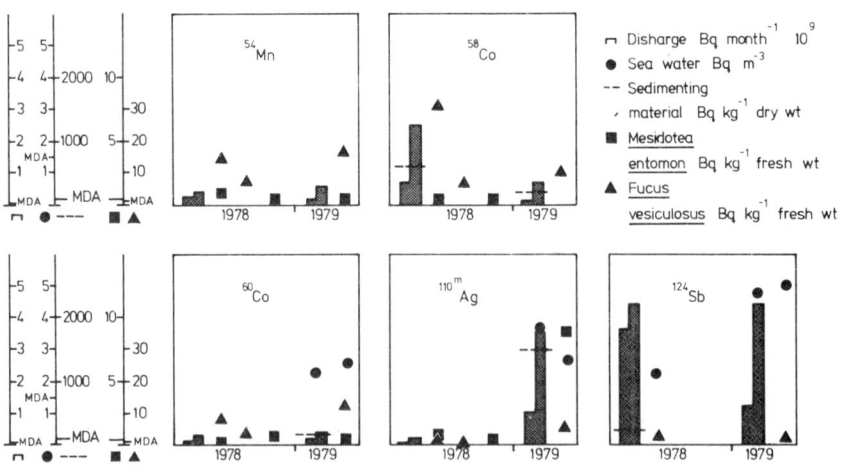

Fig. 5 Short-lived radionuclides in the liquid discharge at the Loviisa nuclear power plant and in the environmental samples in the recipient at 2.5 km from discharge channel. (Ojala and Sjöblom, unpublished).

11. Angeletti, L., Ancellin, L. and Bittel, R.:"Aspects pratiques du comportement du plutonium dans l'environment", Radioprotection 12 (1), 3-26 (1977).

12. Bagge, P. and Salo, A.:"Biological detectors of radioactive contamination in the Baltic", Report SFL-A9, Institute of Radiation Physics, Helsinki (1967).

13. Ojala, J. and Sjöblom, K-L., personal communication.

Discussion

H. KAUTSKY, Federal Republic of Germany

Have you ever made any calculations of the ratio between Sr-90 and Cs-137 ? In the South Western Baltic Sea we have observed, during the bomb tests, the normal ratio of about 1:1.4. After the bomb tests stopped, this ratio changed to 1:1 and it has been nearly constant for many years now. We believe this change is due to the better adsorbing capability of Cs-137 onto sediments. This - for the moment - fairly constant ratio enables us to observe any possible influx of water with a higher content of Cs-137 into the Baltic Sea. In other words, because of the higher content of Cs-137 in the North Sea water entering the Baltic Sea, we hope it might become possible to calculate the amount of in-flowing water by calculating the change of the ratio between Cs-137 and Sr-90. Have you found a similar change in the ratio ?

L.A. SALO, Finland

Yes, we have calculated the Cs-137/Sr-90 ratios for the different water masses in the Baltic Sea since 1960. The ratios vary with sampling area (i.e. from one water mass to another). In the Gulf of Bothnia and Gulf of Finland the ratios have been lower than in the Baltic proper. This is due to the greater influence of the run-off in which Cs-137 is very much more reduced than Sr-90 compared to the deposition. The ratios in all areas have decreased slightly since the time of maximum deposition. The ratios can be used to trace the origin of the water masses.

A.M. ORTINS DE BETTENCOURT, Portugal

I have noticed in your table that you have obtained different concentration factors for stable strontium and strontium-90. Are these differences due to statistical errors or do you have another explanation ?

L.A. SALO, Finland

The differences are partly due to statistical errors (in the case of fish, because of the varying bone content) and partly due to other factors, e.g. the physical-chemical form (in the case of benthic animals Sr-90 and Sr fix to particles to a different extent).

A. AARKROG, Denmark

Concentration factors based upon stable Sr are, in the case of Fucus vesiculosus, usually lower than those obtained from Sr-90 fallout because the latter contaminates the sea weed not only through the water pathway but also by deposition directly from the air, when the seaweed is exposed to the air at low tide.

W.L. TEMPLETON, United States

There is an additional factor which explains the variations in your concentration factors. Your sampling extends from freshwater to sea water, and in such a situation you will find that the calculated concentration factor will vary depending on the Sr/Ca and Cs/K ratios in the water in which the organisms reside. Templeton and Brown published a paper on the variation in the Sr/Ca ratios, and Preston and Jefferies on Cs/K in the early 1960's.

L.A. SALO, Finland

Yes, although the Sr/Ca ratio in the Baltic Sea is constant according to our studies (Salo, Voipio - 1966).

BEHAVIOUR OF ACTIVATION PRODUCTS RELEASED FROM A NUCLEAR
POWER REACTOR INTO COASTAL WATERS

M. Nilsson and S. Mattsson
Radiation Physics Department, University of Lund
Lasarettet, 221 85 LUND, Sweden

ABSTRACT

The macroalgae Fucus vesiculosus has been used as a bioindicator for mapping the release of radionuclides from a nuclear power station into the marine environment. Both the time variation of radionuclide concentration in Fucus and its variation by distance from the power plant have been studied. The uptake and retention of different activation products in Fucus are compared.
The concentration of ^{60}Co in Idothea viridis , Idothea baltica and Gammarus oceanicus living in the Fucus plants is also given.

1.   INTRODUCTION

The handling and processing of nuclear fuel and waste give rise throughout the fuel cycle to releases of radioactive substances into the environment. An important source at present and undoubtedly the most important in the future for the irradiation of man is the marine environment.

We have therefore considered it of great importance to study the transport of radionuclides released under controlled conditions from a nuclear power plant into the marine environment. This applies to both the radionuclide content in water and its concentration along food-chains of interest with respect to estimating the dose commitment to the exposed population.

The discharged radioactivity is, however, dispersed into enormous volumes of water. It is therefore necessary to find a bioindicator which concentrates the substances but still, to a certain degree, describes the radionuclide concentration in its immediate surroundings. We have found the brown seaweed Fucus vesiculosus a very suitable bioindicator for such studies /̲ 1 /̲.

The source of the discharged radioactivity is the Barsebäck nuclear power plant which is located on the Öresund sound between Denmark and Sweden. It consists of two boiling water reactors, each of 1700 MW thermal effect. The first reactor (B1) became operational in 1975, and the second (B2) in 1977 (cf. figure 1). The release of radionuclides, mainly activation products, is during normal operation, due to high filtering and waste containment capacity, quite low. During the summer period, when annual overhaul and partial refuelling takes place, the discharge is much higher. This gives a pulsed pattern of released radionuclides, which is fortunate for our studies, since it simplifies the interpretation and mathematical treatment of data regarding the time variation of concentration, uptake and retention of the radionuclides.

Our studies of the radionuclide content in Fucus started in 1967; being at that time mainly studies of the long term variation of ^{137}Cs activity concentration. These studies were carried out at a considerable distance (140 km north) from the power plant. In the summer of 1976 we for the first time found traces of ^{60}Co in Fucus from this collecting station. From that time on, we expanded our programme to regular collection at eight sampling stations.

2.   MATERIAL AND METHODS

All samples of Fucus vesiculosus, firmly rooted to stones on the bottom, were collected from a water depth varying between 0.5 and 1 m. Samples of the crustaceans Idothea, mainly I. baltica and I. viridis, as well as samples of Gammarus found on the collected Fucus plants have been sorted out for analysis. After drying in air at room-temperature for 2 to 3 days the samples were ground and packed in plastic containers of 5,180 or 2 000 ml volume. The measurements were carried out using Ge(Li)-detectors of volumes 46-100 cm^3 with counting times varying between 5 and 100 hours. After measurement the samples were dried at 105°C for 24 hours to get the reference weight. The detector efficiencies were carefully determined using calibration samples of different densities containing accurately known activities of ^{152}Eu, ^{57}Co and ^{22}Na.

3.   RESULTS AND DISCUSSION

3a.   Activation products in Fucus vesiculosus

The location where collection has most frequently been made is situated 2.9 km north of the discharge pipe. Figure 1 gives the activity concentrations (dry weight) for ^{60}Co (T1/2 = 5.26 a), ^{58}Co (71.3 d), ^{54}Mn (303 d) and ^{65}Zn (245 d) in Fucus v. collected at this location between 1976 and 1979. On the abscissae, different operational data for the plant are given. We have also, on occasion, detected ^{110}Ag^m (255 d), ^{57}Co (270 d), ^{51}Cr (27.8 d) and ^{131}I (8.05 d).

The activation products shown in figure 1 all follow a similar pattern, and clearly reflect the increase in discharged activity during the overhaul periods.

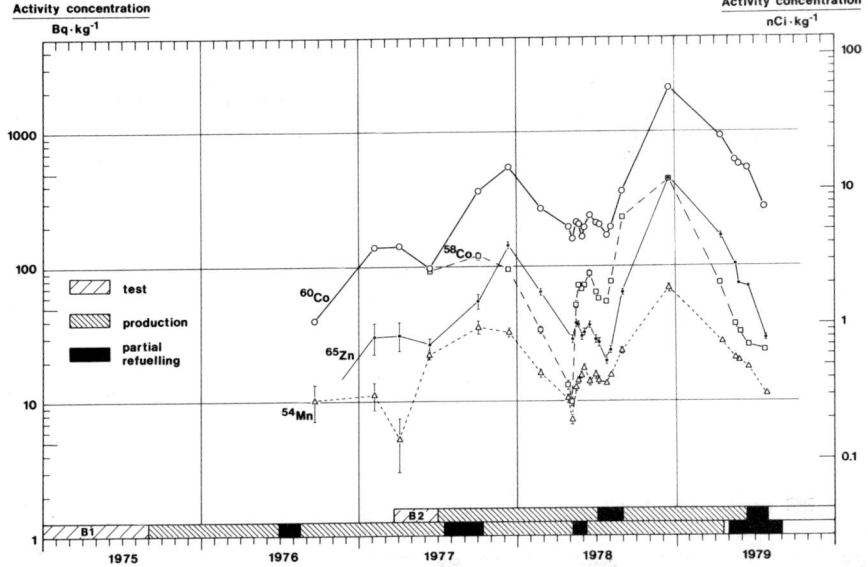

Figure 1. Time variation of the concentration of activation products in Fucus v. 2.9 km north of the Barsebäck discharge pipe.

From these results, we could establish that the increase in activity concentration in Fucus v. per unit activity released from the power plant was about the same for these radionuclides [1]. Furthermore, from the very frequent collection in May-June 1978 we could determine the biological half-life for these nuclides to be (60±15) days [1]. For this purpose we were supplied with data on released activity by the staff of the power plant [2]. The half-lives for the cobalt isotopes are in good agreement with laboratory experiments [3]. For ^{65}Zn, however, our results indicate a lower retention than found in laboratory experiments [4, 5].

We have also followed the ratio of ^{58}Co/^{60}Co activity concentrations in Fucus v. at the same collecting station 2.9 km north of the discharge pipe. This is presented in figure 2. From this figure it is obvious that the power plant releases ^{58}Co and other activation products mainly during the overhaul periods. This figure also clearly shows that the ^{58}Co/^{60}Co activity ratio is an excellent indicator of an increased discharge rate from the plant. The resolution in time is remarkably good although the interval between the two shutdowns in 1978 was only three weeks. The activity ratio still shows a decrease between the shutdowns.

The more distant spread of activity has been investigated at eight collecting stations for Fucus v. along the Swedish west-coast. In figure 3, the variation with distance of the activity concentrations in Fucus v. for ^{60}Co, ^{58}Co, ^{54}Mn, ^{65}Zn, ^{110}Ag^m and ^{57}Co, all activation products, are given. This diagram is based on a collection made on the same day in December 1978.

As discussed in a previous article [1] it is clear that this distance dependence of the activity concentration gives a good fit to a power function of the form:

$$C(z) = \alpha \cdot z^{-\beta}$$

where c(z) = activity concentration at distance z and α and β are constants.

The value of the constant β, as shown in figure 4, is nearly the same for all radionuclides except ^{110}Ag^m, which has a value about 50 % lower. The reason for this is at present not clear, but will be subject to further studies.

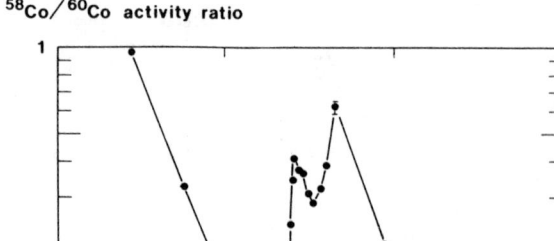

Figure 2
Time variation of the $^{58}Co/^{60}Co$ activity ratio in Fucus v. 2.9 km north of the discharge pipe

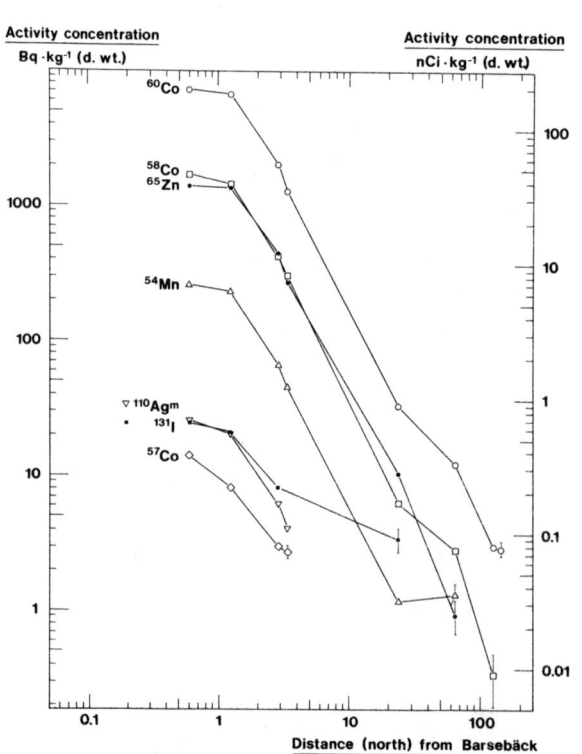

Figure 3.
Variation in activity concentration in Fucus v. by distance northwards from Barsebäck in December 1978

In figure 3, a curve representing ^{131}I is also given. In our previos study $/^-1\_/$ we also had detected ^{131}I, but at that time we were not sure whether the iodine was released from the power plant or from the university hospital in Malmö or Lund (20 km south of the plant) in connection with thyroid therapy. It is now clear that the main source of this iodine is the power plant, since during 1979 we have detected ^{136}Cs in Fucus v., thus indicating that increasing quantities of fission products are being released. The β-value for ^{131}I is only about one-third of the value for the activation products, except ^{110}Ag^m, perhaps illustrating the good solubility of iodine in water. It is also of interest to discuss the variation of the constant β with time for the activation products, except ^{57}Co for which too few data are available. This is shown in figure 4.

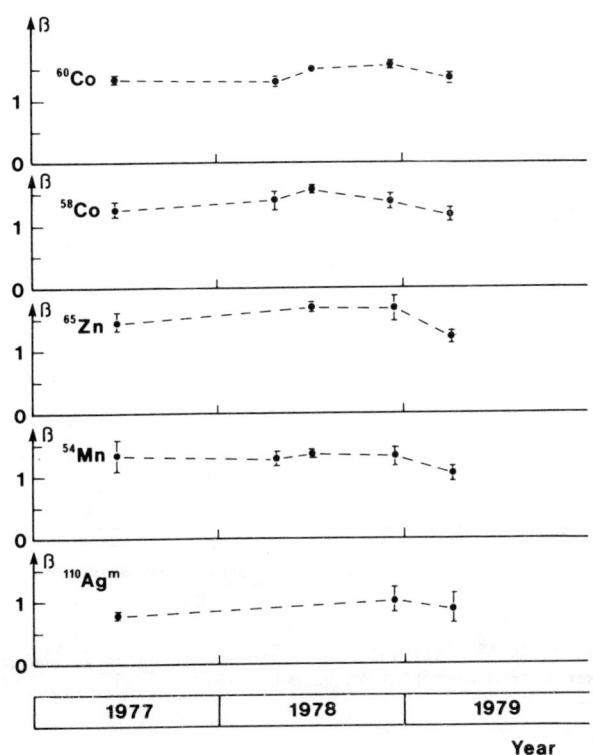

Figure 4. β-values (equation 1) calculated at different times.

It is evident that the β-values for the different radionuclides show a similar time variation; an increase from 1977 to 1978 followed by a decrease in 1979. Since the current situation is very stable in the Öresund. this pattern is probably linked to the discharge rate of activation products (cf. figure 1). The reduced β-values could accordingly be explained by the reduced activity releases mentioned earlier and by there being a considerable time delay between the discharge and established equilibrium conditions between water, sediments and Fucus.

3b.   Cobalt-60 activity concentration in Idothea

Detectable concentrations of ^{60}Co were found in several samples of Idothea. Because of the limited sample weights (0.1-2 g dry weight) the analysis had normally to be restricted to ^{60}Co. Figure 5 shows a comparison of the ^{60}Co activity concentration in Idothea and Fucus v. in December 1978 at various distances from

Barsebäck. The activity concentration in Idothea well reflects the concentration in Fucus v. presumably due to the fact that Idothea is grazing on the seaweed plants. The ratio between the ^{60}Co activity concentration in Idothea and Fucus v.

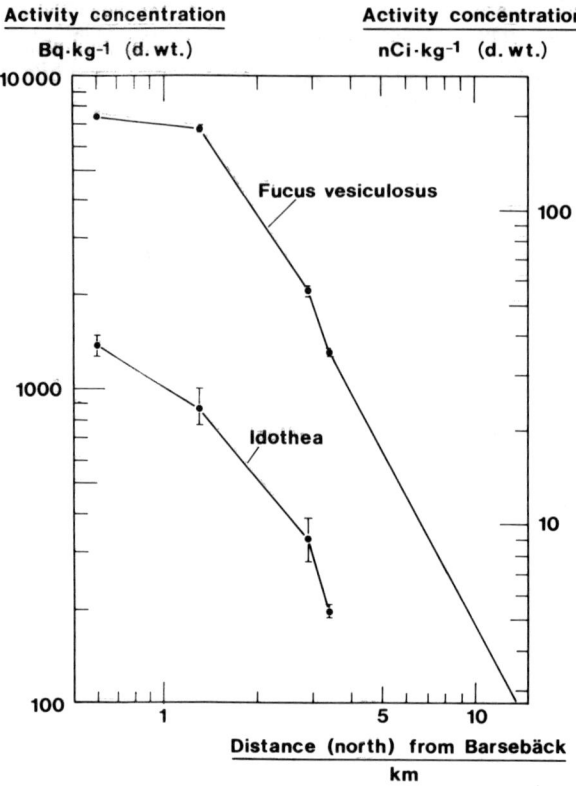

Figure 5. Variation in ^{60}Co activity concentration in Fucus v. and Idothea by distance northwards from Barsebäck in December 1978.

was 0.16±0.04 at that time. Considerably higher ratios, 0.93±0.10, are found in summer. This is illustrated in Figure 6. The explanation for this seasonal variation may be the decreased metabolic activity of Idothea during the winter period. No detectable ^{60}Co activity (< 2 Bq/kg d.wt.) has been found in Gammarus, also living on the Fucus plants. This may indicate a difference in their "food habits". Because Idothea (as well as Gammarus) are eaten by various fish species, these findings may be of radioecological importance.

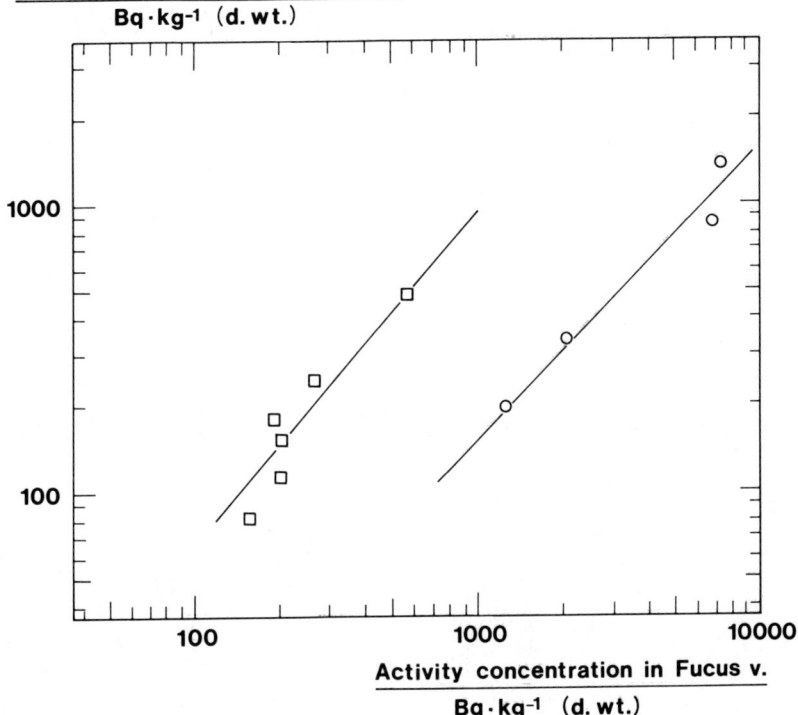

Figure 6. Relation between ^{60}Co activity concentrations in Idothea and Fucus v. in summer (squares) and winter (circles).

4. SUMMARY AND CONCLUSION

The macroalgae Fucus vesiculosus has proven to be a very suitable bioindicator for mapping the release of activation and fission products from a nuclear power plant into the marine environment.
Studies of the ^{60}Co activity concentration in animals living on the Fucus plants have shown that species of Idothea have activity concentrations varying between 0.16 (winter) and 0.93 (summer) of the activity concentration in Fucus v. Because Idothea are eaten by various fish species this may be of importance for the transport of radioactive substances to man.

REFERENCES

(1) Mattsson, S., Finck, R., and Nilsson, M.:
"Temporal and spatial variations in the distribution of activation products from Barsebäck Nuclear Power Plant (Sweden) in the marine environment as established by seaweed". Environmental Pollution (In Press)

(2) Sydkraft (South Swedish Power Board):
"Monthly report over liquid releases from Barsebäck May 1975-July 1979" (in Swedish)

(3) Nakahara, M., Koyanagi, T., and Saiki, M.:
"Concentration of radioactive cobalt by seaweeds in the food chain"
Impacts of nuclear releases into aquatic environment (Proc. Symp. Otaniemi 30 June - 4 July 1975) pp. 301-312. IAEA, Vienna, 1975

(4) Gutknecht, J.:
"Uptake and retention of cesium-137 and zinc-65 by seaweeds". Limnology and Oceanography 10, 58-66 (1975)

(5) Young, M.L.:
"The transfer of ^{65}Zn and ^{59}Fe along a Fucus serratus (L.) - Littorina obtusata (L.) food chain". J. mar. biol. Ass. U.K. 55, 583-610 (1975)

SELECTION OF CRITICAL GROUP IN RELATION TO THE RELEASE OF RADIONUCLIDES FROM NUCLEAR SPENT FUEL REPROCESSING PLANT

Y. Ohmomo
Division of Radioecology, National Institute of
Radiological Sciences. Nakaminato, Ibaraki, Japan

ABSTRACT

In respect of internal radiation due to the coastal release of radionuclides, survey on marine food consumption is most useful for the selection of critical group. Species of marine organisms they usually eat is fully over 100 in the coastal area of Ibaraki prefecture where the fuel reprocessing plant is located. Though it gives only a spot datum, one day's consumption survey a season is of convenience to obtain cooperation from housewives and is of use to pick up critical organisms and those who eat much of them. However, long-term survey is required to estimate ordinary intake of the critical foods or those who are supposed critical people. One day's consumption survey makes it easy to perform the subsequent long-term one.

Short contributions

The nuclear fuel reprocessing plant is in Tokai-mura, about 90 miles north from Tokyo. Low-level liquid radioactive waste is discharged into the Pacific Ocean from the exit nozzle through pipeline. The marine fish and the other sea foods are still the most important protein resources for the Japanese people. So the radioactive contamination of the coastal sea is of great concern for inhabitants, specifically for inshore fishermen and their families there. In respect of internal radiation due to the coastal release, it is considered that the critical group is almost equal to those who eat much of marine foods caught in the sea area significantly contaminated by radionuclides and will be exposed to the highest internal radiation. Species of marine organisms they usually eat is fully over 100 around there. Since it is hard work for housewives to keep record on the consumption without any omission specially for a long-term, it is quite difficult to obtain cooperation from them. Though it gives only a spot datum, one day's consumption survey is of convenience to obtain cooperation from them and is of use to select critical marine organisms and those who eat much of them. Consequently, subsequent longterm consumption survey is concerned only with limited numbers of organisms. In this meaning, one day's consumption survey makes it easy to perform the subsequent investigation. There is another problem in relation to make a record on marine food consumption. Amount of consumption must be given as fresh weight of the materials. However, housewives do not always write in the fresh weight. It is often given as cooked weight or only size or price of the materials. When given by the cooked weight, it is necessary to collect the materials, to cook in the same way as they do and to obtain the conversion factors to fresh weight.

SOME COMMENTS IN CONNECTION WITH JAPANESE EXPERIENCE
ON PRE-ESTIMATION OF RADIATION DOSES TO MAN DUE TO
THE COASTAL RELEASE OF LIQUID RADIOACTIVE WASTES

M. Saiki
Division of Radioecology, National Institute
of Radiological Sciences. Nakaminato-shi,
Ibaraki-ken, Japan

ABSTRACT

Research programme and experience, in connection with coastal discharge of radioactive liquid wastes, during past fifteen years, were summarized. In the first stage, i. e. period obtained only limited practical scientific information, pre-estimation of radiation doses to man with theoretical and assumed factors, weighted enough safety, was carried out. During the second stage, further practical estimation was obtained by use of experimental results and survey data. At the third stage, environmental monitoring data are going to prepare for scientific analysis and safety control of the radioactive effluent. At the same time, radioecological studies on the aquatic organisms in coastal waters and also technical studies on the waste treatment are investigated continuously.

SHORT CONTRIBUTIONS

When construction of the first nuclear spent fuel reprocessing plant at Tokai-mura was planned by Power Reactor and Nuclear Fuel Development Corporation (PNC), about fifteen years ago, the effects of radiation and radioactive materials to man and his environment were discussed by many of scientists in various fields of sciences. The amounts of radioactive wastes, discharged into the coastal waters, are estimated as maximum 65 Ci/90 days, excluding tritium. It will be composed as follows : 1% Sr-89, 1% Sr-90, 6% Ru-103 and 103m, 48% Ru-106 and Rh-106, 4% Cs-137, 1% Ce-141, 20% Ce-144 and Pr-144, 10% Zr-95 and Nb-95 and 9% of other radionuclides. Concentration of radioactivity in the waste is estimated as $2.4 \times 10^{-3} \mu Ci/cm^3$.

For evaluation of the effects of low-level liquid waste, discharged into coastal seas, the internal exposure caused by ingestion of radioactive substances in sea-foods, and the external exposure caused by radiation from the seawater and coastal area, were considered. In order to ensure the health of the whole population, the group, which is considered to have received the highest dose, was taken into account. Therefore, radiation exposures by fishermen who will be irradiated by beta-ray on hands and by gamma-ray on whole body, due to radioactive contaminated fishing nets, ropes and boats, are estimated. On the internal radiation burder in man, irradiation due to oral intake might be the most noticeable problem for Japanese, since the remarkable biological concentration of some radionuclides by marine organisms and because of that Japanese diet consists of considerable amounts of sea-foods of various kinds.

A fish "shirasu", immature anchovy, engraulis japonica, was chosen as a noticeable food, since local inhabitants around the reprocessing plant are fond of this fish which is used to be eaten with whole body. Furthermore, ruthenium, zirconium, cerium and strontium, to be appeared in the effluent, might be concentrated in skins, gills and intestinal contents. For extreme poor swimming abilities of this fish, it was assumed that it will moves with the effluent from the discharge outlet. Therefore, radioactive concentration in immature anchovy were roughly estimated by the following method.

$$\frac{dm}{dt} = - Km + ac$$

a : the accumulation rate of radioactivity ($cm^3/g \cdot sec$).
k : the excretion rate of radioactivity (sec^{-1}). [ex. Ce : 2.67×10^{-6}, Ru : 1.60×10^{-5}]
c : the radioactivity concentration in the seawater along the flow ($\mu Ci/cm^3$). (concentration factor = a/k)
m(t) : the radioactivity concentration in the fish at the time t ($\mu Ci/g$).

In addition to above estimation, the concentration of Ru-106 by algae, collected about 4Km from the discharge outlet, where is a inhabitable craggy clift for seaweeds, was taken into consideration.

The estimation was carried out by PNC and other scientists during 1965 to 1969 for safety evaluation by Specialists Group for safety on Nuclear Fuel Reprocessing plant, Japan AEC. Further detailed information on the method will be obtained in the paper entitled "Environmental considerations and public acceptance aspects on nuclear plant site selection in coastal areas in Japan" by M. Saiki et. al. (Environmental Aspects of Nuclear Power Stations, p. 821-846, STI/PUB/261, IAEA, Vienna, 1971). The Specialists Group AEC recognized that the radiation dose due to coastal disposal from the reprocessing plant is less than one tenth of the dose limit for members of the public. However, the Specialists Group recommended that further practical evaluation should be held in future, before the disposal for full operation of the plant, and, if necessarry, "permissible discharge limit" should be revised, on the basis of the latest scientific results of extensive research and surveys.

Under the auspieces of Japan AEC and Science and Technology Agency, Special Investigation Committee for Coastal

Discharge of Liquid Radioactive wastes was initiated in 1966, in the frame of Nuclear Safety Research Association. The Committee was managed by Prof. Y. Miyake, Prof. E. Tajima, Prof. Y. Hiyama, Dr. K. Tsukamoto, Dr. K. Misono and other Directors General of several national institutes and it was consisted of sub-committees on discharge method with pipe-line (Head : H. Honma), on biology (Head : M. Saiki), on chemistry (Head : N. Yamagata) and on oceanography (Head : S. Sakagishi). During 1967 to 1971, following items were investigated by the Biological Sub-Committee : (1) movement and transfer of radionuclides in coastal waters, (2) distribution of marine biota and production of fisheries foods at around the discharge area, (3) concentration of radionuclides by marine organisms, (4) marine foods consumption by inhabitants around the discharge area, (5) radioactivity background survey, (6) estimation of radiation body burden by critical inhabitants. For investigations on above items, Marine Radioecological Research Station, National Institute of Radiological Sciences (NIRS) was initiated in 1969 at Nakaminato city where is located in center of Tokai-Ooarai area of atomic energy establishments. (Note : Recently, the Research Station was renamed as Nakaminato Laboratory, NIRS, since terrestrial radioecological research group joined there.)

According to the results obtained in the above described activities, the Specialists Group, AEC, re-evaluated the radiation doses in 1977. And it was found that the radiation dose estimated with the latest information was lower than the value estimated in 1969. It shows that factors, used in the first estimation, lean to too much safety side, chiefly depend upon limited information. For example, continuous daily consumption of contaminated immature anchovy was assumed as 200g/day which is far higher than the data obtained by practical survey carried out by Y. Ohmomo and M. Sumiya. Furthermore, it was also assumed that all of immature anchovy moves with the effluent from discharge outlet along the axis which has the highest concentration of activity. In my understanding, during the first stage (until 1969), cautious pre-estimation was done allowing a sufficient margin of safety, thereafter, some practical estimation was followed in the second stage (1970 to 1976), since experimental results and field survey data became available. And the environmental monitoring data on radioactivity due to test operation of the plant have expected in the third stage (after 1977).

The monitoring data might be evaluated periodically by experts of Nuclear Safety Commission, on scientific reasonable basis. If necessarry, the permissible discharge limit could be revised, depending upon recommendation of Nuclear Safety Commission. (Note : Japan AEC was divided in two commissions, i. e. Atomic Energy Commission and Nuclear Safety Commission, in 1978.) On the other hand, the amounts and constitution of radionuclides in the wastes may vary in future, because of that PNC is going to investigate new techniques on the waste treatment.

As was shown in several presentations to the Seminar, research works are continuously performing as follows : on diffusion of coastal waters by PNC and JAERI (Japan Atomic Energy Research Institute), on distribution of marine biota by Ibaraki Prefectural Institute of fisheries, on radioactivity monitoring by PNC and Ibaraki Prefectural Centre on Public Hazard, on transfer and concentration of radionuclides, consumption of sea-foods and constitution of minerals in human body for estimation of radiation dose to man by Nakaminato Branch-Laboratory, NIRS, on radiation effects on aquatic organisms by Chiba Laboratory, NIRS and Zoological Institute, University of Tokyo.

Finally, I would like to add that valuable information are also obtaining by laboratory works, chiefly RI tracer experiments on aquarium fish. Mechanisms on uptake of various physico-chemical forms of radionuclides, ex. sediment-bound radionuclides, by fish have investigated. Variety of uptake rate, effected by various environmental conditions, i. e. temperature, pH etc. of seawater, have also observed. In addition to results obtained in-situ, those experimental data might be valuable for assessment of radiation dose to man and also to fish themselves.

RADIOCESIUM IN INNER DANISH WATERS FROM WINDSCALE

A. Aarkrog
Risø National Laboratory, Health Physics Department
DK-4000 Roskilde, Denmark

ABSTRACT

From measurements of ^{134}Cs and ^{137}Cs in seawater collected in 1978, the transport time of these radionuclides from Windscale to inner Danish waters was estimated at 3 years.

Fig. 1.

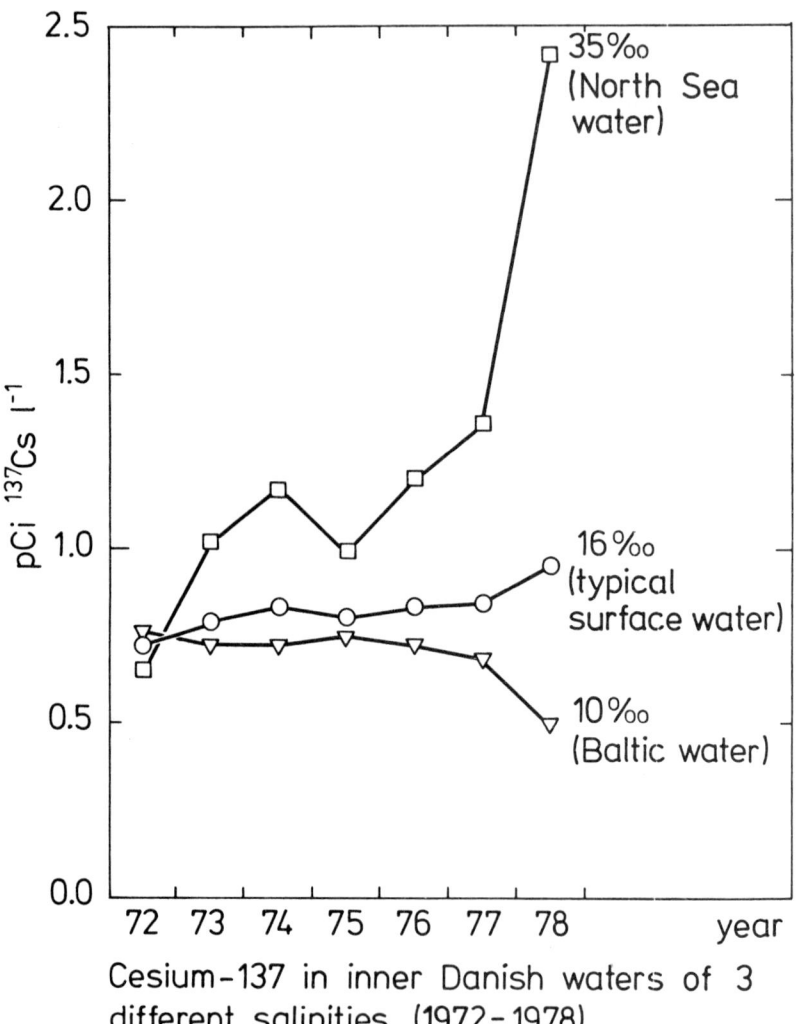

Cesium-137 in inner Danish waters of 3 different salinities (1972-1978)

The inner Danish waters connect the North Sea with the Baltic Sea. The indlow from the North Sea is a southward current of high salinity (~ 35°/oo) bottom water. The outlow from the Baltic Sea through the Danish sounds and belts is a northward surface current of light water.

Since 1972[1]) the ^{137}Cs concentrations of the high salinity component of inner Danish waters have been increasing (fig. 1).

From the comprehensive studies by Kautsky[2]) of ^{137}Cs in seawater we know that the North Sea has received radiocesium from the reprocessing plant at windscale in increasing amounts since the beginning of the seventies. We further know that ^{134}Cs is released from Windscale along with ^{137}Cs.

Table 1 shows the decay-corrected $^{134}Cs/^{137}Cs$ ratio in Windscale releases at August 1978, which is the month of reference for the Danish 1978 seawater samples. It appears that $^{134}Cs/^{137}Cs$ *) in Danish sewater in 1978 corresponded to the release ratio of 1975 and was significantly different from the release ratios in 1974 and 1976. Hence we may conclude that the transport time from Windscale to inner Danish water was approximately 3 years. This estimate is in good agreement with the transport times from Windscale suggested by Kautsky[2]) (2 years to the North Sea and 4-5 years to the Baltic Sea).

In fig. 2 the ^{137}Cs concentrations in high salinity Danish seawater are related to the releases from Windscale 3 years prior to the seawater sampling. The regression line suggests that a release of ,e.g., 1 MCi ^{137}Cs from Windscale in the year (i-3) results in a seawater concentration in the bottom water of the Danish sounds and belts in the year (i) of approximately 10 pCi ^{137}Cs l^{-1}.

REFERENCES:

[1] Risø Reports in the series: Environmental Radioactivity in Denmark in 1972,1973, 1974,1975,1976,1977 and 1978. Nos. 291,305,323,345,361,386 and 403.

[2] Kautsky H. The North Sea region taken as example for the behaviour of artificial radioactive isotopes om nearshore sea areas. Seminar on Marine Radioecology, Tokyo 1-5 Oct. 1979.

Table 1

Cesium-134/Cesium-137 ratios

at August 1978

Windscale releases from 1974 : 0.066

       -          -   - 1975 : 0.077

       -          -   - 1976 : 0.086

Inner Danish waters, 1978 (27 samples):

0.077 (1 SE:0.0025)

Transport time approx. 3 years

---

*) The contributions of ^{137}Cs from fallout was subtracted before the ratios were calculated.

Fig. 2.

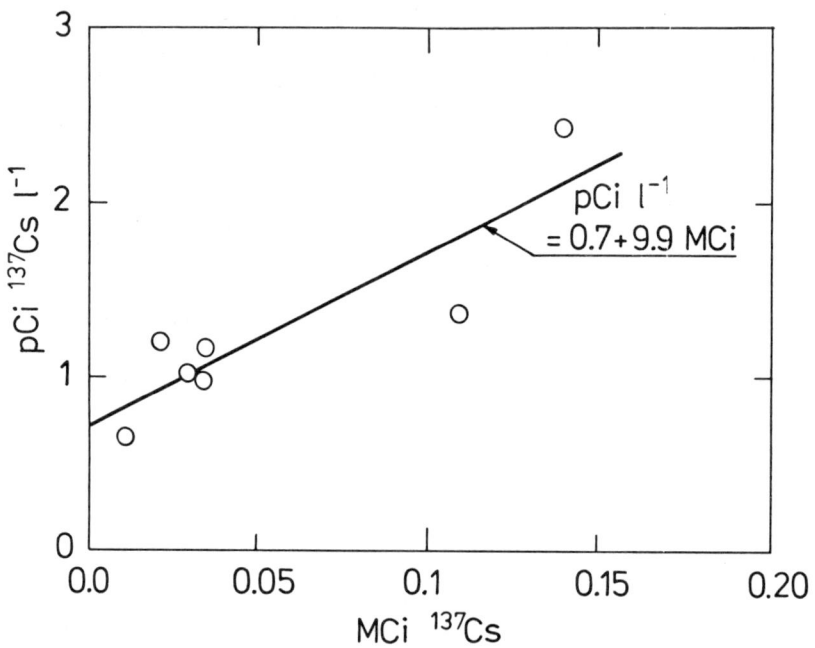

Cesium-137 in Danish bottom seawater (salinity 35‰) 1972-78 as a function of the releases from Windscale 3 years prior to the seawater sampling.

Session 8

CONCLUSIONS OF THE SEMINAR

This summary report has been prepared by
Messrs Templeton and Pentreath and covers the conclusions
of the different sessions presented by the Chairmen,
and a general discussion

Chairman - Président
Mr. W.L. TEMPLETON
(United States)

Séance 8

CONCLUSIONS DU COLLOQUE

Ce texte a été résumé par MM. Templeton et Pentreath ;
il comporte les conclusions des différentes séances
présentées par les Présidents ainsi qu'une discussion générale

In the first deep sea session, Mr. Webb (United Kingdom) drew attention to the recommendations of ICRP 26 and the conditions to be applied to waste management operations i.e. justification, optimisation of protection, and limitation. The key condition is the second, and the technique suggested is that of differential cost benefit analysis, which includes the use of the concept of "dose commitment", the infinite time integral of the dose rate. The third condition is an overriding one that doses shall not exceed the limits. However, to achieve doses below the limits it is necessary to apply an optimization procedure which will indicate whether or not further effort to reduce doses, or dose commitments, is warranted.

Mr. Webb also discussed two forms of radiological assessment. The most usual is a pathway analysis using models, parameter values, and assumptions designed to maximise the dose estimate. This maximum dose estimate is the appropriate one to compare with the ICRP dose limits for compliance with the third condition mentioned above. This was the approach used by the IAEA when defining high-level wastes unsuitable for dumping, as detailed in a paper given by Mr. Templeton (United States). However, these calculations rightly over-estimate the doses to man and should not be taken as appropriate for assessing actual, or most likely, consequences. In order to compare different disposal options i.e. optimization studies, Mr. Webb emphasized that it is also essential to assess, as realistically as possible, the most probable distribution of doses from which to calculate collective dose and/or dose commitments. The latter has not been carried out in the past and Mr. Webb hoped that it would be recognized that data are required to make realistic estimates of dose from specific operations, as well as the need for extreme values to derive maximum estimates. Mr. Webb stated that pathway analysis, and radiological assessment, must be continuing reiterative processes until there is sufficient confidence in both the models, and the parameter values, for decision making on statutory regulations. He emphasized that from his viewpoint, as a modeller, he would like to see research on the following : the rate of release from packaged waste, and its physical and chemical form ; on the local processes at the benthic boundary layer, particularly removal processes by sedimentation ; on physical and biological transport processes, laterally and vertically ; on the stability of long term oceanographic basin models ; and on biological processes in the water column. The remaining two papers in this session described the research being conducted in the deep sea as part of a U.S. program on seabed emplacement of high-level wastes ; and on monitoring conducted by Portugal of seawater, and fishes, adjacent to the Madeira Islands in the Atlantic.

In the second deep sea session the oceanographic research that has been conducted by Japan in the deep oceans in the Western North Pacific, in connection with proposed dumping sites for solid radioactive wastes, was discussed. Five papers described the physical oceanography of the sites, deep bottom current measurements, and the effects of steady flow and upwelling. Two further papers described a biological transport model - comparing the concentration factor approach with a predator/prey model - and details of pre-operational biological surveys. These studies, although incomplete, do provide some further information on deep sea processes hitherto unknown, and are examples of the steps needed to increase our confidence in this methodology.

The third deep sea session included papers on the behavior of radionuclides and stable elements in the deep ocean. Dr. Nagaya (Japan) and Dr. Kautsky (Federal Republic of Germany) presented data of radionuclide distributions in the water column, while Dr. Noshkin (United States) presented evidence for the remobilisation of plutonium at Enewetok and Bikini atolls in the Central Pacific. Mrs Vaz Carreiro (Portugal) discussed, in a short paper, some considerations that need to be made when discussing food chains of deep sea organisms.

The next session, the fourth, dealt with shallow water studies and their possible application to the deep sea. Mr. Ancellin (France) and Dr. Pentreath (United Kingdom) summarized studies conducted at two European reprocessing plants, La Hague in France and Windscale in the United Kingdom. In the first, Mr. Ancellin discussed the research conducted on the distribution, transport and biological concentrations of relatively short-lived fission products that was necessary for a validation of the initial radiological assessment. Dr. Pentreath, on the other hand, described research on the long-lived nuclides, particularly the transuranium elements. He emphasized that these studies were necessary to make realistic estimates of dose commitment from the Windscale operations, and that they also provided data relevant to the deep sea disposal option. In order to extrapolate to the deep sea environment, however, Dr. Pentreath considered that more data were required on the comparative behaviour between the transuranium nuclides and the naturally-occurring actinides, particularly uranium and thorium.

The fifth session continued on shallow water studies with six papers which complemented those of Session 4. Dr. Fukuda (Japan) described recent research on the coastal dispersion of tritium, Mr. Ueda (Japan) discussed the radioecology of Co-60 in a bay system, and Dr. Honda (Japan) detailed experimental work on the formation of organic complexes in seawater. Dr. Smith (Canada) and Mr. Aarkrog (Denmark) discussed studies on the distribution of nuclides in core profiles, and continuing observations on plutonium in the Thule area, respectively. Dr. Persson (Sweden) described measurements of natural (Th, U) and artificial actinides (Pu, Am) in algae from coastal waters.

The sixth session continued with descriptions of shallow water research. Dr. Kautsky described the transport of Cs-137 in the English Channel and North Sea and emphasized that little dilution occurs despite extensive transport. Dr. Pentreath argued that, in order to keep a sense of perspective, it was important to consider the concentrations of natural radionuclides when dose assessments to marine organisms are made. He presented some measurements and calculations that indicated that the radiation dose to deep sea and shallow water organisms from natural radionuclides was very similar. Mr. Van Weers (Netherlands), Mr. Nakahara and Mr. Kurabayashi (Japan) presented data on calculated concentration factors derived from measurements of stable elements in coastal and deep sea marine organisms. Dr. Koyanagi (Japan) presented some data on the accumulation of Co-57 and Co-60 in some demersal fish tissues. The final paper in this session dealt with diet and consumption surveys of marine products around Tokai-Mura, and Nakaminato, as part of the pathway analysis for coastal discharges. Maximum values were about three times the average consumption rate of 200-300 g/day.

In the last technical session, five papers were presented. Three of these dealt with monitoring studies in the Baltic Sea [Mrs Salo (Finland), Dr. Mattsson (Sweden) and Mr. Aarkrog (Denmark)]. Of particular interest was the observation by Mr. Aarkrog that Windscale effluent (Cs-134/Cs-137) appeared to take at least three years to reach the entrance to the Baltic Sea. Dr. Saiki (Japan) reviewed the extensive field and laboratory studies that have been conducted over the last 15 years in connection with the coastal discharges from the Tokai-Mura works.

Following these summaries, the subsequent discussions considered the difficulties and costs of deep sea physical and biological oceanography. A request was made for improved international co-operation in the use of deep sea ocean research vessels, and in the communication of the research data obtained. Dr. Kautsky suggested that, in addition to measurements by anchored instrumentation, there is a need for measurements of real transport processes by buoyancy floats and research submarines. Mr. Webb addressed the need for research priorities. He emphasized that there were large differences and needs between the coastal discharges of low level liquid wastes, the dumping of solid low-level wastes, and the seabed emplacement of high level wastes. In the case of coastal discharges the pathway analyses and radiological assessments can easily be validated during the early years of the discharge, be refined, and produce dose values that are very realistic. In this case it is not difficult to have confidence in ones predictions. In the case of dumping low-level packages wastes on the bottom of the ocean, at depths in excess of 4000 m, Mr. Webb felt that the overall conditions were very much better in terms of limiting the dose to man. However, not only will it be extremely difficult to measure radionuclides at the concentrations that may exist, even in the vicinity of the dumped packages, after tens of years, but their significance in terms of radiation dose to man will be virtually insignificant. Mr. Webb stated that it is not very efficient to expend large sums of money, and the research efforts of valuable people, on something which is intrinsically not very hazardous. However, he stated that we did need to improve our knowledge of water movements, and of biological processes, in order to build models in which one can have reasonable confidence. But Mr. Webb felt that when is comes to the potential disposal of high-level waste the situation is very different. In this case, the packaging itself, or its emplacement in sediment, could result in there being no release of radioactivity for hundreds, or even thousands, of years - and even then the rate of release is likely to be at a very slow rate. For dose assessments, long-term models are required, models which cannot be validated by feedback from actual measurements, except from limited experiments. Such models must therefore be robust, and insensitive to short-term changes in their parameters.

With regard to high-level waste disposal, Dr. Pentreath drew attention to our great lack of knowledge on the longer-lived nuclides, and said that it was here that the radioecologists must provide a greater input. He stated that we are effectively entering a new era, in that fission-product nuclides, and neutron activation nuclides generally, are fairly well characterised ; although one still requires some refinements for local situations. The contribution of the radioecologists should now be to complement the other research being made, such as on the deep sea oceanography, the sedimentalogical aspects, and so on.

Dr. Yayanos said a great deal depended on the efficiency of the "barriers" presumed to exist between the high-level waste and the sea water : whether the barriers be the cannister, the sediment, the waste-form, or even some other type which has not yet been considered ; perhaps even a biological one. A great deal of confidence in such barriers was required because once in the deep sea it would prove to be very expensive to move the waste yet again. And with regard to expenditure, although some people considered that the high cost of the necessary biological and oceanographic research diminished the cost/benefit ratio of the deep sea disposal option, he considered that such research was justified, in any case, because of its fundamental importance. In other words it was essential to understand the deep sea in order to develop its resources, i.e. manganese nodule mining, deep sea commercial fisheries, high-level waste disposal or whatever. He also recommended a commitment by governments to sustain a continuing monitoring programme should the deep sea disposal option be used - in order to validate the initial decision.

Dr. Noshkin noted that emphasis had been given to the marine food chain as the major pathway to man for radionuclides in the sea, and that indeed this was correct ; but there were additional pathways in the sea that should possibly be considered in the context of disposal of waste in the ocean. One of these, which is being currently investigated at Windscale as well as in the Marschall Islands, is the inhalation pathway. In the Pacific, elevated levels of plutonium in the marine aerosols were generated on the ocean reef. This was not a re-suspension of salt spray, which could not account for the excess plutonium in the aerosols. Concentrations of plutonium in algae and sand did not correspond with the amounts measured in the aerosols. Thus he thought there may be some surface water enrichment of elements like plutonium, in near shore areas, which subsequently become airborne.

Mr. Ancellin said that although there was a considerable amount of data on concentration factors for radionuclides in the literature, there appeared to be some problem in agreeing values. For some nuclides the available data do not differ too much, and mean values and ranges could be agreed upon. But for other nuclides the values sometimes differed by factors of 10 to a 100. Some of this was probably due to the fact that the measured water values used were not those of the integrated exposure, and that the values then became quite arbitrary. A consensus of the experts for each nuclide, and each major group of organisms, was required.

Mr. Webb raised a fairly fundamental question concerning the concentration factors - that these are an expression of a very simple equilibrium model of food chain transfer. One of the reasons, perhaps, as to why it is difficult to settle on a good number for a concentration factor is that a simple equilibrium model is not the best description of the real process. In other aspects of food-chain modelling, with which he was concerned - for instance, terrestrial food chains, modelling of transfer through animals like cows and sheep - the tendency is to make a dynamic model of the animal itself. For example in modelling the transport of iodine from grass to cow to milk a dynamic model was constructed which consisted of gut and bowel, blood and milk. This is not done for marine food chains and he therefore questioned the value of equilibrium models.

Dr. Pentreath commented on this. He said that one must appreciate that the term concentration factor is a term of convenience for health physicists, or those concerned with dose assessment. The radioecologist was well aware of the fact that an absolute value, of course, does not exist. However, he thought it was of value in the aquatic environment because the common medium, water, was usually amenable to the derivation of a common value to which others could be related. Not that this necessarily implied that an aquatic organism obtained radionuclides directly from the water, it also obtained them from food ; the relative importance of the two pathways differed considerably from one organism to another, and from nuclide to nuclide. For terrestrial animals, however, accumulation was more usually only by ingestion ; although there are exceptions - such as Pu absorbed across the lungs. He continued that what he would like to see, for the low-level discharges to coastal waters, was a move away from concentration factor data, which are used in the initial assessments, to relating concentrations in terms of per unit input. What one realy wants to know is that a nuclide is going into the environment at a certain rate and that it is coming back to man at a certain rate. Once one has made ones initial calculations, and appreciating all the errors which are inherent in using concentration factor data, one can then move towards an estimation of the concentrations which will obtain in samples taken at a certain point, or even in fish landed at a certain port, for a certain rate of discharge. This approach was frequently more useful, and in many cases one does not even need to understand all the complexities obtaining between the nuclide entering the environment and its return to man, providing one can adequately relate the two. He thought this sort of expression was also useful in terms of high-level waste disposal, in view of the IAEAs concept of release rates.

Mr. Van Weers commented that there was a big difference between a model based on concentrations in marine materials related to unit discharge, and applying that concept to models for the disposal of high-level or low-level waste into deep waters. The concept of the relation between concentration in an organism and unit release could be checked when releasing in near-shore waters - one could from experience, learn of the relation between release rate and concentration in environmental materials. But in the case of the dumping of radioactive materials one cannot check, by measurement, the relation between dumped or released activity and the concentration in biological materials in the environment. One has to rely on models, and the concept of concentration factors cannot be avoided in such a situation.

Dr. Pentreath added that a further aspect which was being brought out in the meeting pertaining to concentration factors was one that had been made previously, and which he thought should be reiterated : that the required accuracy of the values differs somewhat from whether one was dealing with per caput dose, collective dose or committed dose, with a decreasing emphasis on the accuracy of the data required.

Dr. Ichikawa made two points concerning concentration factors. Firstly, in his opinion, the concentration factor of radionuclides between the water and the marine organisms was not constant in principle, but one single value was used as a constant value for practical purpose, even though the value of the concentration factor was changed by environmental factors, physiological factors, and other, various, factors. Secondly, in considering the radiological impact, one uses the annual dose for the unit of impact - in which case one realises that most of the people do not consume the same fish obtained from the same place. He prefered to use the average value of the concentration factors obtained.

Dr. Aarkrog stated that in his view the use of concentration factors in connection with disposal of waste on the bottom of the ocean was especially problematical. For instance, considering radionuclides such as plutonium, and its uptake in benthic organisms, one realises that the applications of concentration factors from water to the organism may lead to wrong conclusions because these benthic organisms may take up most of the activity from the sediments and thus were not dependent on concentrations in the water.

Dr. Pentreath responded that although this was true - the route to an organism may come from many sources, including sediment - it was surprising, nevertheless, that one still found a reasonable relationship between the water and the organisms because there was also a relationship between the water and the sediment. Mr. Van Weers considered that will apply in any case to situations where one can rely on some sort of equilibrium being obtained.

Mr. Webb said that one of the modelling problems he is engaged in at the moment is to try to build a model of the transfer from high-level waste which has been placed in the sediment layer, back through food chains, to man. And in this situation it is waste which has come from containment, and entered the sediment matrix, which is the prime source term. One of the problems that he has encountered was to find the appropriate way of modelling the transfer of this material to the biota. He did not feel that it was actually the right way to go via the water to the benthic boundary layer, and then back, using concentration factors, to model the benthic organisms. There should be some way of modelling the more direct transfer of the contaminated sediment into the organism and he wondered if anyone could give any comments on what would be the appropriate way to set up this kind of system ? Dr. Saiki responded that he thought in experimental studies the transfer from the sediment is usually quite limited.

Dr. Pentreath said that Mr. Webb's problem area had been recognised, and appreciated, by marine radioecologists and there have been attempts to estimate the retention of radionuclides from sediment which has been ingested by marine organisms. In fact he thought that this also illustrated the complexity with which we were dealing, and would imagine that the pathway which could be of importance from deep sea sediments might nevertheless still be via water for some radionuclides. He thought that in the case of a complex element like plutonium it may well be a function of chemical speciation which effects the rate of release back into the water from the sediment, and that this in turn provided a pathway. For example he was of the opinion that demersal fish in the Irish Sea derived plutonium primarily from direct uptake from water, despite their ingestion of highly contaminated sediment ; whereas the principal route of uptake of the majority of radionuclides into fish was through the food. Marine invertebrates, however, were more likely to obtain their plutonium by ingestion. He emphasized that this was another area of research which was still wide open, particularly for those long-lived elements which are of particular concern, and which have a rather complex chemistry.

Mr. Webb said that it was interesting to consider that of the radionuclides which were currently thought to be important in the long term dose assessments of high-level waste disposal, even the plutonium isotopes were short-lived in comparison to nuclides such as $Np-237$, $I-129$, $Sn-126$ and so on. He agreed with Dr. Pentreath's previous remarks that we know very little about such long-lived nuclides, even in shallow coastal waters.

Dr. Ichikawa stated that he was of the opinion that a food chain model must be used. He felt that the IAEA should have included such a deep sea model rather than avoided it, and hoped that it would be included in the future. As far as research was concerned, he felt that further research on concentration factors for deep sea organism was required ; both for stable elements and radionuclides. He would also like to see more studies done on the availability of radionuclides incorporated into marine products.

Dr. Pentreath speculated as to how long our option for very deep water disposal will remain open politically. He wondered if we will not eventually go towards improving our initial containment and gradually move back towards less deep waters which remain within a country's 200 mile territorial limits. Admittedly there are very deep waters within such 200 mile territorial limits, but not in all cases. Therefore, perhaps, some of our concern over our lack of knowledge of the very deep oceans may eventually not fully be needed. Mr. Templeton wondered if it will ever be possible for engineers to design a container that could take care of plutonium, let alone radionuclides with longer half lifes.

Mr. Webb, in response to what Dr. Ichikawa had mentioned earlier about oceanographic models, said that it was not very fair to call the sort of things that are used for this kind of assessment "oceanographic models" ; they were really only mathematical techniques for estimating maximum doses in such a way that it was extremely unlikely that one underestimated the doses. This was really all they were trying to do. The assumption made by the IAEA that bottom water be used as the source term for certain living organisms was not really a silly assumption. It cannot under estimate the effect. He thought that it was quite important to recognise that the purpose of these types of models was only to make sure that the relationship between the calculated doses, and the input of radioactivity, had been calculated in such a way that one was not underestimating the doses. He stressed the point that these models are sufficient for that purpose - that is what they are made for, and they do do that job reasonably satisfactorily. However, they do not

describe what could actually happen, so it was a mistake to imply that because the models had been adopted, and blessed by various international gatherings, they actually represented reality. They are purely mathematical techniques to over-estimate doses.
Dr. Ichikawa, while agreeing with Mr. Webb, said that when one calculated the dose in this way many people assumed that the calculated dose was that to which man was exposed in reality. That was very often the situation in Japan.

LIST OF PARTICIPANTS

LISTE DES PARTICIPANTS

CANADA

SMITH, J.N., Dr., Department of Fisheries and Oceans, Chemical Oceanography Division, Bedford Institute of Oceanography, P.O. Box 1006, Dartmouth NS, B2Y 4A2

DENMARK - DANEMARK

AARKROG, A., Risø National Laboratory, Postbox 49, DK-4000 Roskilde

FINLAND - FINLANDE

SALO, L.A., Dr., Institute of Radiation Protection, P.O. Box 268, 00101 Helsinki 10

FRANCE

ANCELLIN, J., Commissariat à l'Energie Atomique, Département de Protection, Institut de Protection et de Sûreté Nucléaire, B.P. n° 6, 92260 Fontenay-aux-Roses

BRESSON, G., Commissariat à l'Energie Atomique, Département de Protection, Institut de Protection et de Sûreté Nucléaire, B.P. n° 6, 92260 Fontenay-aux-Roses

KIEFFER, J.Y.M., 47 rue Perronet, 92200 Neuilly-sur-Seine

FEDERAL REPUBLIC OF GERMANY - REPUBLIQUE FEDERALE D'ALLEMAGNE

KAUTSKY, H., Dr., Deutsches Hydrographisches Institut, Postfach 220, 2000 Hamburg 4

JAPAN - JAPON

AMANO, H., Dr., Division of Environmental Safety Research, Tokai Research Establishment, Japan Atomic Energy Research Institute, Tokai-mura, Naka-gun, Ibaraki-ken

AOYAGI, I., Nippon Kyodo Hogei Kaisha, Ltd., 3-2-4, Kasumigaseki, Chiyoda-ku, Tokyo

ARAI, M., Association of Radiation Effects, 1-9-16, Kajicho, Chiyoda-ku, Tokyo

DIDA, H., Dr., Oceanographical Division, Marine Department, Meteorological Agency, 1-3-4, Otemachi, Chiyoda-ku, Tokyo

FUJIOKA, J., Head, Administrative Service, National Institute of Radiological Sciences, 9-1, 4-Chome, Anagawa, Chiba 260

FUKUDA, M., Division of Environmental Safety Research, Tokay Research Establishment, Japan Atomic Energy Research Institute, Tokai-mura, Naka-gun, Ibaraki-ken

FUKUDA, S., Dr., Health and Safety Office, Power Reactor and Nuclear Fuel Development Corporation, 9-13, 1-Chome, Akasaka, Minato-ku, Tokyo

FUSHIMI, K., Dr., Geochemical Laboratory, Meteorological Research Institute, Koenji-kita 4-35-8, Suginami, Tokyo 166

HASHIMOTO, T., Radiation Control, Plant Management Department, The Japan Atomic Power Co., 1-6-1, Ohtemachi, Chiyoda-ku, Tokyo

HIRANO, S., Division of Marine Radioecology, National Institute of Radiological Sciences, 3609 Isozaki, Nakaminato, Ibaraki 311-12

HIROSE, K., Dr., Geochemical Laboratory, Meteorological Research Institute, Koenji-kita 4-35-8, Suginami, Tokyo 166

HONDA, Y., Dr., Department of Nuclear Reactor Engineering, Faculty of Science and Technology, Kinki University, 3-4-1 Kowakae, Higashi-Osaka, Osaka 577

HONJO, K., Dr., Seikai Regional Fisheries Research Laboratory, Kokubu-cho, Nagasaki

HORI, K., Dr., National Institute of Radiological Sciences, 3609 Isozaki, Nakaminato, Ibaraki 311-12

ICHIKAWA, R., Dr., Division of Environmental Health, National Institute of Radiological Sciences, 9-1, 4-Chome, Anagawa, Chiba 260

IMAI, K., Division of Environmental Safety Research, Tokai Research Establishment, Japan Atomic Energy Research Institute, Tokai-mura, Naka-gun, Ibaraki-ken

IMAWAKI, S., Geophysical Institute, Kyoto University, Kyoto 606

INOUE, Y., Dr., Department of Sanitary Engineering, Kyoto University, Yoshida hon-Machi, Sakyo-ku, Kyoto City 606

INOUYE, S., Deputy Manager, First Testing Division, Radioactive Waste Management Center, N° 15 Mori Building, 2-8-10, Toranomon, Minato-ku, Tokyo 105

ISHIHARA, T., Managing Director, Radioactive Waste Management Center, N° 15 Mori Building, 2-8-10, Toranomon, Minato-ku, Tokyo 105

ISHII, T., Division of Marine Radioecology, National Institute of Radiological Sciences, 3609 Isozaki, Nakaminato, Ibaraki 311-12

ISHIKAWA, M., Dr., Division of Marine Radioactivity, National Institute of Radiological Sciences, 3609 Isozaki, Nakaminato, Ibaraki 311-12

ISHIYAMA, T., Dr., Radiation Center of Osaka Prefecture, Shinke-cho, Sakai, Osaka 593

IZAWA, M., Chief, Section for Industrial Program and Technology, Japan Atomic Industrial Forum, Inc., 1-13, 1-Chome, Shinbashi, Minato-ku, Tokyo

IZUMO, Y., The Institute of Public Health, 6-1, 4-Chome, Shirokanedai, Minato-ku, Tokyo 108

KADOYA, S., Dr., Nuclear Specialist, Director Rank, Office of
   Engineers-in-chief, Ebara Corporation, 11-1, Haneda Asa i-cho,
   Ota-ku, Tokyo 144

KANAZAWA, T., Mrs., Geochemical Laboratory, Meteorological Research
   Institute, Koenji-kita 4-35-8, Suginami, Tokyo 166

KANEDA, H., Radiation Safety Division, The Chubu Electric Power
   Co. Inc., N° 1, Higashi Shinmachi-cho, Higashi-ku, Nagoya

KASAI, A., Division of Environmental Safety Research, Tokai Research
   Establishment, Japan Atomic Energy Research Institute, Tokai-
   mura, Naka-gun, Ibaraki-ken

KASUKAWA, H., Nuclear Safety Research Association, 1-2-2, Uchisaiwai-
   cho, Chiyoda-ku, Tokyo

KIDACHI, T., Tokai Regional Fisheries Research Laboratory, 5-5
   Kachidoki, Chuo-ku, Tokyo

KIMURA, K., Division of Environmental Health, National Institute of
   Radiological Sciences, 9-1, 4-Chome, Anagawa, Chiba 260

KINOSHITA, M., Health and Safety Office, Power Reactor and Nuclear
   Fuel Development Corporation, 9-13, 1-Chome, Akasaka, Minato-ku,
   Tokyo

KISHIMOTO, Y., Health and Safety Office, Power Reactor and Nuclear
   Fuel Development Corporation, 9-13, 1-Chome, Akasaka, Minato-ku,
   Tokyo

KITAHARA, Y., Health and Safety Office, Power Reactor and Nuclear
   Fuel Development Corporation, 9-13, 1-Chome, Akasaka, Minato-ku,
   Tokyo

KOYANAGI, T., Dr., Division of Marine Radioecology, National
   Institute of Radiological Sciences, 3609 Isozaki, Nakaminato,
   Ibaraki 311-12

KUMATORI, T., Dr., Director, National Institute of Radiological
   Sciences, 9-1, 4-Chome, Anagawa, Chiba 260

KURABAYASHI, M., Health and Safety Office, Power Reactor and Nuclear
   Fuel Development Corporation, 9-13, 1-Chome, Akasaka, Minato-ku,
   Tokyo

KURODA, T., Japan Fisheries Resource Conservation Association,
   1-11-35, Nagata-cho, Chiyoda-ku, Tokyo

KUROKAWA, Y., Dr., Health and Safety Office, Power Reactor and
   Nuclear Fuel Development Corporation, 9-13, 1-Chome, Akasaka,
   Minato-ku, Tokyo

KUROSU, T., Nuclear Power Operations Department, The Kansai Electric
   Power Co. Inc., 3-3-22, Nakanoshima, Kita-ku, Osaka

KUWABARA, T., National Federation of Fisheries Cooperative Association, 1-1-12, Uchikanda, Chiyoda-ku, Tokyo

MACHIDA, C., Super Technical Advisor, Radioactive Waste Management
   Center, N° 15 Mori Building, 2-8-10, Toranomon, Minato-ku,
   Tokyo 105

MATSUMURA, T., Dr., Radiation Center of Osaka Prefecture, Shinke-cho,
   Sakai, Osaka 593

MATSUYAMA, S., Mrs., Centre de Calcul, Université de Hôsei,
   Koganeishi 184

MIYAKE, Y., Dr., Geochemistry Research Association, Koenji-kita, 4-29-2-217, Suginami, Tokyo 166

MORISAWA, M., Fisheries Mutual Insurance Fund, 1-1-12, Uchikanda, Chiyoda-ku, Tokyo

MORITA, S., Environmental Pollution Research Center of Ibaraki-ken, 4043-36, Ishikawa, 1-Chome, Mito-shi 310

NAGAKURA, T., Dr., Nuclear Power Research, Head Quarters, Central Research Institute of Electric Power Industry, 6-1, 1-Chome, Otemachi, Chiyoda-ku, Tokyo

NAGAYA, Y., Dr., Division of Marine Radioecology, National Institute of Radiological Sciences, 3609 Isozaki, Nakaminato, Ibaraki 311-12

NAKAHARA, M., Division of Marine Radioecology, National Institute of Radiological Sciences, 3609 Isozaki, Nakaminato, Ibaraki 311-12

NAKAMURA, K., Division of Marine Radioecology, National Institute of Radiological Sciences, 3609 Isozaki, Nakaminato, Ibaraki 311-12

NAKAMURA, R., Division of Marine Radioecology, National Institute of Radiological Sciences, 3609 Isozaki, Nakaminato, Ibaraki 311-12

NARITA, T., National Salmon Drift-Net Fisheries Association, 2-7-2, Hirakawa-cho, Chiyoda-ku, Tokyo

OHIWA, R., Nuclear Power Division, the Federation of Electric Power Companies, 1-9-4 Ohtemachi, Chiyoda-ku, Tokyo

OHMOMO, Y., Dr., Division of Radioecology, Nakaminato Branch Laboratory, National Institute of Radiological Sciences, 3609 Isozaki, Nakaminato-shi, Ibaraki-ken

OHTA, T., Radiation Safety Division, Atomic Power Department, The Tokyo Electric Power Co. Inc., 1-1-3 Uchisaiwai-cho, Chiyoda-ku, Tokyo

OKUTANI, T., Dr., National Science Museum (Natural History Institute), 3-23-1 Hyakunin-cho, Shinjuku-ku, Tokyo

ONO, K., Managing Director, Radioactive Waste Management Center, N° 15 Mori Building, 2-8-10, Toranomon, Minato-ku, Tokyo 105

SAIKI, M., Dr., Division of Radioecology Nakaminato Laboratory, National Institute of Radiological Sciences, 3609 Isozaki, Nakaminato-shi, Ibaraki-ken

SAKAMOTO, S., Fishing Group Preservation Division, Ministry of Agriculture, Forestry and Fisheries, 1-2-1, Kasumigaseki, Chiyoda-ku, Tokyo

SAKATA, S., Super Technical Advisor, Radioactive Waste Management Center, N° 15 Mori Building, 2-8-10 Toranomon, Minato-ku, Tokyo 105

SARUHASHI, K., Dr., Geochemical Laboratory, Meteorological Research Institute, Koenji-kita, 4-35-8, Suginami, Tokyo 166

SATO, H., Nuclear Industry Division, Agency of Natural Resources and Energy, 3-1, 1-Chome, Kasumigaseki, Chiyoda-ku, Tokyo

SHIGEMUNE, H., Nippon Kyodo Hogei Kaisha, Ltd., 3-2-4, Kasumigaseki, Chiyoda-ku, Tokyo

SHIMIZU, M., Dr., Faculty of Agriculture, University of Tokyo,
1-1-1 Yayoi, Bunkyo-ku, Tokyo

SHIMIZU, S., Deputy Manager, First Testing Division, Radioactive
Waste Management Center, N° 15 Mori Building, 2-8-10 Toranomon,
Minato-ku, Tokyo 105

SUDO, H., Dr., Tokai Regional Fisheries Research Laboratory, 5-1
Kachidoki, 5-Chome, Chuo-ku, Tokyo 104

SUGIMURA, Y., Dr., Geochemical Laboratory, Meteorological Research
Institute, Koenji-kita, 4-35-8, Suginami, Tokyo 166

SUGIURA, Y., Dr., Department of General Education, University of
Kagoshima, Korimoto-cho 1-21-35, Kagoshima-shi 890

SUMIYA, M., Mrs., Division of Radioecology, Nakaminato Laboratory,
National Institute of Radiological Sciences, 3609 Isozaki,
Nakaminato-shi, Ibaraki-ken

SUMITA, N., Federation of Japan Tuna Fisheries Co-operative Association, 2-3-22, Kudankita, Chiyoda-ku, Tokyo

SUYAMA, I., Division of Environmental Health, National Institute of
Radiological Sciences, 9-1, 4-Chome, Anagawa, Chiba 260

SUZUKI, Y., Dr., Division of Marine Radioecology, National Institute
of Radiological Sciences, 3609 Isozaki, Nakaminato, Ibaraki 311-12

SUZUKI, Y., Geochemical Laboratory, Meteorological Research Institute,
Koenji-kita 4-29-2-217, Suginami, Tokyo 166

TAIRA, K., Ocean Research Institute, 1-15-1, Minamidai, Nakano,
Tokyo

TAKAHASHI, M., Japan Fisheries Association, 1-9-13, Akasaka,
Minato-ku, Tokyo

TAKANO, K., Dr., Rikagaku kenkyusho, Wako-shi, Hirosawa, Saitama-ken 351

TAKEUCHI, S., Nippon Kyodo Hogei Kaisha, Ltd., 3-2-4, Kasumigaseki,
Chiyoda-ku, Tokyo

TERASIMA, T., Dr., National Institute of Radiological Sciences,
9-1, 4-Chome, Anagawa, Chiba 260

TSUCHIYA, M., Manager, First Testing Division, Radioactive Waste
Management Center, N° 15 Mori Building, 2-8-10 Toranomon,
Minato-ku, Tokyo 105

TSURUGA, H., Dr., Tokai Regional Fisheries Research Laboratory,
5-1 Kachidoki, 5-5-1, Chuo-ku, Tokyo 104

UEDA, T., Division of Marine Radioecology, National Institute of
Radiological Sciences, 3609 Isozaki, Nakaminato, Ibaraki 311-12

WADA, A., Dr., Hydrolics Department, Civil Engineering Laboratory,
Central Research Institute of Electric Power Industry, 1646,
Abiko-city, Chiba Prefecture

WATANABE, Y., Dr., Tokai Regional Fisheries Research Laboratory,
5-5, Kachidoki, Chuo-ku, Tokyo

YAMADA, R., Deputy Manager, Planning Division Radioactive Waste
Management Center, N° 15 Mori Building, 2-8-10, Toranomon,
Minato-ku, Tokyo 105

YAMAGATA, N., Dr., Department of Radiological Health, Institute of
    Public Health, 4-6-1 Shirokanedai, Minato-ku, Tokyo 108

YAMAMOTO, M., Engineer, Second Testing Division, Radioactive Waste
    Management Center, N° 15 Mori Building, 2-8-10 Toranomon,
    Minato-ku, Tokyo 105

YAMAZAKI, F., Dr., Japan Radioisotope Association, 28-45, 2-Chome,
    Motokomagome, Bunkyo-ku, Tokyo

THE NETHERLANDS - PAYS-BAS

VAN WEERS, A.W., Energieonderzoek Centrum Nederland, P.B. 1, 1755 ZG
    Petten

WEBER, J., Dr., Ministry of Public Health and Environmental Hygiene,
    10 Dr. Reyersstraat, Leidschendam

PORTUGAL

VAZ CARREIRO, M.C., Mrs., LNETI Departamento de Proteccao e Seguranca
    Radiólogica, Estrada Nacional N° 10, 2685 Sacavém

ORTINS DE BETTENCOURT, A., Dr., LNETI Departamento de Proteccao e
    Seguranca Radiológica, Estrada Nacional n° 10, 2685 Sacavém

SWEDEN - SUEDE

PERSSON, B.R.R., Dr., Radiation Physics Department, Lasarettet,
    S-22185 Lund

UNITED KINGDOM - ROYAUME-UNI

WEBB, G.A.M., National Radiological Protection Board, Building 566,
    Harwell, Oxon OX11 ORQ

PENTREATH, R.J., Dr., Ministry of Agriculture, Fisheries and Food,
    Fisheries Radiobiological Laboratory, Hamilton Dock, Lowestoft,
    Suffolk NR32 1DA

ATHERTON, R.S., Dr., Health and Safety Directorate, Rutherford
    House R 102, British Nuclear Fuels Ltd., Risley, Warrington
    WA3 6AS, Cheshire

UNITED STATES - ETATS-UNIS

NOSHKIN, Jr., W.E., Dr., Lawrence Livermore Laboratory, P.O. Box 5507,
    Livermore, California 94550

TEMPLETON, W.L., Battelle-Pacific Northwest Laboratories,
    P.O. Box 999, Richland, WA 99352

YAYANOS, A.A., Dr., University of California, Scripps Institute of
   Oceanography, A-002 3115 MBIR SIO, La Jolla, California 92093

SCIENTIFIC SECRETARIAT - SECRETARIAT SCIENTIFIQUE

RÜEGGER, B., Dr., OECD Nuclear Energy Agency, 38 boulevard Suchet,
   F-75016 Paris, France

LOCAL ARRANGEMENTS - ORGANISATION LOCALE

TANABE, M., Nuclear Safety Division, Nuclear Safety Bureau, Science
   and Technology Agency, 2-2-1 Kasumigaseki, Chiyoda-ku, Tokyo,
   Japan

SOME NEW PUBLICATIONS OF NEA

QUELQUES NOUVELLES PUBLICATIONS DE L'AEN

ACTIVITY REPORTS

RAPPORTS D'ACTIVITE

Activity Reports of the OECD Nuclear Energy Agency (NEA)
- 6th Activity Report (1977)
- 7th Activity Report (1978)

Rapports d'activité de l'Agence de l'OCDE pour l'Energie Nucléaire (AEN)
- 6ème Rapport d'Activité (1977)
- 7ème Rapport d'Activité (1978)

Free on request - Gratuit sur demande

Annual Reports of the OECD HALDEN Reactor Project
- 18th Annual Report (1977)
- 19th Annual Report (1978)

Rapports annuels du Projet OCDE de réacteurs de HALDEN
- 18ème Rapport annuel (1977)
- 19ème Rapport annuel (1978)

Free on request - Gratuit sur demande

Twentieth Anniversary of the OECD Nuclear Energy Agency
- Proceedings on the NEA Symposium on International Co-operation in the Nuclear Field : Perspectives and Prospects

Vingtième Anniversaire de l'Agence de l'OCDE pour l'Energie Nucléaire
- Compte rendu du Symposium de l'AEN sur la coopération internationale dans le domaine nucléaire : bilan et perspectives

Free on request - Gratuit sur demande

NEA at a Glance

Coup d'oeil sur l'AEN

Free on request - Gratuit sur demande

SCIENTIFIC AND TECHNICAL PUBLICATIONS	PUBLICATIONS SCIENTIFIQUES ET TECHNIQUES

NUCLEAR FUEL CYCLE / LE CYCLE DU COMBUSTIBLE NUCLEAIRE

Reprocessing of Spent Nuclear Fuels in OECD Countries	Retraitement du combustible nucléaire dans les pays de l'OCDE

1977
£ 2.50, US$ 5.00, F 20.00

Nuclear Fuel Cycle Requirements and Supply Considerations, Through the Long-Term	Besoins liés au cycle du combustible nucléaire et considérations sur l'approvisionnement à long terme

1978
£ 4.30, US$ 8.75, F 35.00

World Uranium Potential - An International Evaluation	Potentiel mondial en uranium - Une évaluation internationale

1978
£ 7.80, US$ 16.00, F 64.00

Uranium - Resources, Production and Demand	Uranium - Ressources, Production et Demande

1979
£ 8.70, US$ 19.50, F. 78.00

RADIATION PROTECTION / RADIOPROTECTION

Radon Monitoring (Proceedings of the NEA Specialist Meeting, Paris)	Surveillance du radon (Compte rendu d'une réunion de spécialistes de l'AEN, Paris)

1978
£ 8.00, US$ 16.50, F 66.00

Iodine-129 (Proceedings of an NEA Specialist Meeting, Paris)	Iode-129 (Compte rendu d'une réunion de spécialistes de l'AEN, Paris)

1977
£ 3.40, US$ 7.00, F 28.00

Recommendations for Ionization Chamber Smoke Detectors in Implementation of Radiation Protection Standards	Recommandations relatives aux détecteurs de fumée à chambre d'ionisation en application des normes de radioprotection

1977
Free on request - Gratuit sur demande

Management, Stabilisation and Environmental Impact of Uranium Mill Tailings (Proceedings of the Albuquerque Seminar, United States)	Gestion, stabilisation et incidence sur l'environnement des résidus de traitement de l'uranium (Compte rendu du Séminaire d'Albuquerque, Etats-Unis)

1978
£ 9.80, US$ 20.00, F 80.00

Exposure to Radiation from the Natural Radioactivity in Building Materials (Report by an NEA Group of Experts)	Exposition aux rayonnements due à la radioactivité naturelle des matériaux de construction (Rapport établi par un Groupe d'experts de l'AEN)

1979
Free on request - Gratuit sur demande

Marine Radioecology (Proceedings of the Tokyo Seminar)	Radioécologie marine (Compte rendu du Colloque de Tokyo)

1980
in preparation - en préparation

RADIOACTIVE WASTE MANAGEMENT — GESTION DES DECHETS RADIOACTIFS

Bituminization of Low and Medium Level Radioactive Wastes (Proceedings of the Antwerp Seminar)	Conditionnement dans le bitume des déchets radioactifs de faible et de moyenne activités (Compte rendu du Séminaire d'Anvers)

1976
£ 4.70, US$ 10.00, F 42.00

Objectives, Concepts and Strategies for the Management of Radioactive Waste Arising from Nuclear Power Programmes (Report by an NEA Group of Experts)	Objectifs, concepts et stratégies en matière de gestion des déchets radioactifs résultant des programmes nucléaires de puissance (Rapport établi par un Groupe d'experts de l'AEN)

1977
£ 8.50, US$ 17.50, F 70.00

Treatment, Conditioning and Storage of Solid Alpha-Bearing Waste and Cladding Hulls (Proceedings of the NEA/IAEA Technical Seminar, Paris)	Traitement, conditionnement et stockage des déchets solides alpha et des coques de dégainage (Compte rendu du Séminaire technique AEN/AIEA, Paris)

1977
£ 7.30, US$ 15.00, F 60.00

Storage of Spent Fuel Elements (Proceedings of the Madrid Seminar)	Stockage des éléments combustibles irradiés Compte rendu du Séminaire de Madrid)

1978
£ 7.30, US$ 15.00, F 60.00

In Situ Heating Experiments in Geological Formations (Proceedings of the Ludvika Seminar, Sweden)	Expériences de dégagement de chaleur in situ dans les formations géologiques (Compte rendu du Séminaire de Ludvika, Suède)

1978
£ 8.00, US$ 16.50, F 66.00

Migration of Long-lived Radionuclides in the Geosphere (Proceedings of the Brussels Workshop)	Migration des radionucléides à vie longue dans la géosphère (Compte rendu de la réunion de travail de Bruxelles)

1979
£ 8.30, US$ 17.00, F 68.00

Low-Flow, Low-Permeability Measurements in Largely Impermeable Rocks (Proceedings of the Paris Workshop)	Mesures des faibles écoulements et des faibles perméabilités dans des roches relativement imperméables (Compte rendu de la réunion de travail de Paris)

1979
£ 7.80, US$ 16.00, F 64.00

On-Site Management of Power Reactor Wastes (Proceedings of the Zurich Symposium)	Gestion des déchets en provenance des réacteurs de puissance sur le site de la centrale (Compte rendu du Colloque de Zurich)

1979
£ 11.00, US$ 22.50, F 90.00

Recommended Operational Procedures for Sea Dumping of Radioactive Waste	Recommandations relatives aux procédures d'exécution des opérations d'immersion de déchets radioactifs en mer

1979
Free on request - Gratuit sur demande

Guidelines for Sea Dumping Packages of Radioactive Waste (Revised version)	Guide relatif aux conteneurs de déchets radioactifs destinés au rejet en mer (Version révisée)

1979
Free on request - Gratuit sur demande

Use of Argillaceous Materials for the Isolation of Radioactive Waste (Proceedings of the Paris Workshop)	Utilisation des matériaux argileux pour l'isolation des déchets radioactifs (Compte rendu de la réunion de travail de Paris)

1980
in preparation - en préparation

SAFETY

Safety of Nuclear Ships
(Proceedings of the Hamburg
Symposium)

SURETE

Sûreté des navires nucléaires
(Compte rendu du Symposium de
Hambourg)

1978
£ 17.00, US$ 35.00, F 140.00

Nuclear Aerosols in Reactor
Safety
(A State-of-the-Art Report by
a Group of Experts)

Les aérosols nucléaires dans la
sûreté des réacteurs
(Rapport sur l'état des connaissances établi par un Groupe
d'Experts)

1979
£ 8.30, US$ 18.75, F 75.00

Plate Inspection Programme
(Report from the Plate Inspection
Steering Committee - PISC - on
the Ultrasonic Examination of
Three Test Plates)

Programme d'inspection des plaques
(Rapport du Comité de Direction de
l'inspection des plaques - PISC -
sur l'examen par ultrasons de trois
plaques d'essai)

1980

SCIENTIFIC INFORMATION

Neutron Physics and Nuclear
Data for Reactors and other
Applied Purposes
(Proceedings of the Harwell
International Conference)

INFORMATION SCIENTIFIQUE

La physique neutronique et les
données nucléaires pour les
réacteurs et autres applications
(Compte rendu de la Conférence
Internationale de Harwell)

1978
£ 26.80, US$ 55.00, F 220.00

| LEGAL | PUBLICATIONS |
| PUBLICATIONS | JURIDIQUES |

| Convention on Third Party Liability in the Field of Nuclear Energy - Incorporating the provisions of the Additional Protocol of January 1964 | Convention sur la responsabilité civile dans le domaine de l'énergie nucléaire - Texte incluant les dispositions du Protocole Additionnel de janvier 1964 |

1960
Free on request - Gratuit sur demande

| Nuclear Legislation, Analytical Study : "Nuclear Third Party Liability" (revised version) | Législations nucléaires, étude analytique : "Responsabilité civile nucléaire" (version révisée) |

1976
£ 6.00, US$ 12.50, F 50.00

| Nuclear Law Bulletin (Annual Subscription - two issues and supplements) | Bulletin de Droit Nucléaire (Abonnement annuel - deux numéros et suppléments) |

£ 5.60, US$ 12.50, F 50.00

| Index of the first twenty issues of the Nuclear Law Bulletin | Index des vingt premiers numéros du Bulletin de Droit Nucléaire |

Free on request - Gratuit sur demande

| Licensing Systems and Inspection of Nuclear Installations in NEA Member Countries (two volumes) | Régime d'autorisation et d'inspection des installations nucléaires dans les pays de l'AEN (deux volumes) |

Free on request - Gratuit sur demande

| NEA Statute | Statuts de l'AEN |

Free on request - Gratuit sur demande

OECD SALES AGENTS
DÉPOSITAIRES DES PUBLICATIONS DE L'OCDE

ARGENTINA – ARGENTINE
Carlos Hirsch S.R.L., Florida 165, 4° Piso (Galería Guemes)
1333 BUENOS-AIRES, Tel. 33-1787-2391 Y 30-7122

AUSTRALIA – AUSTRALIE
Australia & New Zealand Book Company Pty Ltd.,
23 Cross Street, (P.O.B. 459)
BROOKVALE NSW 2100 Tel. 938-2244

AUSTRIA – AUTRICHE
Gerold and Co., Graben 31, WIEN 1. Tel. 52.22.35

BELGIUM – BELGIQUE
LCLS
44 rue Otlet, B1070 BRUXELLES .Tel. 02-521 28 13

BRAZIL – BRÉSIL
Mestre Jou S.A., Rua Guaipà 518,
Caixa Postal 24090, 05089 SAO PAULO 10. Tel. 261-1920
Rua Senador Dantas 19 s/205-6, RIO DE JANEIRO GB.
Tel. 232-07. 32

CANADA
Renouf Publishing Company Limited,
2182 St. Catherine Street West,
MONTREAL, Quebec H3H 1M7 Tel. (514) 937-3519

DENMARK – DANEMARK
Munksgaards Boghandel,
Nørregade 6, 1165 KØBENHAVN K. Tel. (01) 12 85 70

FINLAND – FINLANDE
Akateeminen Kirjakauppa
Keskuskatu 1, 00100 HELSINKI 10. Tel. 65-11-22

FRANCE
Bureau des Publications de l'OCDE,
2 rue André-Pascal, 75775 PARIS CEDEX 16. Tel. (1) 524.81.67
Principal correspondant :
13602 AIX-EN-PROVENCE : Librairie de l'Université.
Tel. 26.18.08

GERMANY – ALLEMAGNE
OECD Publications and Information Centre
4 Simrockstrasse
5300 BONN Tel. 21 60 46

GREECE – GRÈCE
Librairie Kauffmann, 28 rue du Stade,
ATHÈNES 132. Tel. 322.21.60

HONG-KONG
Government Information Services,
Sales and Publications Office, Beaconsfield House, 1st floor,
Queen's Road, Central. Tel. 5-233191

ICELAND – ISLANDE
Snaebjörn Jönsson and Co., h.f.,
Hafnarstraeti 4 and 9, P.O.B. 1131, REYKJAVIK.
Tel. 13133/14281/11936

INDIA – INDE
Oxford Book and Stationery Co.:
NEW DELHI, Scindia House. Tel. 45896
CALCUTTA, 17 Park Street. Tel.240832

ITALY – ITALIE
Libreria Commissionaria Sansoni:
Via Lamarmora 45, 50121 FIRENZE. Tel. 579751
Via Bartolini 29, 20155 MILANO. Tel. 365083
Sub-depositari:
Editrice e Libreria Herder,
Piazza Montecitorio 120, 00 186 ROMA. Tel. 674628
Libreria Hoepli, Via Hoepli 5, 20121 MILANO. Tel. 865446
Libreria Lattes, Via Garibaldi 3, 10122 TORINO. Tel. 519274
La diffusione delle edizioni OCSE è inoltre assicurata dalle migliori
librerie nelle città più importanti.

JAPAN – JAPON
OECD Publications and Information Center
Akasaka Park Building, 2-3-4 Akasaka, Minato-ku,
TOKYO 107. Tel. 586-2016

KOREA - CORÉE
Pan Korea Book Corporation,
P.O.Box n°101 Kwangwhamun, SÉOUL. Tel. 72-7369

LEBANON – LIBAN
Documenta Scientifica/Redico,
Edison Building, Bliss Street, P.O.Box 5641, BEIRUT.
Tel. 354429–344425

MALAYSIA – MALAISIE
University of Malaya Co-operative Bookshop Ltd.
P.O. Box 1127, Jalan Pantai Baru
KUALA LUMPUR Tel. 51425, 54058, 54361

THE NETHERLANDS – PAYS-BAS
Staatsuitgeverij
Verzendboekhandel
Chr. Plantijnstraat
'S-GRAVENHAGE Tel. nr. 070-789911
Voor bestellingen: Tel. 070-789208

NEW ZEALAND – NOUVELLE-ZÉLANDE
The Publications Manager,
Government Printing Office,
WELLINGTON: Mulgrave Street (Private Bag),
World Trade Centre, Cubacade, Cuba Street,
Rutherford House, Lambton Quay, Tel. 737-320
AUCKLAND: Rutland Street (P.O.Box 5344), Tel. 32.919
CHRISTCHURCH: 130 Oxford Tce (Private Bag), Tel. 50.331
HAMILTON: Barton Street (P.O.Box 857), Tel. 80.103
DUNEDIN: T & G Building, Princes Street (P.O.Box 1104),
Tel. 78.294

NORWAY – NORVÈGE
J.G. TANUM A/S
P.O. Box 1177 Sentrum
Karl Johansgate 43
OSLO 1 Tel (02) 80 12 60

PAKISTAN
Mirza Book Agency, 65 Shahrah Quaid-E-Azam, LAHORE 3.
Tel. 66839

PORTUGAL
Livraria Portugal, Rua do Carmo 70-74,
1117 LISBOA CODEX.
Tel. 360582/3

SPAIN – ESPAGNE
Mundi-Prensa Libros, S.A.
Castelló 37, Apartado 1223, MADRID-1. Tel. 275.46.55
Libreria Bastinos, Pelayo, 52, BARCELONA 1. Tel. 222.06.00

SWEDEN – SUÈDE
AB CE Fritzes Kungl Hovbokhandel,
Box 16 356, S 103 27 STH, Regeringsgatan 12,
DS STOCKHOLM. Tel. 08/23 89 00

SWITZERLAND – SUISSE
Librairie Payot, 6 rue Grenus, 1211 GENÈVE 11. Tel. 022-31.89.50

TAIWAN – FORMOSE
National Book Company,
84-5 Sing Sung Rd., Sec. 3, TAIPEI 107. Tel. 321.0698

THAILAND – THAILANDE
Suksit Siam Co., Ltd.
1715 Rama IV Rd.
Samyan, Bangkok 5
Tel. 2511630

UNITED KINGDOM – ROYAUME-UNI
H.M. Stationery Office, P.O.B. 569,
LONDON SE1 9 NH. Tel. 01-928-6977, Ext. 410 or
49 High Holborn, LONDON WC1V 6 HB (personal callers)
Branches at: EDINBURGH, BIRMINGHAM, BRISTOL,
MANCHESTER, CARDIFF, BELFAST.

UNITED STATES OF AMERICA – ÉTATS-UNIS
OECD Publications and Information Center, Suite 1207,
1750 Pennsylvania Ave., N.W. WASHINGTON, D.C.20006.
Tel. (202)724-1857

VENEZUELA
Libreria del Este, Avda. F. Miranda 52, Edificio Galipàn,
CARACAS 106. Tel. 32 23 01/33 26 04/33 24 73

YUGOSLAVIA – YOUGOSLAVIE
Jugoslovenska Knjiga, Terazije 27, P.O.B. 36, BEOGRAD.
Tel. 621-992

Les commandes provenant de pays où l'OCDE n'a pas encore désigné de dépositaire peuvent être adressées à :
OCDE, Bureau des Publications, 2 rue André-Pascal, 75775 PARIS CEDEX 16.
Orders and inquiries from countries where sales agents have not yet been appointed may be sent to:
OECD, Publications Office, 2 rue André-Pascal, 75775 PARIS CEDEX 16.

OECD PUBLICATIONS, 2 rue André-Pascal, 75775 Paris Cedex 16 - No. 41 421 1980
PRINTED IN FRANCE
(2150 HS 66 80 05 3) ISBN 92-64-02053-5